contemporary
accelerator
physics

contemporary accelerator physics

Stephan I Tzenov
Università degli Studi di Salerno, Italy

NEW JERSEY • LONDON • SINGAPORE • BEIJING • SHANGHAI • HONG KONG • TAIPEI • CHENNAI

Published by

World Scientific Publishing Co. Pte. Ltd.
5 Toh Tuck Link, Singapore 596224
USA office: Suite 202, 1060 Main Street, River Edge, NJ 07661
UK office: 57 Shelton Street, Covent Garden, London WC2H 9HE

British Library Cataloguing-in-Publication Data
A catalogue record for this book is available from the British Library.

CONTEMPORARY ACCELERATOR PHYSICS

ISBN 981-238-900-8

Printed in Singapore by World Scientific Printers (S) Pte Ltd

To my wife Anastassia with gratitude

Preface

At present, we attest to the rapid development of the physics of accelerators and storage rings in a number of aspects, from purely scientific ones to particular applications in industry, medicine and even everyday life. This inevitably stimulates the progress of the theory of charged particle beams as well. This progress is apparently due to the fact that the physics of charged particle beams is at the border of a wide spectrum of theoretical disciplines ranging from analytical mechanics to elements of quantum mechanics and quantum field theory. It is natural that the contemporary methods of classical mechanics, plasma physics, statistical physics and quantum mechanics merge to find their application in the physics of beams. This was the basic premise I have been guided by, during the work on the book. In addition, the selection and the manner of exposition of the material was to a large extent influenced by my own taste as a theorist, as well as by the aspiration to make the book useful to researchers from other fields. I apologize in advance for many important applications of accelerator theory have not been covered here, hoping that the gap could be filled in a possible subsequent edition.

There are many excellent references on the physics of particle accelerators both introductory and advanced ones, which cover various aspects of the theory, as well as issues more experimentally oriented. Amongst these, I would like to mention the seminal book by *Kolomensky* and *Lebedev* [Kolomensky (1966)] and the one by *Bruck* [Bruck (1966)]. From the more recent editions, the books by *Chao* [Chao (1993)], *Bryant* and *Johnsen* [Bryant (1993)], *Edwards* and *Syphers* [Edwards (1993)], *Humphries* [Humphries (1986)], *Lee* [Lee (1999)], *Wiedemann* [Wiedemann (1993)] and *Wilson* [Wilson (2001)] are mandatory to be pointed out. A couple of options were available to me at the beginning of the work on the book – either to write a standard introductory text on accelerator physics, or try to create a reference containing material on specific issues with an emphasis on contemporary methods of classical mechanics, statistical physics and plasma physics, that are relevant to the physics of charged particle beams. Well aware of the risk I run, I took the second opportunity being motivated by two reasons. Since a vast bibliography on the principles of particle accelerators and storage rings exists, I considered it unnecessary to add more confusion to the potential reader's choice. Secondly, modern methods and mathematical tools widely used in accelerator theory are rather scattered in the literature, so that it appeared to me a good idea to bring them together in a single volume. It is, of course, the reader's irreversible right to judge whether I have succeeded or not. In this aspect, all comments and criticism are most welcome.

In the course of the work on the book some of the material has been discussed with many colleagues. I would like to thank them all, as well as all my collaborators for the joint efforts, without which many of the results described in the subsequent chapters would not have been obtained. It is my great pleasure to thank *Alex Chao* and *Michel Dubois-Violette* for their constant interest and encouragement, and for many valuable comments and suggestions on the early version of the draft. Numerous illuminating discussions with *S. De Martino* and *S. De Siena* together with their support is gratefully acknowledged. Many thanks are due also to *Y. Cai, P. Colestock, V. Danilov, R. Davidson, D. Dinev, M. Furman, N. Goldenfeld, D. Kaltchev, V. Mikhailov, Y. Oono, F. Ruggiero, R. Ruth, E. Startsev* and *Y. Yan.* I am greatly indebted to my first teacher, colleague and friend *Ivan Enchevich*, who introduced me to the fascinating field of charged particle beams.

I am grateful to my wife *Anastassia* and to my son *Theodore* for their confidence and patience.

Stephan Tzenov

University of Salerno
November, 2003

Contents

Chapter 1

Hamiltonian Formulation of Single Particle Dynamics

1.1 Introduction

The motion of charged particles in accelerators and storage rings is extremely complicated. Complicated is the motion even of a single individual particle in the external confining and accelerating electromagnetic fields, not to mention about the collective behaviour of the beam particles as an ensemble. Complete description of beam propagation can be achieved only by numerical integration of the equations of motion (*Hamilton*'s equations), however, as we will see in chapter 7 even such procedure is not feasible.

Fortunately, however for accelerator theorists, the roles of various factors influencing the trajectory of particles can be in some sense classified in a hierarchical structure by order of magnitude. Thus taking into account more and more intrinsic features, one can achieve increasingly detailed description of particle orbits order by order. Throughout this and the subsequent chapters, we will show that the design particle orbit comprises the zero order approximation. This is the closed orbit of a synchronous particle with strictly prescribed momentum (or energy) matching the confining dipole magnetic field. The deviation from the design orbit due to the finite energy spread with respect to the synchronous particle can be regarded as a first order correction. The accelerating electric field is usually small in comparison with the guiding magnetic field, so that the particle energy does not vary noticeably in the course of one revolution. This fact simplifies considerably the treatment in higher orders, because it provides a possibility to study the longitudinal (synchrotron) motion separately from the transversal (betatron) motion. Synchrotron and betatron oscillations come into play next and can be regarded as a second order superstructure. One can continue further and consider the nonlinear terms.

It should be emphasized that such step-by-step description of particle motion is in good agreement with experimental observations and is preferable for its physical clearness.

1.2 Hamiltonian Formalism

It is well-known that the Lagrangian of a particle with rest mass m_0 and charge e moving in an electromagnetic field, specified by the scalar potential φ and the vector potential $\mathbf{A}$ can be written as [Goldstein (1980)]

$$\mathcal{L} = -m_0 c^2 \sqrt{1-\beta^2} + e\mathbf{v}\cdot\mathbf{A} - e\varphi, \tag{1.1}$$

where $\mathbf{v} = \mathrm{d}\mathbf{r}/\mathrm{d}t$ is the particle velocity, $\beta = v/c$ is the relative particle velocity, and c is the speed of light in vacuum.

The first step in the description of the dynamics of particles in accelerators and storage rings consists in defining the design particle orbit $\mathbf{r}_0(s)$ [Kolomensky (1966)], where s is the path length along the trajectory. Generally enough, it is possible to show [Courant (1958)] that in a magnetic field possessing a plane of symmetry (median plane) and perpendicular to that plane in every point, there exists at least one design orbit. Moreover, the design orbit is a plane curve lying entirely in the median plane, and is determined by the curvature $K(s)$ and the center of the curve in every point.

In the natural coordinate system associated with the design orbit the position vector of a particle $\mathbf{r}(s)$ is

$$\mathbf{r}(x,z,s) = \mathbf{r}_0(s) + x\mathbf{n}(s) + z\mathbf{b}(s), \tag{1.2}$$

where x is the deviation from the design orbit in the direction of the unit normal vector $\mathbf{n}(s)$, and z is the deviation in the direction of the unit binormal vector $\mathbf{b}(s)$. The vector triple $(\mathbf{n},\mathbf{b},\boldsymbol{\tau})$ ($\boldsymbol{\tau}$ being the unit tangent vector) defines a local coordinate system along with the design orbit. It satisfies the equations [Sternberg (1964)]

$$\boldsymbol{\tau}(s) = \frac{\mathrm{d}\mathbf{r}_0}{\mathrm{d}s}, \qquad \frac{\mathrm{d}\boldsymbol{\tau}}{\mathrm{d}s} = -K(s)\mathbf{n}(s), \tag{1.3}$$

$$\frac{\mathrm{d}\mathbf{n}}{\mathrm{d}s} = K(s)\boldsymbol{\tau}(s) + \kappa(s)\mathbf{b}(s), \qquad \frac{\mathrm{d}\mathbf{b}}{\mathrm{d}s} = -\kappa(s)\mathbf{n}(s), \tag{1.4}$$

known as *Fresnet-Serret* formulas. Here $\kappa(s)$ is the torsion ($\kappa \equiv 0$ for plane curves). Differentiating equation (1.2) with respect to the time we obtain

$$\mathbf{v} = v_x \mathbf{n} + v_z \mathbf{b} + (1 + xK) v_s \boldsymbol{\tau}, \tag{1.5}$$

where

$$v_x = \frac{\mathrm{d}x}{\mathrm{d}t}, \qquad v_z = \frac{\mathrm{d}z}{\mathrm{d}t}, \qquad v_s = \frac{\mathrm{d}s}{\mathrm{d}t},$$

Taking into account

$$\sqrt{1-\beta^2} = \sqrt{1 - \left[\beta_x^2 + \beta_z^2 + (1+xK)^2 \beta_s^2\right]},$$

$$\mathbf{v} \cdot \mathbf{A} = v_x A_x + v_z A_z + (1 + xK) v_s A_s,$$

as well as [Goldstein (1980)]

$$\mathcal{H} = v_x p_x + v_z p_z + v_s p_s - \mathcal{L}, \tag{1.6}$$

relating the particle's Hamiltonian $\mathcal{H}$ to the Lagrangian $\mathcal{L}$, we can write the Hamiltonian as

$$\mathcal{H} = \frac{m_0 c^2}{\sqrt{1-\beta^2}} + e\varphi, \tag{1.7}$$

or

$$\mathcal{H} = c\sqrt{m_0^2 c^2 + (p_x - eA_x)^2 + (p_z - eA_z)^2 + \left(\frac{p_s}{1+xK} - eA_s\right)^2} + e\varphi. \tag{1.8}$$

Here the canonical momenta (p_x, p_z, p_s) are given by the expressions

$$p_u = \frac{\partial \mathcal{L}}{\partial v_u} = \frac{m_0 v_u}{\sqrt{1-\beta^2}} + eA_u, \qquad u = (x, z), \tag{1.9}$$

$$p_s = \frac{\partial \mathcal{L}}{\partial v_s} = \frac{m_0 v_s (1+xK)^2}{\sqrt{1-\beta^2}} + e(1+xK)A_s. \tag{1.10}$$

It is well-known [Whittaker (1944)], [Dragt (1982)] that one of the canonical variables, say the path length s along the design orbit can be chosen as an independent variable. Then the inverse time $-t$ and the Hamiltonian $\mathcal{H}$ become canonically conjugate pair of variables, while the new Hamiltonian H is simply the longitudinal canonical momentum, taken with an opposite sign

$$H = -p_s =$$

$$-(1+xK)\sqrt{\frac{(\mathcal{H}-e\varphi)^2}{c^2} - m_0^2c^2 - (p_x - eA_x)^2 - (p_z - eA_z)^2}$$

$$-e(1+xK)A_s. \tag{1.11}$$

To show that we write down the *Hamilton*'s equations of motion derived from the Hamiltonian (1.8)

$$\frac{\mathrm{d}u}{\mathrm{d}t} = \frac{\partial \mathcal{H}}{\partial p_u}, \qquad \frac{\mathrm{d}p_u}{\mathrm{d}t} = -\frac{\partial \mathcal{H}}{\partial u}, \qquad u = (x, z) \tag{1.12}$$

$$\frac{\mathrm{d}s}{\mathrm{d}t} = \frac{\partial \mathcal{H}}{\partial p_s}, \qquad \frac{\mathrm{d}p_s}{\mathrm{d}t} = -\frac{\partial \mathcal{H}}{\partial s}, \tag{1.13}$$

from which it follows

$$\frac{\mathrm{d}u}{\mathrm{d}s} = \frac{\mathrm{d}u/\mathrm{d}t}{\mathrm{d}s/\mathrm{d}t} = \frac{\partial \mathcal{H}/\partial p_u}{\partial \mathcal{H}/\partial p_s}, \qquad \frac{\mathrm{d}p_u}{\mathrm{d}s} = \frac{\mathrm{d}p_u/\mathrm{d}t}{\mathrm{d}s/\mathrm{d}t} = -\frac{\partial \mathcal{H}/\partial u}{\partial \mathcal{H}/\partial p_s}, \tag{1.14}$$

$$\frac{\mathrm{d}t}{\mathrm{d}s} = \frac{1}{\partial \mathcal{H}/\partial p_s}, \qquad \frac{\mathrm{d}\mathcal{H}}{\mathrm{d}s} = \frac{\mathrm{d}\mathcal{H}/\mathrm{d}t}{\mathrm{d}s/\mathrm{d}t} = \frac{\partial \mathcal{H}/\partial t}{\partial \mathcal{H}/\partial p_s}. \tag{1.15}$$

The obvious condition to impose is that the longitudinal particle velocity does not vanish

$$\frac{\mathrm{d}s}{\mathrm{d}t} = \frac{\partial \mathcal{H}}{\partial p_s} \neq 0.$$

This means that the longitudinal particle coordinate should be monotonically increasing or decreasing function of time. Let us now view the longitudinal canonical momentum p_s as a function of x, z, p_x, p_z, $\mathcal{H}$ and $-t$. Comparison of the total differentials of p_s and $\mathcal{H}$ yields

$$\frac{\partial p_s}{\partial u} = -\frac{\partial \mathcal{H}/\partial u}{\partial \mathcal{H}/\partial p_s}, \qquad \frac{\partial p_s}{\partial p_u} = -\frac{\partial \mathcal{H}/\partial p_u}{\partial \mathcal{H}/\partial p_s}, \tag{1.16}$$

$$\frac{\partial p_s}{\partial(-t)} = \frac{\partial \mathcal{H}/\partial t}{\partial \mathcal{H}/\partial p_s}, \qquad \frac{\partial p_s}{\partial \mathcal{H}} = \frac{1}{\partial \mathcal{H}/\partial p_s}. \tag{1.17}$$

Thus, *Hamilton*'s equations can be rewritten

$$\frac{\mathrm{d}u}{\mathrm{d}s} = \frac{\partial(-p_s)}{\partial p_u}, \qquad \frac{\mathrm{d}p_u}{\mathrm{d}s} = -\frac{\partial(-p_s)}{\partial u}, \qquad u = (x, z) \tag{1.18}$$

$$\frac{\mathrm{d}(-t)}{\mathrm{d}s} = \frac{\partial(-p_s)}{\partial \mathcal{H}}, \qquad \frac{\mathrm{d}\mathcal{H}}{\mathrm{d}s} = -\frac{\partial(-p_s)}{\partial(-t)}. \tag{1.19}$$

The canonical conjugate pairs of variables (x, p_x), (z, p_z) and $(-t, \mathcal{H})$ behave as if their motion is governed by a new Hamiltonian:

$$H = -p_s(x, p_x, z, p_z, -t, \mathcal{H}; s), \tag{1.20}$$

where s now plays the role of the new "time" variable. The new set of *Hamilton*'s equations in terms of the new Hamiltonian is

$$\frac{\mathrm{d}u}{\mathrm{d}s} = \frac{\partial H}{\partial p_u}, \qquad \frac{\mathrm{d}p_u}{\mathrm{d}s} = -\frac{\partial H}{\partial u}, \qquad u = (x, z) \tag{1.21}$$

$$\frac{\mathrm{d}(-t)}{\mathrm{d}s} = \frac{\partial H}{\partial \mathcal{H}}, \qquad \frac{\mathrm{d}\mathcal{H}}{\mathrm{d}s} = -\frac{\partial H}{\partial(-t)}. \tag{1.22}$$

Note that $(-t, \mathcal{H})$ now play the role of a third coordinate and conjugate momentum.

1.3 Canonical Transformations

Dynamical systems are described in terms of a set of variables, coordinates and canonically conjugate momenta. In an accelerator these are the horizontal and vertical deviations of an orbiting particle in a plane transverse to the design orbit, supplemented by the corresponding canonical momenta. As we already saw, a third pair of variables consists of the inverse time and the energy of the particle. In a number of applications it is more convenient to express the equations of motion in terms of different variables which are functions of the old ones. It is desirable to have the equations governing the time evolution of the new variables again in Hamiltonian form

$$\frac{\mathrm{d}X_i}{\mathrm{d}s} = \frac{\partial K}{\partial P_i}, \qquad \frac{\mathrm{d}P_i}{\mathrm{d}s} = -\frac{\partial K}{\partial X_i}, \tag{1.23}$$

where $\mathbf{X} = (X_1, X_2, \ldots, X_n)$ and $\mathbf{P} = (P_1, P_2, \ldots, P_n)$ are the new canonically conjugate coordinates and momenta, and $K(\mathbf{X}, \mathbf{P}; s)$ is the new Hamiltonian.

The transformation from the old canonical variables $\mathbf{x}$ and $\mathbf{p}$ to the new ones $\mathbf{X}$ and $\mathbf{P}$ can be expressed by the $2n$ equations

$$X_i = f_i(\mathbf{x}, \mathbf{p}; s), \qquad P_i = g_i(\mathbf{x}, \mathbf{p}; s). \tag{1.24}$$

Obviously, of the $4n$ variables only $2n$ are independent because of the $2n$ functional relations (1.24). Therefore, the canonical transformations we are interested in must involve only $2n$ independent variables.

To find the transformations that meet the above requirements we follow *Goldstein* [Goldstein (1980)]. *Hamilton*'s equations of motion can be derived from a variational principle. Consider the evolution of the system from s_1 to s_2 and the action integral

$$\mathcal{S} = \int_{s_1}^{s_2} \mathrm{d}s \mathcal{L}(\mathbf{x}(s), \mathbf{x}'(s); s), \tag{1.25}$$

where $\mathcal{L}(\mathbf{x}(s), \mathbf{x}'(s); s)$ is the Lagrangian, written in the old coordinates $\mathbf{x} = (x_1, x_2, \ldots, x_n)$ and velocities, and the prime implies differentiation with respect to s. Taking into account equation (1.6), and varying the vector function $\mathbf{x}(s)$ so that the end points remain fixed and the action integral is stationary, we obtain the *Lagrange*'s equations of motion

$$\frac{\mathrm{d}}{\mathrm{d}s}\left(\frac{\partial \mathcal{L}}{\partial x_i'}\right) - \frac{\partial \mathcal{L}}{\partial x_i} = 0, \tag{1.26}$$

which are equivalent to *Hamilton*'s equations

$$\frac{\mathrm{d}(p_i)}{\mathrm{d}s} + \frac{\partial H}{\partial x_i} = 0, \qquad x_i' = \frac{\partial H}{\partial p_i}. \tag{1.27}$$

Hamilton's principle of least action in terms of the new variables $\mathbf{X}$ and $\mathbf{P}$, and with the new Hamiltonian K must again be valid

$$\delta \mathcal{S}_1 = \delta \int_{s_1}^{s_2} \mathrm{d}s \left[\sum_i P_i X_i' - K(\mathbf{X}, \mathbf{P}; s)\right] = 0. \tag{1.28}$$

This suggests that the new and old Lagrangian can differ by the total "time" derivative of some function $\mathcal{F}$, which must be a function of the new and old variables. As an example, let us consider a function depending only on the new and old coordinates

$$\mathcal{F} = F_1(\mathbf{x}, \mathbf{X}; s).$$

Then we have

$$\sum_i p_i x_i' - H = \sum_i P_i X_i' - K + \frac{\mathrm{d}F_1}{\mathrm{d}s}.$$

Expanding the total "time" derivative of F_1 we obtain

$$\sum_i x_i'\left(p_i - \frac{\partial F_1}{\partial x_i}\right) - \sum_i X_i'\left(P_i + \frac{\partial F_1}{\partial X_i}\right) - \left(H - K + \frac{\partial F_1}{\partial s}\right) = 0. \quad (1.29)$$

Equation (1.29) holds identically, provided the coefficients of x' and X' vanish, because $\mathbf{x}$ and $\mathbf{X}$ represent $2n$ independent variables. Thus we must have

$$p_i = \frac{\partial F_1}{\partial x_i}, \qquad P_i = -\frac{\partial F_1}{\partial X_i}, \qquad K = H + \frac{\partial F_1}{\partial s}. \quad (1.30)$$

Equations (1.30) specify the relations between the old and new variables in a canonical transformation defined by the function F_1, called the generating function. If the canonical transformation is not degenerate, the first two of these equations can be solved for $\mathbf{x}$ and $\mathbf{p}$ in terms of $\mathbf{X}$ and $\mathbf{P}$. The new Hamiltonian, given by the third equation in (1.30) is then

$$K(\mathbf{X}, \mathbf{P}; s) = H(\mathbf{x}(\mathbf{X}, \mathbf{P}; s), \mathbf{p}(\mathbf{X}, \mathbf{P}; s); s) + \frac{\partial F_1}{\partial s}(\mathbf{x}(\mathbf{X}, \mathbf{P}; s), \mathbf{X}; s). \quad (1.31)$$

Rather than choosing the old coordinates and new coordinates as variables of the canonical transformation, we could have chosen the old coordinates and new momenta $(\mathbf{x}, \mathbf{P})$. In this case we have a different type of generating function $F_2(\mathbf{x}, \mathbf{P}; s)$, and a different set of equations for the canonical transformation

$$p_i = \frac{\partial F_2}{\partial x_i}, \qquad X_i = \frac{\partial F_2}{\partial P_i}, \qquad K = H + \frac{\partial F_2}{\partial s}. \quad (1.32)$$

The generating functions F_2 and F_1 are related by a *Legendre* transformation of the form

$$F_2(\mathbf{x}, \mathbf{P}; s) = F_1(\mathbf{x}, \mathbf{X}; s) + \sum_i x_i P_i. \quad (1.33)$$

In a similar way it is possible to define two additional generating functions, $F_3(\mathbf{p}, \mathbf{X}; s)$ depending on the old momenta and new coordinates, and $F_4(\mathbf{p}, \mathbf{P}; s)$ depending on the old momenta and new momenta.

1.4 Electric and Magnetic Fields

The accelerating electric **E** and the guiding magnetic **B** fields in an accelerator can be derived from the vector potential **A**. In a machine with separated functions the vector potential includes contributions from all the static multipole elements (magnetic dipoles, quadrupoles, sextupoles, etc.), as well as the accelerating RF field. Without loss of generality the only nonzero component of **A** will be the longitudinal one A_s, which automatically implies that the longitudinal component of the magnetic field B_s vanishes.

It is however instructive to consider the general case, where all three components of the vector potential are non vanishing. Here we utilize the method of *Teng* (see e.g. [Suzuki (1978)]) to obtain the contribution to the vector potential **A**, coming from all multipole elements. The static magnetic field is

$$\mathbf{B} = \nabla \times \mathbf{A}, \tag{1.34}$$

which in the curvilinear coordinates (1.2) can be written as

$$B_x = \frac{1}{h_0}\left[\frac{\partial(h_0 A_s)}{\partial z} - \frac{\partial A_z}{\partial s}\right], \tag{1.35}$$

$$B_z = \frac{1}{h_0}\left[\frac{\partial A_x}{\partial s} - \frac{\partial(h_0 A_s)}{\partial x}\right], \tag{1.36}$$

$$B_s = \frac{\partial A_z}{\partial x} - \frac{\partial A_x}{\partial z}, \tag{1.37}$$

where $h_0 = 1 + xK$. From the gauge condition

$$xA_x + zA_z = 0$$

it follows:

$$A_x = -z\widetilde{F}(x, z, s), \qquad A_z = x\widetilde{F}(x, z, s), \tag{1.38}$$

where $\widetilde{F}(x, z, s)$ is yet unknown function. By substitution of equation (1.38) into equations (1.35) - (1.37) it is straightforward to obtain

$$2\widetilde{F} + \left(x\frac{\partial}{\partial x} + z\frac{\partial}{\partial z}\right)\widetilde{F} = B_s,$$

$$\frac{xK}{1+xK}A_s + \left(x\frac{\partial}{\partial x} + z\frac{\partial}{\partial z}\right)A_s = zB_x - xB_z.$$

Assuming that $\widetilde{F}$ and A_s are sums of homogeneous polynomials in x and z and applying the *Euler*'s theorem (see e.g. [Fleming (1966)]) for homogeneous functions one finds

$$\widetilde{F} = \frac{1}{2}B_s^{(0)} + \frac{1}{3}B_s^{(1)} + \frac{1}{4}B_s^{(2)} + \dots, \tag{1.39}$$

$$\widetilde{G}_{x,z} = \left(1 + \frac{xK}{2}\right)B_{x,z}^{(0)} + \left(\frac{1}{2} + \frac{xK}{3}\right)B_{x,z}^{(1)} + \left(\frac{1}{3} + \frac{xK}{4}\right)B_{x,z}^{(2)} + \dots, \tag{1.40}$$

$$A_s = \frac{z\widetilde{G}_x - x\widetilde{G}_z}{1+xK}, \tag{1.41}$$

where $B^{(0)}$, $B^{(1)}$, $B^{(2)}$, etc. denote homogeneous polynomials in x and z of order 0, 1, 2, etc., respectively. From *Maxwell*'s equations

$$\nabla \times \mathbf{B} = 0, \qquad\qquad \nabla \cdot \mathbf{B} = 0$$

one observes that the magnetic induction vector $\mathbf{B}$ can be expressed as a gradient of a potential function Ψ

$$\mathbf{B} = \nabla\Psi,$$

satisfying the *Laplace* equation

$$\nabla^2\Psi = 0.$$

Taking into account the median symmetry it is easy to check that the potential Ψ is an odd function of z

$$\Psi = \left(a_0 + a_1 x + \frac{a_2 x^2}{2!} + \dots\right)z - \left(b_0 + b_1 x + \frac{b_2 x^2}{2!} + \dots\right)\frac{z^3}{3!}$$

$$+(c_0 + c_1 x + \dots)\frac{z^5}{5!} + \dots, \tag{1.42}$$

where all a's, b's, c's, etc. are functions of s, and the coefficients b_n, c_n, etc. are related to a_n via the expressions

$$b_0 = a_0'' + Ka_1 + a_2, \tag{1.43}$$

$$b_1 = -2Ka_0'' - K'a_0' + a_1'' - K^2 a_1 + Ka_2 + a_3, \tag{1.44}$$

$$b_2 = 6K^2a_0'' + 6KK'a_0' - 4Ka_1'' - 2K'a_1' + a_2'' + 2K^3a_1 - 2K^2a_2 + Ka_3 + a_4, \tag{1.45}$$

$$c_0 = b_0'' + Kb_1 + b_2. \tag{1.46}$$

Here the primes denote differentiation with respect to s. The a's have a simple physical meaning. They represent the vertical component of the magnetic field and its derivatives with respect to x in the median plane

$$a_0 = (B_z)_{x,z=0}, \qquad a_1 = \left(\frac{\partial B_z}{\partial x}\right)_{x,z=0}, \qquad a_2 = \left(\frac{\partial^2 B_z}{\partial x^2}\right)_{x,z=0}. \tag{1.47}$$

From the above considerations it becomes clear that the magnetic field **B** can be in principle computed up to arbitrarily high multipole order, provided its value in the median plane (the vertical component B_z only) is known.

For various accelerator lattice elements in particular we have:

(1) Cavity.
The accelerating electric field **E** is given by the expression:

$$E_s = -\widehat{\mathcal{E}}_0(s)\sin(\omega t - \phi_0), \qquad E_x = E_z = 0,$$

where the minus signs are introduced only for later convenience. Here $\widehat{\mathcal{E}}_0(s)$ is the RF voltage, $\omega = k\omega_s$ is the RF frequency, ϕ_0 is the initial RF phase, ω_s is the angular frequency of the synchronous particle and k is the harmonic acceleration number. Taking into account the relation between the electric field and the vector potential

$$\mathbf{E} = -\frac{\partial \mathbf{A}}{\partial t},$$

we obtain

$$A_s = -\frac{\widehat{\mathcal{E}}_0(s)}{\omega}\cos(\omega t - \phi_0). \tag{1.48}$$

(2) Quadrupole.
Since $K = 0$, for the components of the magnetic field we have

$$B_x = a_1 z, \qquad B_z = a_1 x, \qquad B_s = 0.$$

The vector potential is simply

$$A_s = \frac{a_1}{2}(z^2 - x^2). \tag{1.49}$$

(3) Sextupole.
The components of the magnetic field are

$$B_x = a_2 xz, \qquad B_z = \frac{a_2}{2}\left(x^2 - z^2\right), \qquad B_s = 0.$$

For the vector potential we obtain

$$A_s = -\frac{a_2}{6}\left(x^3 - 3xz^2\right). \tag{1.50}$$

(4) Octupole.
The components of the magnetic field are

$$B_x = -\frac{a_3}{6}\left(z^3 - 3x^2 z\right), \qquad B_z = \frac{a_3}{6}\left(x^3 - 3xz^2\right), \qquad B_s = 0.$$

For the vector potential we obtain

$$A_s = -\frac{a_3}{24}\left(x^4 - 6x^2 z^2 + z^4\right). \tag{1.51}$$

(5) Synchrotron magnet.
In this case for the vector potential we have

$$A_s = -\frac{a_0}{2K}(1 + xK) + \frac{a_1}{2}\left(z^2 - x^2\right). \tag{1.52}$$

1.5 Synchro-Betatron Formalism in Beam Dynamics

The starting point in our exposition of the synchro-betatron formalism in cyclic accelerators and storage rings is the Hamiltonian (1.11). To include collective effects, such as beam-beam interaction and space charge, we keep the scalar potential φ nonzero. Moreover, we assume that apart from terms corresponding to various lattice elements, an additional term in the longitudinal component of the vector potential A_s is present. It is responsible for collective effects, and as we will see, it adds to the scalar potential (with appropriate coefficient) to form the collective part of the Hamiltonian. Following *Tzenov* [Tzenov (1991)], [Tzenov (1997b)] we introduce the new scaled canonically conjugated set of variables

$$u \implies \widetilde{p}_u = \frac{p_u}{p_{0s}}, \qquad -\beta_s ct = \tau \implies h = \frac{\mathcal{H}}{\beta_s^2 E_s}, \tag{1.53}$$

where hereafter $u = (x, z)$, unless stated explicitly, and cast the new Hamiltonian in the form

$$\widetilde{H} = \frac{H}{p_{0s}} =$$

$$-(1 + xK)\sqrt{\left(\beta_s h - \frac{e\varphi}{\beta_s E_s}\right)^2 - \frac{1}{\beta_s^2\gamma_s^2} - \left(\widetilde{p}_x - \frac{eA_x}{p_{0s}}\right)^2 - \left(\widetilde{p}_z - \frac{eA_z}{p_{0s}}\right)^2}$$

$$-(1 + xK)\frac{eA_s}{p_{0s}}. \tag{1.54}$$

Here $p_{0s} = m_0\beta_s\gamma_s c$ and $E_s = m_0\gamma_s c^2$ are the momentum and the energy of the synchronous particle, respectively, $\beta_s = v_s/c$ – its relative velocity and $\gamma_s = \left(1 - \beta_s^2\right)^{-1/2}$ is the *Lorentz* factor. Note that we have implicitly assumed the momentum of the synchronous particle p_{0s} to be constant. The effect of acceleration on the dynamics of particles will be discussed in the next chapter, where we will study the adiabatic damping of betatron oscillations.

Since the energy of collective interaction $e\varphi$ is much smaller than the energy of the synchronous particle, and the scaled transverse momenta $\widetilde{p}_{x,z}$ are small quantities, we can expand the square root in equation (1.54) in a power series in $\widetilde{p}_{x,z}/\Gamma$ and $e\varphi/E_s$, where

$$\Gamma = \sqrt{\beta_s^2 h^2 - \frac{1}{\beta_s^2\gamma_s^2}}. \tag{1.55}$$

Taking into account the expressions for the longitudinal component of the vector potential corresponding to various lattice elements, obtained in the previous paragraph one finds

$$\widetilde{H} = \widetilde{H}_0 + \widetilde{H}_1 + \widetilde{H}_2 + \widetilde{H}_3 + \cdots + \widetilde{H}_{col}, \tag{1.56}$$

where

$$\widetilde{H}_0 = -\Gamma + \frac{e\widehat{\mathcal{E}}_0(s)}{\beta_s E_s}\frac{c}{\omega}\cos\left(\frac{\omega\tau}{c\beta_s} + \phi_0\right), \tag{1.57}$$

$$\widetilde{H}_1 = -(\Gamma - 1)Kx, \tag{1.58}$$

$$\widetilde{H}_2 = \frac{1}{2\Gamma}\left(\widetilde{p}_x^2 + \widetilde{p}_z^2\right) + \frac{1}{2R^2}\left(G_x x^2 + G_z z^2\right), \tag{1.59}$$

$$\widetilde{H}_3 = K\left[\frac{\widetilde{p}_x^2 + \widetilde{p}_z^2}{2\Gamma} + \frac{g_0}{2R^2}\left(x^2 - z^2\right)\right]x + \frac{\lambda_0}{6R^3}\left(x^3 - 3xz^2\right), \tag{1.60}$$

$$\widetilde{H}_{col} = \frac{e}{E_s}(1 + xK)\left(\frac{h\varphi_{col}}{\Gamma} - \frac{v_s A_{col}}{\beta_s^2}\right). \tag{1.61}$$

In equations (1.59) - (1.61) R is the mean machine radius, φ_{col} and A_{col} are the parts of the scalar and vector potentials accounting for collective effects, respectively, and

$$K = \frac{e}{p_{0s}}(B_z)_{x,z=0}, \qquad g_0 = \frac{eR^2}{p_{0s}}\left(\frac{\partial B_z}{\partial x}\right)_{x,z=0}, \tag{1.62}$$

$$\lambda_0 = \frac{eR^3}{p_{0s}}\left(\frac{\partial^2 B_z}{\partial x^2}\right)_{x,z=0}, \tag{1.63}$$

$$G_x = g_0 + R^2K^2, \qquad G_z = -g_0. \tag{1.64}$$

Note that the second term in the square brackets of equation (1.60) is due to the combined effect of the curvature and the quadrupole component of the static magnetic field. In an ideal lattice, since quadrupoles are located in straight sections, this term disappears.

It is convenient to describe the dynamics of particles in cyclic accelerators and storage rings in terms of a new independent variable $\theta = s/R$ – the azimuthal angle along the circumference of the machine. To preserve the canonical structure of the equations of motion one has to multiply the Hamiltonian (1.56) simply by R. We consider a canonical transformation defined by the generating function

$$\mathcal{F}_2\left(u, \widetilde{\widetilde{p}}_u, \tau, \eta; \theta\right) = x\widetilde{\widetilde{p}}_x + z\widetilde{\widetilde{p}}_z + \tau\left(\eta + \frac{1}{\beta_s^2}\right) + \eta R\theta. \tag{1.65}$$

The new canonical variables are

$$\widetilde{u} = u, \qquad \widetilde{\widetilde{p}}_u = \widetilde{p}_u, \qquad \eta = h - \frac{1}{\beta_s^2}, \qquad \sigma = R\theta + \tau. \tag{1.66}$$

The new Hamiltonian can be written as

$$\widetilde{\widetilde{H}}_0 = \frac{R\eta^2}{2\gamma_s^2} + \frac{e\widehat{\mathcal{E}}_0(s)}{\beta_s E_s}\frac{Rc}{\omega}\cos\left(\frac{\omega\sigma}{c\beta_s} - k\theta + \phi_0\right), \tag{1.67}$$

$$\widetilde{\widetilde{H}}_1 = -RK\widetilde{x}\eta, \tag{1.68}$$

$$\widetilde{\widetilde{H}}_2 = \frac{R}{2}\left(\widetilde{p}_x^2 + \widetilde{p}_z^2\right) + \frac{1}{2R}\left(G_x\widetilde{x}^2 + G_z\widetilde{z}^2\right) + \dots, \tag{1.69}$$

$$\widetilde{\widetilde{H}}_3 = RK\left[\frac{\widetilde{p}_x^2 + \widetilde{p}_z^2}{2} + \frac{g_0}{2R^2}\left(\widetilde{x}^2 - \widetilde{z}^2\right)\right]\widetilde{x} + \frac{\lambda_0}{6R^2}\left(\widetilde{x}^3 - 3\widetilde{x}\widetilde{z}^2\right) + \dots, \tag{1.70}$$

$$\widetilde{H}_{col} = \frac{eR}{\beta_s^2 E_s}(1 + \widetilde{x}K)(\varphi_{col} - v_s A_{col}) + \dots, \tag{1.71}$$

where the dots in equations (1.69) - (1.71) stand for terms proportional to η and its higher powers. These terms are responsible for chromatic effects, that will be discussed later.

The next step in the synchro-betatron formalism consists in the definition of the dispersion function D. Let us consider the canonical transformation

$$\mathcal{G}_2(\widetilde{u}, \widehat{p}_u, \sigma, \widehat{\eta}; \theta) = \sigma\widehat{\eta} + \widetilde{z}\widehat{p}_z + \widehat{p}_x(\widetilde{x} - \widehat{\eta}D) + \frac{\widetilde{x}\widehat{\eta}}{R}\frac{\mathrm{d}D}{\mathrm{d}\theta} - \frac{\widehat{\eta}^2}{2R}D\frac{\mathrm{d}D}{\mathrm{d}\theta}, \tag{1.72}$$

defining the new canonical variables

$$\widetilde{x} = \widehat{x} + \widehat{\eta}D, \qquad \widetilde{p}_x = \widehat{p}_x + \frac{\widehat{\eta}}{R}\frac{\mathrm{d}D}{\mathrm{d}\theta}, \qquad \widetilde{z} = \widehat{z}, \qquad \widetilde{p}_z = \widehat{p}_z, \tag{1.73}$$

$$\sigma = \widehat{\sigma} + \widehat{p}_x D - \frac{\widehat{x}}{R}\frac{\mathrm{d}D}{\mathrm{d}\theta}, \qquad \eta = \widehat{\eta}. \tag{1.74}$$

If we now require that the dispersion function satisfies the equation

$$\frac{\mathrm{d}^2 D}{\mathrm{d}\theta^2} + G_x D = R^2 K, \tag{1.75}$$

as can be easily checked the $\widehat{H}_1$ term in the new Hamiltonian vanishes. Thus, we have

$$\widehat{H}_0 = -\frac{R}{2}\left(KD - \frac{1}{\gamma_s^2}\right)\widehat{\eta}^2 + \frac{e\widehat{\mathcal{E}}_0(s)}{\beta_s E_s}\frac{Rc}{\omega}\cos\left(\frac{\omega\sigma}{c\beta_s} - k\theta + \phi_0\right), \tag{1.76}$$

$$\widehat{H}_2 = \frac{R}{2}\left(\widehat{p}_x^2 + \widehat{p}_z^2\right) + \frac{1}{2R}\left(G_x\widehat{x}^2 + G_z\widehat{z}^2\right). \tag{1.77}$$

If the dispersion function and its derivative are equal to zero in the cavities' locations, that is there is no coupling between the longitudinal and the transverse degrees of freedom, we may perform the standard procedure of averaging the Hamiltonian (1.76) over one turn. Defining the momentum compaction factor α_M as

$$\alpha_M = \frac{1}{2\pi} \int_0^{2\pi} \mathrm{d}\theta K(\theta) D(\theta), \tag{1.78}$$

we obtain

$$\widehat{H}_0 = -\frac{R\mathcal{K}}{2}\widehat{\eta}^2 + \frac{\Delta E_0}{\beta_s E_s} \frac{c}{2\pi\omega} \cos\left(\frac{\omega\widehat{\sigma}}{c\beta_s} + \Phi_0\right), \tag{1.79}$$

where $\mathcal{K}$ is the phase slip coefficient, given by the expression

$$\mathcal{K} = \alpha_M - \frac{1}{\gamma_s^2}. \tag{1.80}$$

The RF phase Φ_0 is

$$\Phi_0 = \phi_0 - \Theta_0, \qquad \Theta_0 = \arctan\left(\frac{\mathcal{A}_s}{\mathcal{A}_c}\right), \tag{1.81}$$

where

$$\mathcal{A}_s = \frac{eR}{2\pi} \int_0^{2\pi} \mathrm{d}\theta \widehat{\mathcal{E}}_0(\theta) \sin k\theta, \qquad \mathcal{A}_c = \frac{eR}{2\pi} \int_0^{2\pi} \mathrm{d}\theta \widehat{\mathcal{E}}_0(\theta) \cos k\theta.$$

The quantity

$$\Delta E_0 = 2\pi\sqrt{\mathcal{A}_s^2 + \mathcal{A}_c^2} \tag{1.82}$$

comprises the maximum energy gain per turn.

The Hamiltonian (1.77) governs the linear uncoupled betatron motion. It describes particle oscillations in a plane transverse to the design orbit. These will be studied in more detail in the next chapter. The Hamiltonian (1.79) governs the synchrotron motion. It comprises longitudinal oscillations of the orbiting particles in a circular accelerator, which are in general nonlinear. It is worthwhile to note, that synchrotron and betatron oscillations are, strictly speaking, coupled through various sources of coupling. One of them we already briefly mentioned, that is the nonzero dispersion function and its derivative in the cavities. Another source of synchro-betatron coupling are the terms proportional to the deviation from the energy of the synchronous particle η and its powers, that we have omitted in equations (1.69) - (1.71).

The synchro-betatron formalism is a powerful tool to study the single particle dynamics in cyclic accelerators and storage rings. It has a number of advantages compared to the standard approach [Kolomensky (1966)], [Bruck (1966)] in which transverse and longitudinal dynamics are treated separately. The method provides a unified treatment of synchrotron and betatron motion, and is simple to handle and apply in particular cases of practical interest and importance. After using the fully six-dimensional description to derive the Hamiltonian, the dispersion function is introduced via a canonical transformation so that the *symplectic structure* of the equations of motion is completely preserved. At last, but not least, important accelerator parameters, such as the momentum compaction factor and the phase slip coefficient are introduced in a natural way.

Chapter 2

Linear Betatron Motion

2.1 Introduction

In parallel with the growing complication and the main goal to improve the performance of the accelerators, the analytical theory of particle orbits and their stability passed through various stages of development, as the problems varied greatly with the type and the size of the machine. The basic role of the theory is the prescription of ideal stable orbits and the study of factors which can perturb the confining magnetic and the accelerating electric fields. In addition, an important subject of study are the consequences of a deviation from the ideal stable case such perturbations can cause. One of the most difficult problem is the study of the effect caused by direct interaction between particles within the beam, indirect interaction via the environment, scattering on the residual gas in the vacuum chamber, electromagnetic radiation of the accelerated particles, random external perturbations, etc. As a first step, one can consider the analysis of the stability of motion in an ideal magnetic field configuration provided that the energy gain per turn is negligibly small. This analysis is the mile-stone in the preliminary design of every machine and as a rule, the closer the real guiding and accelerating fields are to the design values, the better the performance of the accelerator is. Sometimes, it is said that the machine is close to the linear one.

As already mentioned in the previous chapter, it is possible in cyclic accelerators and storage rings to indicate a distinguished trajectory (design orbit) which is contingent on the synchronism with the accelerating field. However, from the very beginning of the acceleration process, the beam particles acquire a spread in coordinates and momenta, as well as a spread in energies around the energy of the synchronous particle. In other words, there is a variety of initial conditions for different particles in the beam. To ensure the stability of motion, the guiding magnetic field must have in the neighbourhood of the design orbit a certain configuration. In the present chapter, we will study this configuration, as well as the basic properties of linear stability.

2.2 The Transfer Matrix

The *Hamilton*'s equations of motion following from the Hamiltonian (1.77) can be written as

$$\frac{\mathrm{d}\widehat{x}}{\mathrm{d}\theta} = R\widehat{p}_x, \qquad \frac{\mathrm{d}\widehat{p}_x}{\mathrm{d}\theta} = -\frac{G_x}{R}\widehat{x}, \tag{2.1}$$

and similar equations for $\widehat{z}$ and $\widehat{p}_z$. In terms of $\widehat{x}$ and $\widehat{z}$ these become

$$\frac{\mathrm{d}^2 u}{\mathrm{d}\theta^2} + G_u(\theta)u = 0, \tag{2.2}$$

where u stands for either $\widehat{x}$ or $\widehat{z}$. Equations (2.2) describe the motion of beam particles near the design orbit, usually called betatron oscillations. The focusing strengths G_u, comprised of contributions from quadrupole and bending magnets [see equation (1.64)] at different locations along the circumference of the machine, depend on θ. This dependence is periodic since the design orbit is a closed curve. In circular accelerators and storage rings the magnetic lattice ideally consists of N identical sections, therefore the period of G_u is $2\pi/N$.

We consider now a linear homogeneous second order differential equation

$$\frac{\mathrm{d}^2 u}{\mathrm{d}\theta^2} + F_u(\theta)\frac{\mathrm{d}u}{\mathrm{d}\theta} + G_u(\theta)u = 0, \tag{2.3}$$

where the coefficients $F_u(\theta)$ and $G_u(\theta)$ are generic functions of θ, not necessarily periodic. Define a vector

$$\mathbf{U}(\theta) = \begin{pmatrix} u(\theta) \\ \dot{u}(\theta) \end{pmatrix}, \tag{2.4}$$

associated with a particular integral $u(\theta)$ of equation (2.3), where the dot implies differentiation with respect to θ. An equation of the form

$$\mathbf{U}(\theta_1) = \widehat{\mathcal{M}}(\theta_1\,|\theta_0)\mathbf{U}(\theta_0) \tag{2.5}$$

relates the values of the vector (2.4) in two points θ_0 and θ_1. The matrix $\widehat{\mathcal{M}}(\theta_1\,|\theta_0)$ called the *transfer matrix* does not depend on the particular choice of $u(\theta)$, but only on the functions $F_u(\theta)$ and $G_u(\theta)$ between θ_0 and θ_1.

To prove the above statement we utilize the well-known fact from the theory of linear differential equations of second order, asserting that any such equation has two linearly independent solutions (fundamental solutions). Moreover, any particular solution can be constructed as a linear combination of the fundamental solutions. We select two particular integrals $v(\theta)$ and $w(\theta)$ of equation (2.3) such that at the initial point θ_0 we have

$$v(\theta_0) = 1, \qquad \dot{v}(\theta_0) = 0, \tag{2.6}$$

$$w(\theta_0) = 0, \qquad \dot{w}(\theta_0) = 1. \tag{2.7}$$

Since all solutions of a linear differential equation are represented as a linear combination of two particular integrals we can write

$$\mathbf{U}(\theta) = \widehat{\mathcal{W}}(\theta)\mathbf{U}(\theta_0),$$

where

$$\widehat{\mathcal{W}}(\theta) = \begin{pmatrix} v(\theta) & w(\theta) \\ \dot{v}(\theta) & \dot{w}(\theta) \end{pmatrix}. \tag{2.8}$$

For $\theta = \theta_1$ we have

$$\mathbf{U}(\theta_1) = \widehat{\mathcal{W}}(\theta_1)\mathbf{U}(\theta_0),$$

so that

$$\widehat{\mathcal{M}}(\theta_1\,|\theta_0) = \widehat{\mathcal{W}}(\theta_1). \tag{2.9}$$

It can be easily verified that the transfer matrix for any interval made up of subintervals is simply the product of the matrices for the subintervals, that is,

$$\widehat{\mathcal{M}}(\theta_2\,|\theta_0) = \widehat{\mathcal{M}}(\theta_2\,|\theta_1)\widehat{\mathcal{M}}(\theta_1\,|\theta_0). \tag{2.10}$$

A transfer matrix of particular interest in the case of equation (2.2) is the one describing the motion of particles through a whole period

$$\widehat{\mathcal{M}}(\theta) = \widehat{\mathcal{M}}\left(\theta + \frac{2\pi}{N}\middle|\,\theta\right), \tag{2.11}$$

starting from θ. Since its elements are periodic functions of θ the transfer matrix for a full passage through one revolution can be written as

$$\widehat{\mathcal{M}}(\theta + 2\pi \,|\theta) = \widehat{\mathcal{M}}^N(\theta). \tag{2.12}$$

The transfer matrix for M revolutions is then

$$\widehat{\mathcal{M}}(\theta + 2\pi M \,|\theta) = \widehat{\mathcal{M}}^{MN}(\theta). \tag{2.13}$$

The determinant of the transfer matrix is related to the so-called *Wronskian*

$$W(\theta) = \det \begin{pmatrix} u_1(\theta) & u_2(\theta) \\ \dot{u}_1(\theta) & \dot{u}_2(\theta) \end{pmatrix} \tag{2.14}$$

of the differential equation (2.3). Here $u_1(\theta)$ and $u_2(\theta)$ are two independent integrals. Upon multiplying equation (2.3) corresponding to u_1 by u_2, multiplying the one corresponding to u_2 by u_1, and subtracting them we obtain

$$\frac{\mathrm{d}W}{\mathrm{d}\theta} + F_u(\theta)W = 0.$$

This can be directly integrated to give

$$W(\theta) = W_0 \exp\left[-\int_{\theta_0}^{\theta} \mathrm{d}\tau F_u(\tau)\right], \tag{2.15}$$

which in the important case of $F_u(\theta) = 0$ reduces to

$$W(\theta) = \text{const.} \tag{2.16}$$

Let us now consider the transfer matrix for an arbitrary interval from θ_0 to θ. From equation (2.9) we have

$$\widehat{\mathcal{M}}(\theta \,|\theta_0) = \widehat{\mathcal{W}}(\theta).$$

Since

$$\det \widehat{\mathcal{M}}(\theta_0 \,|\theta_0) = \det \widehat{\mathcal{I}} = 1$$

we obtain

$$\det \widehat{\mathcal{M}}(\theta \,|\theta_0) = \exp\left[-\int_{\theta_0}^{\theta} \mathrm{d}\tau F_u(\tau)\right],$$

where $\widehat{\mathcal{I}}$ is the unit matrix. For $F_u(\theta) = 0$, that is, the case of equation (2.2), we arrive at

$$\det \widehat{\mathcal{M}}(\theta \,|\theta_0) = 1. \tag{2.17}$$

Equation (2.17) is of utmost importance in the theory of Hamiltonian systems and is a direct consequence of the *symplectic character* of Hamiltonian dynamics [Dragt (1982)]. It is also equivalent to the condition for area preservation in phase space (*Liouville's theorem*) [Klimontovich (1986)].

2.3 Hill's Equation and Floquet's Theorem

Equation of the form (2.2), where the coefficient $G_u(\theta)$ is a periodic function of θ with a period Θ

$$G_u(\theta + \Theta) = G_u(\theta) \tag{2.18}$$

is known as *Hill's* equation [Stoker (1950)], [Ince (1944)]. There exists an important result which gives the form of the solution of linear differential equations with periodic coefficients, called the *Floquet's theorem.* It states that an equation of *Hill's* type (2.2) possesses two independent solutions, namely

$$u_1(\theta) = f_1(\theta) e^{i\nu_u \theta}, \qquad\qquad u_2(\theta) = f_2(\theta) e^{-i\nu_u \theta}, \tag{2.19}$$

where the functions $f_1(\theta)$ and $f_2(\theta)$ are periodic with period Θ

$$f_{1,2}(\theta + \Theta) = f_{1,2}(\theta). \tag{2.20}$$

We are looking for integrals of *Hill's* equation called solutions in a "normal" form which satisfy the relation

$$u_i(\theta + \Theta) = \lambda_i u_i(\theta), \qquad\qquad (i = 1, 2). \tag{2.21}$$

Let $w_1(\theta)$ and $w_2(\theta)$ be two independent solutions of *Hill's* equation. Since $G_u(\theta)$ is periodic in θ, the functions $w_1(\theta + \Theta)$ and $w_2(\theta + \Theta)$ are also solutions[1] and can be represented as a linear combination of w_1 and w_2 in the form

$$\mathbf{w}(\theta + \Theta) = \widehat{\mathcal{A}} \mathbf{w}(\theta). \tag{2.22}$$

[1]This can be easily checked by formally shifting the independent variable θ in *Hill's* equation by one period Θ, and taking into account the periodicity condition (2.18).

Here $\mathbf{w}$ is a column vector with components w_1 and w_2, and $\widehat{\mathcal{A}}$ is a 2×2 matrix with constant coefficients.

We now introduce the new functions $u_1(\theta)$ and $u_2(\theta)$ which are also independent particular integrals of *Hill*'s equation. They are related to the w's via the expression

$$\mathbf{u}(\theta) = \widehat{\mathcal{B}}\mathbf{w}(\theta). \tag{2.23}$$

By noting that $u_{1,2}(\theta + \Theta)$ are solutions of *Hill*'s equation as well, we can write

$$\mathbf{u}(\theta + \Theta) = \widehat{\mathcal{C}}\mathbf{u}(\theta). \tag{2.24}$$

Substitution of equations (2.22) and (2.23) into equation (2.24) yields

$$\widehat{\mathcal{C}} = \widehat{\mathcal{B}}\widehat{\mathcal{A}}\widehat{\mathcal{B}}^{-1}. \tag{2.25}$$

Comparing equations (2.21) and (2.24) we conclude that the matrix $\widehat{\mathcal{C}}$ must be diagonal

$$\left(\widehat{\mathcal{C}}\right)_{ij} = \lambda_i \delta_{ij},$$

where δ_{ij} is the *Kronecker*'s delta-symbol. Equivalently, the problem can be formulated in finding a matrix $\widehat{\mathcal{B}}$ which diagonalizes the matrix $\widehat{\mathcal{A}}$ through the similarity transformation (2.25). Such matrix $\widehat{\mathcal{B}}$ exists and its inverse is made up of the eigenvectors $\mathbf{b}_i$, satisfying the equation

$$\widehat{\mathcal{A}}\mathbf{b}_i = \lambda_i \mathbf{b}_i.$$

Furthermore, λ_i are the eigenvalues of the matrix $\widehat{\mathcal{A}}$ which solve the secular equation

$$\det\left(\widehat{\mathcal{A}} - \lambda\widehat{\mathcal{I}}\right) = 0,$$

or

$$\lambda^2 - \lambda \mathrm{tr}\widehat{\mathcal{A}} + \det\widehat{\mathcal{A}} = 0, \tag{2.26}$$

where $\mathrm{tr}\widehat{\mathcal{A}}$ is the trace of the matrix $\widehat{\mathcal{A}}$. It can be shown that the eigenvalues are unique and independent of the choice of the two integrals $w_1(\theta)$ and $w_2(\theta)$ of *Hill*'s equation [Bruck (1966)].

To find the relation between the matrix $\widehat{\mathcal{A}}$ and the transfer matrix we differentiate equation (2.22) with respect to θ

$$\dot{\mathbf{w}}(\theta + \Theta) = \widehat{\mathcal{A}}\dot{\mathbf{w}}(\theta).$$

The last equation put together with equation (2.22) can be written in a compact form as

$$\widehat{\mathcal{W}}_{12}(\theta + \Theta) = \widehat{\mathcal{W}}_{12}(\theta)\widehat{\mathcal{A}}^T, \tag{2.27}$$

where $\widehat{\mathcal{W}}_{12}(\theta)$ is similar to (2.8), but comprised of the functions $w_1(\theta)$ and $w_2(\theta)$ instead of $v(\theta)$ and $w(\theta)$, and $\widehat{\mathcal{A}}^T$ is the transposed of $\widehat{\mathcal{A}}$. On the other hand, taking into account the definition (2.9) of the transfer matrix we have the two equations

$$\mathbf{W}_i(\theta + \Theta) = \widehat{\mathcal{M}}(\theta + \Theta\,|\theta)\mathbf{W}_i(\theta) \qquad\qquad (i = 1, 2),$$

where $\mathbf{W}_i(\theta)$ is now similar to (2.4) and consists of $w_i(\theta)$ and $\dot{w}_i(\theta)$. In a compact form this can be written as

$$\widehat{\mathcal{W}}_{12}(\theta + \Theta) = \widehat{\mathcal{M}}(\theta + \Theta\,|\theta)\widehat{\mathcal{W}}_{12}(\theta). \tag{2.28}$$

By comparing equations (2.27) and (2.28) we obtain the following result

$$\widehat{\mathcal{A}}^T = \widehat{\mathcal{W}}_{12}^{-1}(\theta)\widehat{\mathcal{M}}(\theta + \Theta\,|\theta)\widehat{\mathcal{W}}_{12}(\theta). \tag{2.29}$$

Since the similarity transformation (2.29) leaves the determinant and the trace of the transfer matrix unchanged we have

$$\det \widehat{\mathcal{A}} = \det \widehat{\mathcal{M}} = 1, \qquad\qquad \mathrm{tr}\widehat{\mathcal{A}} = \mathrm{tr}\widehat{\mathcal{M}}.$$

Therefore the solutions of the secular equation (2.26) are

$$\lambda_{1,2} = e^{\pm i\nu_u\Theta}, \tag{2.30}$$

where

$$\cos \nu_u\Theta = \frac{1}{2}\mathrm{tr}\widehat{\mathcal{M}}. \tag{2.31}$$

It is worthwhile to mention that ν_u can be complex. The fact that the determinant of the transfer matrix is equal to unity implies that the eigenvalues come in reciprocal pairs, that is if λ is an eigenvalue $1/\lambda$ is an eigenvalue too. Let us also note that the eigenvalues of the transfer matrix are independent of the reference point θ. From equation (2.10) we have

$$\widehat{\mathcal{M}}\left(\theta_2 + \frac{2\pi}{N}\middle|\theta_1\right) = \widehat{\mathcal{M}}(\theta_2)\widehat{\mathcal{M}}(\theta_2\,|\theta_1),$$

$$\widehat{\mathcal{M}}\left(\theta_2 + \frac{2\pi}{N}\middle|\theta_1\right) = \widehat{\mathcal{M}}\left(\theta_2 + \frac{2\pi}{N}\middle|\theta_1 + \frac{2\pi}{N}\right)\widehat{\mathcal{M}}(\theta_1) = \widehat{\mathcal{M}}(\theta_2\,|\theta_1)\widehat{\mathcal{M}}(\theta_1),$$

so that

$$\widehat{\mathcal{M}}(\theta_2) = \widehat{\mathcal{M}}(\theta_2 \,|\theta_1)\widehat{\mathcal{M}}(\theta_1)\widehat{\mathcal{M}}^{-1}(\theta_2 \,|\theta_1). \tag{2.32}$$

Since the one period transfer matrices $\widehat{\mathcal{M}}(\theta_1)$ and $\widehat{\mathcal{M}}(\theta_2)$ are related by a similarity transformation they have the same trace and therefore the same eigenvalues.

We are now in a position to examine the stability of linear betatron oscillations. Because the eigenvalues of the one period transfer matrix $\widehat{\mathcal{M}}(\theta)$ have the form given by equation (2.30), the eigenvalues of the transfer matrix for M revolutions $\widehat{\mathcal{M}}^{MN}(\theta)$ are

$$\lambda_{1,2}^{(M)} = e^{\pm 2\pi i M \nu_u}. \tag{2.33}$$

For the motion to be stable both $\lambda_{1,2}^{(M)}$ must remain bounded as $M \to \infty$. This implies that ν_u must be a real quantity, and the eigenvalues in this case are located on the unit circle in the complex plane. In other words, the motion is stable provided

$$\mathrm{tr}\widehat{\mathcal{M}}(\theta) < 2, \tag{2.34}$$

and unstable if

$$\mathrm{tr}\widehat{\mathcal{M}}(\theta) > 2. \tag{2.35}$$

If the form (2.19) of the independent solutions of *Hill*'s equation is substituted into equation (2.21) the due account of equation (2.30) yields immediately the periodicity property (2.20) of the *Floquet* amplitudes $f_{1,2}(\theta)$.

2.4 Twiss Parameters and Courant-Snyder Invariant

In order to obtain the explicit form of the solution to *Hill*'s equation (2.2) we employ the method of normal forms [Arnold (1983)]. The main idea is to transform the Hamiltonian (1.77) into a normal form (in the sense of *Poincare-Dulac* [Arnold (1983)])

$$H_2 = \frac{\dot{\chi}_x(\theta)}{2}\left(P_x^2 + X^2\right) + \frac{\dot{\chi}_z(\theta)}{2}\left(P_z^2 + Z^2\right), \tag{2.36}$$

by means of a canonical transformation, where (X, P_x) and (Z, P_z) are the new pairs of canonical variables, and $\chi_{x,z}(\theta)$ are yet unknown functions of

θ. For that purpose we choose a generating function of the second type depending on the old coordinates and the new momenta

$$F_2(\widehat{u}, P_u; \theta) = \sum_{u=(x,z)} \left(\frac{\widehat{u} P_u}{\sqrt{\beta_u}} - \frac{\alpha_u \widehat{u}^2}{2\beta_u} \right), \tag{2.37}$$

where the coefficients $\alpha_u(\theta)$ and $\beta_u(\theta)$ called the *Twiss parameters* are to be determined. Note that in one dimension for instance, we could have chosen the generating function to be a complete bilinear form in $\widehat{x}$ and P_x, including a term proportional to P_x^2. However, there are three equations confining three types of homogeneous monomial terms of second order in $\widehat{x}$ and P_x, that would have involved four coefficients [the function $\dot{\chi}_x(\theta)$ included]. Therefore, one of the coefficients is arbitrary (for instance, the coefficient multiplying P_x^2), and without loss of generality can be set equal to zero.

For the sake of simplicity we confine ourselves to the horizontal degree of freedom only. Since the motion in the horizontal and vertical directions as described by the Hamiltonian (1.77) is decoupled, similar results will hold for the vertical degree of freedom. The old and the new canonical variables are related through the expressions

$$\widehat{x} = X\sqrt{\beta}, \qquad \widehat{p} = \frac{1}{\sqrt{\beta}}(P - \alpha X), \tag{2.38}$$

where for brevity we have dropped the index x referring to the horizontal motion. We substitute now these into the new Hamiltonian given be the last of equations (1.32) and compare it with the Hamiltonian in a normal form (2.36). Equating coefficients of similar terms we obtain

$$\dot{\chi} = \frac{R}{\beta}, \tag{2.39}$$

$$\dot{\alpha} = \frac{G\beta}{R} - R\gamma, \tag{2.40}$$

$$\dot{\beta} = -2R\alpha, \tag{2.41}$$

where a third *Twiss* parameter $\gamma(\theta)$ has been introduced via the expression

$$\beta\gamma - \alpha^2 = 1. \tag{2.42}$$

To derive a single equation for the β-function we differentiate equation (2.41) with respect to θ and make use of equations (2.40) and (2.42). We obtain

$$\frac{1}{2}\beta\ddot{\beta} - \frac{1}{4}\dot{\beta}^2 + G\beta^2 = R^2. \tag{2.43}$$

The *Hamilton*'s equations of motion

$$\dot{X} = \dot{\chi}P, \qquad \dot{P} = -\dot{\chi}X$$

following from the normal form Hamiltonian (2.36) can be easily solved. The result is

$$\mathbf{X}(\theta) = \begin{pmatrix} X(\theta) \\ P(\theta) \end{pmatrix} = \widehat{\mathcal{R}}(\theta, \theta_0)\mathbf{X}(\theta_0), \tag{2.44}$$

where $\widehat{\mathcal{R}}(\theta, \theta_0)$ is an orthogonal matrix

$$\widehat{\mathcal{R}}(\theta, \theta_0) = \begin{pmatrix} \cos\Delta\chi & \sin\Delta\chi \\ -\sin\Delta\chi & \cos\Delta\chi \end{pmatrix}, \tag{2.45}$$

and

$$\Delta\chi(\theta, \theta_0) = \chi(\theta) - \chi(\theta_0) \tag{2.46}$$

is the phase advance between the points θ and θ_0. The change of variables (2.38) can be written in a matrix form as

$$\mathbf{X} = \widehat{\mathcal{L}}(\theta)\begin{pmatrix} \widehat{x} \\ \widehat{p} \end{pmatrix} = \widehat{\mathcal{L}}(\theta)\widehat{\mathbf{x}},$$

where

$$\widehat{\mathcal{L}}(\theta) = \begin{pmatrix} \beta^{-1/2}(\theta) & 0 \\ \alpha(\theta)\beta^{-1/2}(\theta) & \beta^{1/2}(\theta) \end{pmatrix}. \tag{2.47}$$

Thus, equation (2.44) in terms of the old canonical variables can be rewritten as

$$\widehat{\mathbf{x}}(\theta) = \widehat{\mathcal{L}}^{-1}(\theta)\widehat{\mathcal{R}}(\theta, \theta_0)\widehat{\mathcal{L}}(\theta_0)\widehat{\mathbf{x}}(\theta_0). \tag{2.48}$$

Recall now that according to equation (2.5) we must have

$$\widehat{\mathbf{x}}(\theta) = \widehat{\mathcal{M}}(\theta\,|\theta_0)\widehat{\mathbf{x}}(\theta_0),$$

so that

$$\widehat{\mathcal{M}}(\theta\,|\theta_0) = \widehat{\mathcal{L}}^{-1}(\theta)\widehat{\mathcal{R}}(\theta, \theta_0)\widehat{\mathcal{L}}(\theta_0).$$

Performing the matrix multiplication we can write the transfer matrix as

$$\widehat{\mathcal{M}}(\theta\,|\theta_0) = \begin{pmatrix} m_{11} & m_{12} \\ m_{21} & m_{22} \end{pmatrix}, \tag{2.49}$$

where the matrix elements are given by

$$m_{11} = \sqrt{\frac{\beta}{\beta_0}}(\cos\Delta\chi + \alpha_0 \sin\Delta\chi), \tag{2.50}$$

$$m_{12} = \sqrt{\beta\beta_0}\sin\Delta\chi, \tag{2.51}$$

$$m_{21} = \frac{1}{\sqrt{\beta\beta_0}}[(\alpha_0 - \alpha)\cos\Delta\chi - (1 + \alpha\alpha_0)\sin\Delta\chi], \tag{2.52}$$

$$m_{22} = \sqrt{\frac{\beta_0}{\beta}}(\cos\Delta\chi - \alpha\sin\Delta\chi). \tag{2.53}$$

Equation (2.49) together with equations (2.50)–(2.53) is the most general expression for the transfer matrix, characterizing the passage through an arbitrary interval from the point θ_0 to the point θ. Since the *Twiss* parameters are periodic functions of θ with the same period $2\pi/N$ as the focusing strength $G(\theta)$, the transfer matrix for one period acquires a very simple form

$$\widehat{\mathcal{M}}(\theta) = \begin{pmatrix} \cos\mu + \alpha\sin\mu & \beta\sin\mu \\ -\gamma\sin\mu & \cos\mu - \alpha\sin\mu \end{pmatrix}, \tag{2.54}$$

where

$$\mu = R\int\limits_{\theta}^{\theta+2\pi/N} \frac{d\tau}{\beta(\tau)}. \tag{2.55}$$

For a machine with N identical sections the phase advance per one full revolution will be $N\mu$. The quantity

$$\nu = \frac{N\mu}{2\pi} = \frac{R}{2\pi}\int\limits_{\theta}^{\theta+2\pi} \frac{d\tau}{\beta(\tau)}. \tag{2.56}$$

refers to the number of betatron oscillations in one revolution. It is called the *betatron tune*.

It should be noted that the matrix $\widehat{\mathcal{M}}(\theta)$ can be written as

$$\widehat{\mathcal{M}}(\theta) = \widehat{\mathcal{I}} \cos \mu + \widehat{\mathcal{J}} \sin \mu, \tag{2.57}$$

where

$$\widehat{\mathcal{J}}(\theta) = \begin{pmatrix} \alpha & \beta \\ -\gamma & -\alpha \end{pmatrix} \tag{2.58}$$

is a matrix with zero trace and unit determinant. It also satisfies the identity

$$\widehat{\mathcal{J}}^2 = -\widehat{\mathcal{I}}, \tag{2.59}$$

which ensures that the transfer matrix for one period has properties similar to those of complex numbers on the unit circle. In particular, it is easy to see that an expression

$$\left(\widehat{\mathcal{I}} \cos \mu_1 + \widehat{\mathcal{J}} \sin \mu_1\right)\left(\widehat{\mathcal{I}} \cos \mu_2 + \widehat{\mathcal{J}} \sin \mu_2\right) =$$

$$\widehat{\mathcal{I}} \cos (\mu_1 + \mu_2) + \widehat{\mathcal{J}} \sin (\mu_1 + \mu_2), \tag{2.60}$$

analogue to the *Moivre*'s formula holds. The k-th power of the matrix $\widehat{\mathcal{M}}(\theta)$ is thus

$$\widehat{\mathcal{M}}^k(\theta) = \widehat{\mathcal{I}} \cos k\mu + \widehat{\mathcal{J}} \sin k\mu. \tag{2.61}$$

Let the quantity $I(X, P; \theta)$ be a function of the canonical variables X and P, and the "time" θ. For I to be a constant of the motion or invariant, it must satisfy the fundamental defining equation

$$\frac{\mathrm{d}I}{\mathrm{d}\theta} = \frac{\partial I}{\partial \theta} + \frac{\partial I}{\partial X}\dot{X} + \frac{\partial I}{\partial P}\dot{P} = \frac{\partial I}{\partial \theta} + [I,\, H] = 0, \tag{2.62}$$

where

$$[I,\, H] = \frac{\partial I}{\partial X}\frac{\partial H}{\partial P} - \frac{\partial I}{\partial P}\frac{\partial H}{\partial X} \tag{2.63}$$

is the *Poisson bracket* and H is the Hamiltonian of the system. For the Hamiltonian in a normal form (2.36) it is straightforward to check that the invariant acquires the form

$$I(X, P; \theta) = X^2 + P^2. \tag{2.64}$$

In other words, the invariant quantity (2.64) is proportional to the transverse energy of the particle. It is called the *Courant-Snyder invariant* in normal coordinates, and represents a circle in the X-P plane. To transform the *Courant-Snyder* invariant to the original canonical variables $(\widehat{x}, \widehat{p})$ we utilize the equation (2.38). Thus we obtain

$$I(\widehat{x}, \widehat{p}; \theta) = \gamma(\theta)\widehat{x}^2 + 2\alpha(\theta)\widehat{x}\widehat{p} + \beta(\theta)\widehat{p}^2. \tag{2.65}$$

2.5 Action-Angle Variables and Beam Emittance

For the normal form Hamiltonian (2.36) we have seen that a constant of the motion exists, and we have found it explicitly. It is important to construct the action-angle variables for this problem, since they represent a useful technique not only to approach the linear betatron motion from a different point of view, but also for the subsequent exposition. We require that the new Hamiltonian is a linear function of the action variable J alone

$$\overline{H}_2 = \dot{\chi} J. \tag{2.66}$$

Note that the action J plays now the role of the new momentum variable.

To determine the generating function of the first kind $F_1(X, \alpha; \theta)$, where α is the angle variable, we write down the *Hamilton-Jacobi* equation

$$\frac{\partial F_1}{\partial \theta} + \dot{\chi}\frac{\partial F_1}{\partial \alpha} + \frac{\dot{\chi}}{2}\left[\left(\frac{\partial F_1}{\partial X}\right)^2 + X^2\right] = 0, \tag{2.67}$$

following from the last of equations (1.30). The obvious ansatz

$$F_1(X, \alpha; \theta) = \frac{1}{2}X^2 S(\alpha; \theta)$$

yields an equation

$$\frac{\partial S}{\partial \theta} + \dot{\chi}\frac{\partial S}{\partial \alpha} + \dot{\chi}(1 + S^2) = 0$$

for the function $S(\alpha; \theta)$. The latter can be integrated in a straightforward manner to give

$$S(\alpha; \theta) = -\tan\alpha,$$

so that

$$F_1(X, \alpha; \theta) = -\frac{X^2}{2}\tan\alpha. \tag{2.68}$$

The complete set of transformation equations becomes

$$X = \sqrt{2J}\cos\alpha, \qquad\qquad P = -\sqrt{2J}\sin\alpha, \tag{2.69}$$

or

$$J = \frac{1}{2}\left(X^2 + P^2\right), \qquad\qquad \alpha = -\arctan\left(\frac{P}{X}\right). \tag{2.70}$$

In the new action-angle canonical variables the solution to the *Hamilton*'s equations of motion is

$$J = J_0 = \text{const}, \qquad\qquad \alpha(\theta) = \chi(\theta) + \alpha_0. \tag{2.71}$$

Thus, we have constructed explicitly an invariant J, which is half of the *Courant-Snyder* invariant (2.64).

To acquire a clearer notion of the meaning of the action invariant consider a single particle propagating through the periodic focusing structure of an accelerator. At every location along the circumference of the machine its normalized coordinate X and normalized momentum P lie on a circle in normalized phase space, prescribed by equation (2.70). However, the locus of points in $\widehat{x}$-$\widehat{p}$ phase space representing the particle's actual position and momentum (neglecting the dispersion) is an ellipse, specified by equation (2.65). As the particle traverses the accelerator structure, this ellipse also evolves such that the area it encloses remains constant. The invariant J is simply related to that area

$$S_{enclosed} = 2\pi J. \tag{2.72}$$

An important quantity in accelerator and storage ring terminology is the beam *emittance.* Nevertheless closely related to the action invariant, the emittance characterizes an ensemble of particles. It is well-known [Klimontovich (1986)] that the equilibrium distribution of particles in phase space is a generic function of the invariant(s) of the motion. We choose the *grand canonical Gibbs ensemble* [Tzenov (1998a)]

$$f(\widehat{x}, \widehat{p}; \theta) = \frac{1}{2\pi\epsilon}\exp\left[-\frac{I(\widehat{x}, \widehat{p}; \theta)}{2\epsilon}\right] \tag{2.73}$$

to represent the equilibrium phase space density of particles in an accelerator, where $I(\widehat{x}, \widehat{p}; \theta)$ is the *Courant-Snyder* invariant (2.65) and ϵ is the beam emittance. Integrating over $\widehat{p}$ we are left with the particle density distribution in configuration space

$$\rho(\widehat{x};\theta) = \frac{1}{\sqrt{2\pi\epsilon\beta(\theta)}} \exp\left[-\frac{\widehat{x}^2}{2\epsilon\beta(\theta)}\right]. \tag{2.74}$$

Therefore, the emittance ϵ is related to the *rms* beam size as

$$\sigma^2(\theta) = \epsilon\beta(\theta). \tag{2.75}$$

In terms of the action variable J the above equation can be rewritten as

$$\epsilon = \langle J \rangle, \tag{2.76}$$

where the bracket indicates an average over the phase space density (2.73).

It is instructive to write down an equation for the *rms* beam size. To derive this equation we use the definition (2.75) and take into account equation (2.43) satisfied by the β-function. We find

$$\frac{\mathrm{d}^2\sigma}{\mathrm{d}\theta^2} + G\sigma = \frac{R^2\epsilon^2}{\sigma^3}. \tag{2.77}$$

This is the so-called *envelope* equation. It governs the evolution of the *rms* beam size during the beam propagation through the periodic accelerator lattice.

2.6 Adiabatic Damping of Betatron Oscillations

In the last paragraph of the previous chapter the synchro-betatron formalism was developed under the basic assumption that the (design) momentum of the synchronous particle is constant. This was further elaborated in the present chapter to build adequate tools for the description of the linear betatron motion. Particles in accelerators are being accelerated, besides, the acceleration time is much longer compared to the revolution period or the period of betatron oscillations. It is interesting and important for direct applications in practice to study the effect of slow increase of the design momentum p_{0s} together with the slow increase of the confining magnetic fields (proportional to p_{0s}) on the transverse beam dynamics.

To see the effect mentioned above we first recall that the Hamiltonian (1.77) can be written as

$$\widehat{H}_2 = \frac{R}{2p_{0s}}\widehat{p}^2 + \frac{p_{0s}}{2R}G\widehat{x}^2, \tag{2.78}$$

where now $\widehat{H}_2$ is the corresponding part of the original Hamiltonian (1.11) and $\widehat{p}$ is the actual horizontal canonical momentum without the scaling (1.53). To make things simple we neglect in what follows the effect of slow acceleration on the dispersion function[2]. In the *Hamilton*'s equations of motion derived from the Hamiltonian (2.78) we eliminate the momentum $\widehat{p}$ and obtain

$$\frac{\mathrm{d}^2\widehat{x}}{\mathrm{d}\theta^2} + \frac{\dot{p}_{0s}}{p_{0s}}\frac{\mathrm{d}\widehat{x}}{\mathrm{d}\theta} + G\widehat{x} = 0. \tag{2.79}$$

Due to the presence of a first derivative term, equation (2.79) describes a damped oscillator with periodically varying frequency. For that reason the effect of acceleration on the betatron motion is called adiabatic damping of betatron oscillations.

A few remarks are now in order. First of all, note that the coefficient $\dot{p}_{0s}/p_{0s}$ multiplying the first derivative term in equation (2.79) is positive in the case of acceleration, so that indeed a process of damping takes place. A natural question therefore arises; where does the damping come from, in spite of the fact that we are dealing with a Hamiltonian system. It is however known that dissipative systems can be equivalently described by Hamiltonian systems with *time-dependent mass* entering the Hamiltonian. This is the so-called *Caldirola-Kanay Hamiltonian* which is not the energy of the original system. Careful inspection of the Hamiltonian (2.78) shows that we have a similar case here with the only difference that our Hamiltonian is in fact, the energy of the system. More extensive discussion and related references can be found in [Tzenov (1997a)].

To proceed further we perform a canonical transformation specified by the generating function

$$\overline{F}_2(\widehat{x}, \overline{p}; \theta) = \widehat{x}\overline{p}\sqrt{p_{0s}} - \frac{\dot{p}_{0s}\widehat{x}^2}{4R} \tag{2.80}$$

and cast the Hamiltonian (2.78) into the form

$$\overline{H}_2 = \frac{R}{2}\overline{p}^2 + \frac{\overline{G}}{2R}\overline{x}^2, \tag{2.81}$$

similar to the Hamiltonian given by equation (1.77). Now the focusing

[2]It is possible to reformulate the synchro-betatron formalism taking into account acceleration effects [Tzenov (1990)]. This is crucial in particular to study the so-called *accelerated orbits* in isochronous cyclotrons.

strength $\overline{G}$ is modified as

$$\overline{G} = G + \frac{\dot{p}_{0s}^2}{4p_{0s}^2} - \frac{\ddot{p}_{0s}}{2p_{0s}}, \tag{2.82}$$

and the transformation equations are

$$\widehat{x} = \frac{\overline{x}}{\sqrt{p_{0s}}}, \qquad \widehat{p} = \overline{p}\sqrt{p_{0s}} - \frac{\dot{p}_{0s}\overline{x}}{2R\sqrt{p_{0s}}}. \tag{2.83}$$

Note that if we assume a nearly linear increase of the design momentum [so that the last term in equation (2.82) can be neglected] the acceleration (or deceleration) has an effect of increasing the focusing strength.

We can in principle repeat the analysis of the previous two paragraphs with a starting point the Hamiltonian (2.81), taking into account only small constant contribution (which is not essential, but simplest for a direct parallel) to the focusing strength that comes from the second and the third terms in equation (2.82). What we will find finally is that the *rms* beam size (in the original $\widehat{x}$, $\widehat{p}$ coordinates) is inversely proportional to the square root of the design momentum, that is

$$\sigma^2(\theta) = \frac{\epsilon\beta(\theta)}{p_{0s}(\theta)}. \tag{2.84}$$

Equation (2.84) provides another reason to call the effect we are considering here the adiabatic damping of betatron oscillations. Due to this variation it is convenient to introduce an auxiliary quantity

$$\epsilon_N = \beta_s\gamma_s\epsilon, \tag{2.85}$$

called the *normalized emittance*, which is constant.

Chapter 3

Nonlinear Resonances of Betatron Oscillations

3.1 Introduction

Nonlinear effects of charged particle dynamics in accelerators and storage rings are contingent on the presence of nonlinear terms in the equations of motion. In general, sources of such terms are not only various types of nonlinearities in the magnetic structure of the machine, but also the relativistic form of the equations of motion itself. From these considerations a very important conclusion follows, that the dynamics of particles in modern high-energy accelerators and storage rings is essentially *nonlinear*. Yet, a third source of nonlinear behavior are the self and image fields driven by high intensity beams, as well as the electromagnetic interaction between colliding bunches in the interaction region of $p\overline{p}$ and $e\overline{e}$ colliders.

The concept of resonance is usually related to the phenomenon of interaction between two oscillation modes with equal or close eigenfrequencies. In the linear case resonance phenomena are intuitively clear, because of their simple properties and physical demonstrativeness. If the linear part of a nonlinear Hamiltonian system is stable, it is always possible to define the oscillation eigenfrequencies[1]. Subsequently, the scenario according to which the nonlinear resonance develops reminds to a large extent the linear case, except for two major particulars. First, a specific nonlinear term enters so to speak seriously into play, whenever a certain condition between the eigenfrequencies is fulfilled. Sometimes it is said that there exists a one to one correspondence between a particular nonlinear term and a specific nonlinear resonance, once the eigenfrequencies in linear approximation are known. The second specific feature consists in the fact that the resonance

[1] This assertion follows from the symplecticity of the transfer matrix for Hamiltonian systems.

frequency *depends* on the oscillation amplitude. This however is one of the basic characteristics of a nonlinear resonance.

The motion of particles in cyclic accelerators and storage rings is in general accomplished in three degrees of freedom. Therefore, there exist three types of resonances: one-dimensional, two-dimensional and three-dimensional (or synchro-betatron) resonances. If the nonlinear resonance involves only the two transverse degrees of freedom, it is usually said that a nonlinear resonance of betatron oscillations takes place. If in addition the longitudinal degree of freedom is included in the process the resonance is called *synchro-betatron resonance.* As we already briefly mentioned in previous chapters synchro-betatron resonances are excited: i) in cavities when the dispersion function and its derivative with respect to the azimuth are nonzero simultaneously, ii) in the case of interaction between beam particles and self and image fields, iii) in the case of beam-beam interaction with a nonzero dispersion at the interaction point, as well as in the case of beam-beam interaction at a crossing angle.

3.2 General Description and Basic Properties of a Nonlinear Resonance

In the present chapter we consider mainly nonlinear resonances of betatron oscillations. In chapter 1 we derived the Hamiltonian governing the betatron motion

$$\widehat{H} = \widehat{H}_2 + \widehat{H}_p, \tag{3.1}$$

where $\widehat{H}_2$ is the Hamiltonian (1.77) describing the linear betatron oscillations. The part $\widehat{H}_p$ collects all nonlinear terms and as we already saw it can be written as a sum of homogeneous polynomials in the canonical variables

$$\widehat{H}_p = \sum_{I=2}^{\infty} \sum_{\substack{k,l,m,n=0 \\ k+l+m+n=I}}^{I} a_{klmn}^{(I)}(\theta) \widehat{x}^k \widehat{p}_x^l \widehat{z}^m \widehat{p}_z^n, \tag{3.2}$$

where $a_{klmn}^{(I)}(\theta)$ are periodic functions of θ with period 2π. Note that we have included in $\widehat{H}_p$ quadratic terms (summation over I starts from $I = 2$) that conventionally belong to $\widehat{H}_2$. However, these are non structural terms due to quadrupole errors of various character, which have periodicity 2π (unlike the periodicity $2\pi/N$ of structural terms in $\widehat{H}_2$) and are usually small.

It will prove convenient for later purposes to remove the explicit "time" dependence in $\widehat{H}_2$. To achieve that, instead of a canonical transformation specified by the generating function (2.68) we consider the following one

$$F_1(U, a_u; \theta) = -\frac{1}{2} \sum_{u=(x,z)} U^2 \tan \left[a_u + \chi_u(\theta) - \nu_u \theta\right]. \tag{3.3}$$

Here again u refers to either x or z, a_u is the angle variable, $\chi_u(\theta)$ and ν_u are the phase advance (2.46) and the betatron tune (2.56), respectively. Combining equations (2.38) and (2.69), we obtain

$$\widehat{u} = \sqrt{2\beta_u J_u} \cos \psi_u, \qquad \widehat{p}_u = -\sqrt{\frac{2J_u}{\beta_u}} (\sin \psi_u + \alpha_u \cos \psi_u), \tag{3.4}$$

where

$$\psi_u(a_u; \theta) = a_u + \chi_u(\theta) - \nu_u \theta. \tag{3.5}$$

Substitution of equation (3.4) into equation (3.2) yields

$$\widehat{H}_p = \sum_{I=2}^{\infty} \sum_{\substack{K,L,M,N=0 \\ K+L+M+N=I}}^{I} h_{KLMN}^{(I)}(\theta)(2J_x)^{(K+L)/2}(2J_z)^{(M+N)/2}$$

$$\times e^{i[(K-L)a_x + (M-N)a_z]}, \tag{3.6}$$

where

$$h_{KLMN}^{(I)}(\theta) = \frac{1}{2^I} \sum_{k,l,m,n} (-1)^{l+n} a_{klmn}^{(I)}(\theta) \beta_x^{(k-l)/2} \beta_z^{(m-n)/2}$$

$$\times \sum_{k'=0}^{k} \binom{k}{k'} \sum_{m'=0}^{m} \binom{m}{m'} \sum_{l'=0}^{l} \sum_{n'=0}^{n} \binom{l}{l'} \binom{n}{n'} (\alpha_x - i)^{l-l'} (\alpha_x + i)^{l'}$$

$$\times (\alpha_z - i)^{n-n'} (\alpha_z + i)^{n'} e^{i[(K-L)(\chi_x - \nu_x\theta) + (M-N)(\chi_z - \nu_z\theta)]}, \tag{3.7}$$

and the summation indices are related via the expressions

$$K+L = k+l, \qquad M+N = m+n, \qquad k'+l' = L, \qquad m'+n' = N. \tag{3.8}$$

Since the coefficients $h^{(I)}_{KLMN}(\theta)$ are periodic functions of θ, they can be expanded in *Fourier* series. Then equation (3.6) can be rewritten as

$$\widehat{H}_p = \sum_{I=2}^{\infty} \sum_{\substack{K,L,M,N=0 \\ K+L+M+N=I}}^{I} \sum_{q=-\infty}^{\infty} h^{(I)}_{KLMNq} (2J_x)^{(K+L)/2} (2J_z)^{(M+N)/2}$$

$$\times e^{i[(K-L)a_x + (M-N)a_z - q\theta]}, \tag{3.9}$$

where

$$h^{(I)}_{KLMNq} = \frac{1}{2\pi} \int_0^{2\pi} d\theta h^{(I)}_{KLMN}(\theta) e^{iq\theta}. \tag{3.10}$$

Noting that $h^{(I)}_{KLMN-q} = h^{(I)*}_{KLMNq}$, where $h^{(I)*}_{KLMNq}$ denotes the complex conjugated of $h^{(I)}_{KLMNq}$, we cast equation (3.9) into the form:

$$\widehat{H}_p = 2\sum_{I=2}^{\infty} \sum_{\substack{K,L,M,N=0 \\ K+L+M+N=I}}^{I} \sum_{q=0}^{\infty} \left| h^{(I)}_{KLMNq} \right| (2J_x)^{(K+L)/2} (2J_z)^{(M+N)/2}$$

$$\times \cos\left[(K-L)a_x + (M-N)a_z - q\theta + \arg\left(h^{(I)}_{KLMNq} \right) \right], \tag{3.11}$$

A closer inspection of the above expression shows that the perturbing Hamiltonian $\widehat{H}_p$ consists of two parts:

(1) Stabilizing part of the perturbing Hamiltonian

$$\widehat{H}_{ps} = \sum_{I} \sum_{r+s=I} h^{(2I)}_{rrss0} (2J_x)^r (2J_z)^s. \tag{3.12}$$

(2) Driving resonant part of the perturbing Hamiltonian including all resonance terms in equation (3.11).

Let us now assume that for some values of K, L, M and N the betatron tunes satisfy exactly, or are close to satisfying the equation $(K-L)\nu_x + (M-N)\nu_z = q$. The rest of the terms in (3.11) are fast oscillating, and do not contribute to the dynamics to the lowest order of approximation. If only the largest term for which the above equation holds is retained in $\widehat{H}_p$, we thereby *isolate* the corresponding resonance. Thus, the Hamiltonian for an isolated resonance can be written as

$$\mathcal{H} = \mathcal{H}_0 + \mathcal{H}_s + \mathcal{H}_1, \tag{3.13}$$

where

$$\mathcal{H}_0 = \nu_1 J_1 + \nu_2 J_2, \qquad \mathcal{H}_s = \widehat{H}_{ps}, \tag{3.14}$$

$$\mathcal{H}_1 = \mathcal{D}(2J_1)^{p_1}(2J_2)^{p_2} \cos(2p_1\alpha_1 \pm 2p_2\alpha_2 - m\theta + \delta_{\mathcal{D}}). \tag{3.15}$$

For convenience of notation hereafter we use indices 1 and 2 instead of x and z respectively, and denote the action-angle variables by (α_1, J_1) and (α_2, J_2). Moreover $p_i = 1/2, 1, 3/2, \ldots$ $(i = 1, 2)$, the "+" sign corresponds to *sum resonances*, while the "–" sign corresponds to *difference resonances.* The relation

$$2p_1\nu_1 \pm 2p_2\nu_2 = m + \varepsilon \tag{3.16}$$

is usually called the *resonance condition*, where ε is the *resonance detuning.* The quantity $N^* = 2p_1 + 2p_2$ determines the order of the resonance.

Intuitively it is clear that the further a dynamical system is from a certain resonance (in other words, the larger the resonance detuning is), the weaker the system feels that resonance. Besides that, the smaller the amplitude $\mathcal{D}$ of the driving term is, the weaker is its influence on the system. Therefore, a condition relating the quantities ε and $\mathcal{D}$ must exist. Once this condition is fulfilled the motion of the system described by the Hamiltonian (3.13) is stable. Formulated in a different way, these considerations imply that the motion is periodic and *invariant tori* exist (KAM theorem [Lichtenberg (1982)]), if the amplitude of the driving resonance term is sufficiently small and the system is far enough from the resonance.

In most of the early papers on the subject [Schoch (1957)], [Hagedorn (1957)], [Guignard (1978)] the basic idea to study the stability region of an isolated nonlinear resonance consists in finding invariant(s) in *involution*[2] with the Hamiltonian, from which the so-called *resonance bandwidth* is obtained. The bandwidth is the lower limit of the resonance detuning beyond which the motion is unstable. The main drawback of the resonance bandwidth is the fact that it is written in a non invariant form, depending on the initial amplitudes of betatron oscillations. Since the amplitudes of

[2]Two dynamic variables A and B are said to be in involution if their *Poisson* bracket is equal to zero.

betatron oscillations change as time passes, at some later instant of time (chosen now as the initial one) we would have completely different stability criterion. Therefore, the correct procedure is to define a stability region in the space of control parameters, nevertheless sometimes in practice the resonance bandwidths yield satisfactory stability estimates.

In order to eliminate the explicit time dependence in equation (3.15) we perform the canonical transformation

$$\widetilde{E}_2\left(\alpha_1, \alpha_2, \widetilde{J}_1, \widetilde{J}_2; \theta\right) = [\alpha_1 + (\varepsilon_1 - \nu_1)\theta + \delta_1]\widetilde{J}_1$$
$$+[\alpha_2 + (\pm\varepsilon_2 - \nu_2)\theta \pm \delta_2]\widetilde{J}_2, \tag{3.17}$$

where

$$\varepsilon_i = \frac{\varepsilon}{4p_i}, \qquad \delta_i = \frac{\delta_{\mathcal{D}}}{4p_i}. \qquad (i = 1, 2) \tag{3.18}$$

We obtain

$$\widetilde{\mathcal{H}}_0 = \varepsilon_1 J_1 \pm \varepsilon_2 J_2, \qquad \widetilde{\mathcal{H}}_s = \mathcal{H}_s, \tag{3.19}$$

$$\widetilde{\mathcal{H}}_1 = \mathcal{D}(2J_1)^{p_1}(2J_2)^{p_2} \cos(2p_1\widetilde{\alpha}_1 \pm 2p_2\widetilde{\alpha}_2). \tag{3.20}$$

The subsequent canonical transformation given by the generating function [Lichtenberg (1982)]

$$S_2\left(\widetilde{\alpha}_1, \widetilde{\alpha}_2, \overline{J}_1, \overline{J}_2\right) = (2p_1\widetilde{\alpha}_1 \pm 2p_2\widetilde{\alpha}_2)\overline{J}_1 \mp p_2\widetilde{\alpha}_2\overline{J}_2 \tag{3.21}$$

specifies the resonant canonical variables

$$\overline{\alpha}_1 = 2p_1\widetilde{\alpha}_1 \pm 2p_2\widetilde{\alpha}_2, \qquad \overline{\alpha}_2 = \mp p_2\widetilde{\alpha}_2, \tag{3.22}$$

$$J_1 = 2p_1\overline{J}_1, \qquad J_2 = \pm p_2\left(2\overline{J}_1 - \overline{J}_2\right), \tag{3.23}$$

and the resonant Hamiltonian

$$\overline{\mathcal{H}} = \varepsilon\overline{J}_1 - \frac{\varepsilon\overline{J}_2}{4} + \mathcal{H}_s\left(\overline{J}_1, \overline{J}_2\right) + \mathcal{D}\left(4p_1\overline{J}_1\right)^{p_1}\left[\pm 2p_2\left(2\overline{J}_1 - \overline{J}_2\right)\right]^{p_2} \cos\overline{\alpha}_1. \tag{3.24}$$

Inspecting the resonant Hamiltonian (3.24) it is straightforward to observe that it does not depend on the angle $\overline{\alpha}_2$. Therefore, $\overline{J}_2$ is an integral of motion

$$\overline{J}_2 = \frac{J_1}{p_1} \mp \frac{J_2}{p_2} = C_\pm = \text{const.} \tag{3.25}$$

The existence of the invariant (3.25) actually reduces the analysis of an isolated nonlinear resonance of betatron oscillations to a problem in one degree of freedom. As far as the Hamiltonian (3.24) does not depend explicitly on θ, the dynamical system it governs is integrable. The invariant (3.25) ensures that in the case of difference resonances the amplitudes J_1 and J_2 cannot grow infinitely. If one of them, say J_1 decreases by a certain amount, the other one J_2 increases by the same amount multiplied by p_2/p_1, so that their sum remains constant. For the sum resonances there is no such mechanism of amplitude redistribution, and for that reason they are more dangerous compared to the difference resonances.

3.3 The Method of Effective Potential

Let us first write down the *Hamilton*'s equations of motion following from the resonant Hamiltonian (3.24)

$$\frac{\mathrm{d}\overline{\alpha}_1}{\mathrm{d}\theta} = \frac{\partial \overline{\mathcal{H}}}{\partial \overline{J}_1} = \varepsilon + \frac{\partial \mathcal{H}_s}{\partial \overline{J}_1} + \frac{\partial F_{\mathcal{D}}}{\partial \overline{J}_1} \cos \overline{\alpha}_1, \tag{3.26}$$

$$\frac{\mathrm{d}\overline{J}_1}{\mathrm{d}\theta} = -\frac{\partial \overline{\mathcal{H}}}{\partial \overline{\alpha}_1} = F_{\mathcal{D}} \sin \overline{\alpha}_1, \tag{3.27}$$

where to simplify notations we have introduced the quantity

$$F_{\mathcal{D}}(\overline{J}_1) = \mathcal{D}(4p_1\overline{J}_1)^{p_1} \left[\pm 2p_2(2\overline{J}_1 - C_{\pm})\right]^{p_2}. \tag{3.28}$$

Differentiating equation (3.27) with respect to θ and making use of equations (3.26) and (3.24) to eliminate the dependence on the angle variable $\overline{\alpha}_1$ we obtain [Tzenov (1992b)]

$$\frac{\mathrm{d}^2\overline{J}_1}{\mathrm{d}\theta^2} = \frac{1}{2}\frac{\partial F_{\mathcal{D}}^2}{\partial \overline{J}_1} + \left(\varepsilon + \frac{\partial \mathcal{H}_s}{\partial \overline{J}_1}\right)\left(\mathcal{A}_{\pm} - \varepsilon \overline{J}_1 - \mathcal{H}_s\right), \tag{3.29}$$

where the new invariant $\mathcal{A}_{\pm}$ is a linear combination of $C_{\pm}$ and $\overline{\mathcal{H}}$ (being also an invariant of motion) and has the form

$$\mathcal{A}_{\pm} = \overline{\mathcal{H}} + \frac{\varepsilon C_{\pm}}{4}. \tag{3.30}$$

It is easy to verify that equation (3.29) can be formally derived from the *effective Hamiltonian*

$$\mathcal{H}_{eff}(\overline{J}_1, p_0) = \frac{p_0^2}{2} + \mathcal{V}_{eff}(\overline{J}_1), \tag{3.31}$$

where

$$\mathcal{V}_{eff}(\overline{J}_1) = -\frac{F_{\mathcal{D}}^2 - \mathcal{H}_s^2}{2} + \frac{\varepsilon^2 \overline{J}_1^2}{2} - \varepsilon \overline{J}_1(\mathcal{A}_\pm - \mathcal{H}_s) - \mathcal{A}_\pm \mathcal{H}_s \tag{3.32}$$

is the *effective potential.* Note that $(\overline{J}_1, p_0)$ is a pair of canonical variables, where the action $\overline{J}_1$ plays the role of the canonical coordinate. Using the *Hamilton*' s equations

$$\frac{d\overline{J}_1}{d\theta} = \frac{\partial \mathcal{H}_{eff}}{\partial p_0} = p_0, \qquad \frac{dp_0}{d\theta} = -\frac{\partial \mathcal{H}_{eff}}{\partial \overline{J}_1} = -\frac{\partial \mathcal{V}_{eff}}{\partial \overline{J}_1}$$

for the system governed by (3.31) it is not difficult to check that the "effective energy" is non-positive, namely:

$$\mathcal{H}_{eff} = -\frac{\mathcal{A}_\pm^2}{2} \leq 0. \tag{3.33}$$

This result will be of major importance in what follows.

The basic condition we will further impose on the effective potential $\mathcal{V}_{eff}$ concerns the possibility of forming a sufficiently deep potential well in which finite motion occurs. In the case of $\mathcal{H}_s = 0$ from equations (3.28) and (3.32) it is evident that the effective potential is a polynomial of $\overline{J}_1$ of order N^*. The analysis of the stability of betatron oscillations, that is the conditions for potential well formation is carried out in a relatively simple way for nonlinear resonances of third and fourth order. This will be presented in the next paragraph. However, the study of fifth and higher order resonances encounters certain difficulties of computational character, nevertheless in some special cases an exact treatment is still possible.

3.4 Stability Analysis of Third and Fourth Order Resonances

In this paragraph we study the most unfavorable case of a pure resonance without stabilization ($\mathcal{H}_s = 0$). Let us consider the third order resonance $\nu_1 + 2\nu_2 = m + \varepsilon$. The effective potential takes the explicit form

$$\mathcal{V}_{eff}(\overline{J}_1) = -16\mathcal{D}^2\overline{J}_1^3 + \frac{1}{2}(\varepsilon^2 + 32C_+\mathcal{D}^2)\overline{J}_1^2 - (\varepsilon\mathcal{A}_+ + 4C_+^2\mathcal{D}^2)\overline{J}_1. \tag{3.34}$$

It is well-known that the potential function (3.34) is a *fold catastrophe* [Thom (1975)] up to a linear coordinate transformation. Its structural stability is achieved if the discriminant of the equation

$$\frac{\partial \mathcal{V}_{eff}}{\partial \overline{J}_1} = 0$$

is positive. This gives:

$$a = \frac{1}{4}\left(\varepsilon^2 + 32C_+\mathcal{D}^2\right)^2 - 48\mathcal{D}^2\left(\varepsilon\mathcal{A}_+ + 4C_+^2\mathcal{D}^2\right) > 0. \tag{3.35}$$

The condition (3.35) however, turns out to be insufficient for the stability of the motion on the whole. Since, $\mathcal{H}_{eff} \leq 0$ [see equation (3.33)] we must have

$$\mathcal{V}_{eff}\left(\overline{J}_1^{(2)}\right) > 0, \tag{3.36}$$

where

$$\overline{J}_1^{(1,2)} = \frac{1}{48\mathcal{D}^2}\left[\frac{1}{2}\left(\varepsilon^2 + 32C_+\mathcal{D}^2\right) \mp \sqrt{a}\right] \tag{3.37}$$

are the extremal points of the effective potential. In other words, if the effective potential is positive for the value of $\overline{J}_1$ for which it has a maximum, the right turning point does exist, provided the relation (3.33) is valid. It is easy to check that equations (3.35) and (3.36) hold simultaneously if

$$a > \frac{1}{16}\left(\varepsilon^2 + 32C_+\mathcal{D}^2\right)^2. \tag{3.38}$$

The condition (3.38) determines the stability domain in the control parameter space $(\varepsilon, \mathcal{D}, C_+, \mathcal{A}_+)$.

The next step is to cast the effective potential (3.34) to the canonical form [Gilmore (1981)] of the fold catastrophe by means of the simple canonical transformation

$$\mathcal{G}_2\left(\overline{J}_1, P_\xi\right) = \frac{1}{\xi_1}\left(\overline{J}_1 + \xi_2\right)P_\xi,$$

where

$$\xi_1^3 = \frac{1}{48\mathcal{D}^2}, \qquad \xi_2 = -\frac{\varepsilon^2 + 32C_+\mathcal{D}^2}{96\mathcal{D}^2}. \tag{3.39}$$

The new effective Hamiltonian can be written as

$$\mathcal{H}_\xi(\xi, P_\xi) = \frac{P_\xi^2}{2\xi_1^2} + \mathcal{V}_\xi(\xi), \qquad \mathcal{V}_\xi(\xi) = -\frac{\xi^3}{3} + \frac{a\xi_1}{48\mathcal{D}^2}\xi. \tag{3.40}$$

The oscillation period of finite motion determined by the condition (3.38) is found to be

$$\Theta = 2\xi_1 \int\limits_{\xi_0}^{\xi^{(1)}} \frac{d\xi}{\sqrt{2[\mathcal{H}_\xi - \mathcal{V}_\xi(\xi)]}} = 2\xi_1 \sqrt{\frac{6}{\xi^{(2)} - \xi_0}} \mathbf{K}(k), \tag{3.41}$$

where

$$\xi_0 = \frac{\overline{J}_{10} + \xi_2}{\xi_1}, \qquad \xi^{(1,2)} = -\frac{\xi_0}{2} \mp \frac{1}{2}\sqrt{\frac{a\xi_1}{4\mathcal{D}^2} - 3\xi_0^2}, \tag{3.42}$$

and

$$k = \sqrt{\frac{\xi^{(1)} - \xi_0}{\xi^{(2)} - \xi_0}}. \tag{3.43}$$

In the above equation (3.41) $\mathbf{K}(k)$ denotes the complete elliptic integral of the first kind [Abramowitz (1984)] and $\overline{J}_{10}$ is the initial value of the action $\overline{J}_1$. The amplitude of resonance beating is given by the expression

$$\text{Amp}(\overline{J}_1) = \frac{\xi_1}{2}\left(\xi^{(1)} - \xi_0\right) = -\frac{\xi_1}{4}\left(3\xi_0 + \sqrt{\frac{a\xi_1}{4\mathcal{D}^2} - 3\xi_0^2}\right). \tag{3.44}$$

At this point we introduce a rather important characteristic of the nonlinear resonance. We define the *evolution time* of the nonlinear resonance as the number of revolutions, indispensable for the amplitude to grow to infinity in the case of degeneration (i.e. $a = 0$) of the effective potential (3.40). The solution to the *Hamilton*'s equations following from the Hamiltonian (3.40) when $a = 0$ is readily obtained in the form

$$\theta(\xi) = -\frac{\xi_1}{3^{1/4}}\sqrt{\frac{3}{2|\xi_0|}}[\mathbf{F}(\phi, k_1) + \mathbf{F}(1, k_1)], \tag{3.45}$$

where now

$$k_1 = \sin\left(\frac{5\pi}{12}\right) = \frac{\sqrt{2+\sqrt{3}}}{2}, \qquad \phi = \frac{\xi + (1-\sqrt{3})|\xi_0|}{\xi + (1+\sqrt{3})|\xi_0|}. \tag{3.46}$$

For the number of revolutions determining the evolution time one obtains

$$N_\infty = \frac{1}{2\pi}\lim_{\xi\to\infty}|\theta(\xi)| = \frac{1}{2\pi}\frac{\xi_1}{3^{1/4}}\sqrt{\frac{6}{|\xi_0|}}\mathbf{F}(1, k_1). \tag{3.47}$$

In the above expressions (3.45) and (3.47) $\mathbf{F}(\phi, k_1)$ denotes the incomplete elliptic integral of the first kind [Abramowitz (1984)]. Although the evolution time thus defined is a rather idealized characteristic of the nonlinear resonance, it can be regarded as a good and useful in practice estimate to assess how dangerous a particular resonance is.

It is worth noting that the result (3.38) can be obtained if one imposes the condition for the effective potential (3.34) to cross the zero level of the effective energy. This means that the algebraic equation

$$\mathcal{V}_e(\overline{J}_1) = 0 \tag{3.48}$$

must possess three real roots. One of them is always equal to zero, and therefore the problem of finding the other two can be reduced to the analysis of a simple quadratic equation. It is not difficult to show that the condition for the two nonzero roots of equation (3.48) to be real is exactly the same as the inequality (3.38). This observation simplifies considerably the treatment of higher order nonlinear resonances. Taking into account equation (3.33) and the shape of the effective potential curve, one can see from simple geometric considerations that stable motion for a nonlinear resonance of order N^* occurs if equation (3.48) has in general $N^* - 1$ real nonzero roots.

Not going into details for the resonances of fourth order we find the stability domain in the form:

(1) For the resonance $\nu_1 + 3\nu_2 = m + \varepsilon$

$$\left(\frac{\varepsilon^2}{36}\right)^3 > \frac{\varepsilon^2 \mathcal{D}^2}{4}\left(\mathcal{A}_+ - \frac{\varepsilon C_+}{4}\right)^2, \tag{3.49}$$

(2) For the resonance $2\nu_1 + 2\nu_2 = m + \varepsilon$

$$\frac{1}{27}\left(\frac{C_+^2}{3} + \frac{\varepsilon^2}{64\mathcal{D}^2}\right)^3 > \left(\frac{C_+^3}{27} + \frac{\varepsilon \mathcal{A}_+}{32\mathcal{D}^2} - \frac{\varepsilon^2 C_+}{192\mathcal{D}^2}\right)^2. \tag{3.50}$$

In quite the same way one can proceed further with the analysis of the resonances of fifth order. It appears that the generalization of the above procedure for nonlinear resonances of higher orders is a difficult (if not impossible) task. One faces the problem of involving more sophisticated tools of catastrophe theory, which sometimes lead to very complicated results inconvenient for practical use.

3.5 The Method of Successive Linearization

As we stated before, the analysis of the stability domain in the case of higher order nonlinear resonances encounters serious computational difficulties. Therefore, a necessity arises in finding an alternative way to obtain a criterion that yields satisfactory stability estimates and is simple and effective to apply for practical purposes. A good candidate for such a technique is the *method of successive linearization* [Guignard (1986)], which gives as we will see below excellent results.

First of all, let us demonstrate the method of successive linearization on the third order resonance $\nu_1 + 2\nu_2 = m + \varepsilon$, examined already in the previous paragraph. Equation (3.29) becomes

$$\frac{\mathrm{d}^2 \overline{J}_1}{\mathrm{d}\theta^2} = 48\mathcal{D}^2 \overline{J}_1^2 - \left(\varepsilon^2 + 32C_+\mathcal{D}^2\right)\overline{J}_1 + \varepsilon\mathcal{A}_+ + 4C_+^2\mathcal{D}^2. \tag{3.51}$$

In practice the nonlinear (in $\overline{J}_1$) part on the right-hand-side of equation (3.51) can be regarded as small compared to the other terms, and an approximate solution by means of the first linearization

$$\frac{\mathrm{d}^2 x^{(0)}}{\mathrm{d}\theta^2} = -\left(\varepsilon^2 + 32C_+\mathcal{D}^2\right)x^{(0)} + \varepsilon\mathcal{A}_+ + 4C_+^2\mathcal{D}^2 \tag{3.52}$$

can be found. If the expression in the brackets on the right-hand-side of equation (3.52) is positive[3] we can write the particular solution of the linearized equation as

$$x^{(0)} = \frac{\varepsilon\mathcal{A}_+ + 4C_+^2\mathcal{D}^2}{\varepsilon^2 + 32C_+\mathcal{D}^2}. \tag{3.53}$$

The general solution of equation (3.51) can be written as

$$\overline{J}_1(\theta) = u(\theta) + x^{(0)},$$

where the function $u(\theta)$ satisfies the equation

$$\frac{\mathrm{d}^2 u}{\mathrm{d}\theta^2} = -\left(\varepsilon^2 + 32C_+\mathcal{D}^2 - 96\mathcal{D}^2 x^{(0)}\right)u + 48\mathcal{D}^2\left(u^2 + x^{(0)2}\right). \tag{3.54}$$

Performing the second linearization we obtain

$$\frac{\mathrm{d}^2 u^{(0)}}{\mathrm{d}\theta^2} = -\left(\varepsilon^2 + 32C_+\mathcal{D}^2 - 96\mathcal{D}^2 x^{(0)}\right)u^{(0)} + 48\mathcal{D}^2 x^{(0)2}. \tag{3.55}$$

[3] This requirement ensures that the general solution of equation (3.52) is stable. Note that it represents the stability criterion to the lowest order.

As a result of the two linearizations the approximate solution of equation (3.51) is simply

$$\overline{J}_1(\theta) = x^{(0)} + u^{(0)}.$$

It is clear that the stability of motion on the whole depends on the condition for bounded n-th approximation. Proceeding further we find

$$u^{(0)} = \frac{48\mathcal{D}^2 x^{(0)2}}{\varepsilon^2 + 32C_+\mathcal{D}^2 - 96\mathcal{D}^2 x^{(0)}}, \tag{3.56}$$

$$v^{(0)} = \frac{48\mathcal{D}^2 u^{(0)2}}{\varepsilon^2 + 32C_+\mathcal{D}^2 - 96\mathcal{D}^2\left(x^{(0)} + u^{(0)}\right)}, \tag{3.57}$$

and so on. The n-th approximation is bounded if

$$\varepsilon^2 + 32C_+\mathcal{D}^2 - 96\mathcal{D}^2\Big(x^{(0)} + u^{(0)} + v^{(0)} + \dots\Big) > 0. \tag{3.58}$$

The continued fraction representation (3.58) tends, of course, to the exact stability criterion (3.38), however in practice the corresponding expressions derived after performing the second linearization provide sufficiently good estimates.

Let us now consider the general case of a N^*-th order isolated nonlinear resonance. Equation (3.29) takes the form

$$\frac{\mathrm{d}^2\overline{J}_1}{\mathrm{d}\theta^2} = \pm 8p_1p_2\mathcal{D}^2\left(4p_1\overline{J}_1\right)^{2p_1-1}\left[\pm 2p_2\left(2\overline{J}_1 - C_\pm\right)\right]^{2p_2-1}$$

$$\times\left(N^*\overline{J}_1 - p_1C_\pm\right) - \varepsilon^2\overline{J}_1 + \varepsilon\mathcal{A}_\pm. \tag{3.59}$$

As it was mentioned above, it is sufficient to carry out two successive linearizations of equation (3.59). Skipping the technical details of the calculation (we leave these to the reader as an exercise) we can write down the stability criterion in the form

$$|\varepsilon| > 8p_1p_2\mathcal{D}\left(\frac{4p_1\mathcal{A}_\pm}{\varepsilon}\right)^{p_1-1}\left[\pm 2p_2\left(\frac{2\mathcal{A}_\pm}{\varepsilon} - C_\pm\right)\right]^{p_2-1}$$

$$\times\sqrt{\frac{2N^*(N^*-1)\mathcal{A}_\pm^2}{\varepsilon^2} - \frac{4p_1(N^*-1)\mathcal{A}_\pm C_\pm}{\varepsilon} + p_1(2p_1-1)C_\pm^2}. \tag{3.60}$$

In the case of $p_1 = 1/2$ the stability criterion (3.60) reduces to

$$|\varepsilon| > 4p_2\mathcal{D}\left[\pm 2p_2\left(\frac{2\mathcal{A}_\pm}{\varepsilon} - C_\pm\right)\right]^{p_2-1}\sqrt{(N^* - 1)\left(\frac{N^*\mathcal{A}_\pm}{\varepsilon} - C_\pm\right)}. \quad (3.61)$$

Note that the expression (3.61) is valid also in the case of $p_2 = 1/2$ after redefinition of the degrees of freedom (the vertical degree of freedom becomes horizontal with index 1 and vice versa). It has correct behavior in the limit of vanishing initial amplitude of betatron oscillations as well.

3.6 Adiabatic Crossing of a Nonlinear Resonance

To derive the resonant Hamiltonian (3.13) in paragraph 3.2 we assumed that the mean radius of the machine R is constant and the focusing coefficients $G_{x,z}(\theta)$ are purely periodic functions of the azimuth θ. In other words, the effect of acceleration as well as other perturbing effects have been neglected. In isochronous cyclotrons for instance, acceleration is essentially present even in the quadratic part of the Hamiltonian [Tzenov (1990)], because the radius of the accelerated orbit (which is the analogue of the design orbit in synchrotrons and storage rings) and the energy of the particle considerably change in the course of one revolution. This leads to a slow change of the betatron tunes $\nu_{x,z}$ with θ. We already saw in paragraph 2.6 that for synchrotrons and storage rings such a change of the betatron tunes is also caused by acceleration, chromatic effects and other perturbing factors.

Taking into account the dependence of the betatron tunes $\nu_{1,2}(\theta)$ on the azimuth and performing the canonical transformation (3.3) we obtain

$$\mathcal{H}(\mathbf{J}, \boldsymbol{\alpha}; \theta) = \mathcal{H}_0(\mathbf{J}; \theta) + \Delta\mathcal{H}_1(\mathbf{J}, \boldsymbol{\alpha}; \theta), \quad (3.62)$$

where $\mathbf{J} = (J_1, J_2)$ and $\boldsymbol{\alpha} = (\alpha_1, \alpha_2)$ are the action-angle variables in vector notation, and Δ is a formal small parameter accounting for the weak perturbative character of the resonance driving term. Moreover the expressions (3.14) and (3.15) acquire the form

$$\mathcal{H}_0(\mathbf{J}; \theta) = \overline{\nu}_1(\theta)J_1 + \overline{\nu}_2(\theta)J_2 + \mathcal{H}_s(\mathbf{J}; \theta), \quad (3.63)$$

$$\mathcal{H}_1(\mathbf{J}, \boldsymbol{\alpha}; \theta) = \mathcal{D}(\theta)F_{\mathcal{D}}(\mathbf{J})\cos\Phi(\boldsymbol{\alpha}; \theta), \quad (3.64)$$

where

$$\overline{\nu}_i(\theta) = \frac{\mathrm{d}}{\mathrm{d}\theta}[\theta\nu_i(\theta)] = \nu_i(\theta) + \theta\frac{\mathrm{d}\nu_i}{\mathrm{d}\theta}, \qquad (i = 1, 2) \quad (3.65)$$

$$F_{\mathcal{D}}(\mathbf{J}) = (2J_1)^{p_1}(2J_2)^{p_2}, \tag{3.66}$$

$$\Phi(\boldsymbol{\alpha}; \theta) = 2p_1\alpha_1 \pm 2p_2\alpha_2 - m\theta + \delta_{\mathcal{D}}(\theta). \tag{3.67}$$

We can write now the resonance condition (3.16) as

$$2p_1\overline{\nu}_1(\theta) \pm 2p_2\overline{\nu}_2(\theta) = m + \overline{\varepsilon}(\theta), \tag{3.68}$$

where the resonance detuning $\overline{\varepsilon}$ depends on the azimuth θ. The canonical transformation (3.17) becomes

$$\widetilde{\mathcal{E}}_2\left(\widetilde{\mathbf{J}}, \boldsymbol{\alpha}; \theta\right) = (\alpha_1 + \varepsilon_1 - \nu_1\theta)\widetilde{J}_1 + (\alpha_2 \pm \varepsilon_2 - \nu_2\theta)\widetilde{J}_2, \tag{3.69}$$

where

$$2p_1\varepsilon_1 + 2p_2\varepsilon_2 = \varepsilon + \delta_{\mathcal{D}}, \qquad \varepsilon(\theta) = \int \mathrm{d}\theta\overline{\varepsilon}(\theta), \tag{3.70}$$

while the new Hamiltonian can be written as

$$\widetilde{\mathcal{H}}_0(\mathbf{J}; \theta) = \dot{\varepsilon}_1 J_1 \pm \dot{\varepsilon}_2 J_2 + \mathcal{H}_s(\mathbf{J}; \theta), \tag{3.71}$$

$$\widetilde{\mathcal{H}}_1(\mathbf{J}, \widetilde{\boldsymbol{\alpha}}; \theta) = \mathcal{D}(\theta)F_{\mathcal{D}}(\mathbf{J}) \cos\widetilde{\Phi}(\widetilde{\boldsymbol{\alpha}}), \tag{3.72}$$

$$\widetilde{\Phi}(\widetilde{\boldsymbol{\alpha}}) = 2p_1\widetilde{\alpha}_1 \pm 2p_2\widetilde{\alpha}_2. \tag{3.73}$$

The canonical transformation specified by the generating function (3.21) remains the same without modification, so that the new resonant canonical variables coincide with those, defined by equations (3.22) and (3.23). The resonant Hamiltonian can be written as

$$\overline{\mathcal{H}}_0\left(\overline{\mathbf{J}}; \theta\right) = \left(\overline{\varepsilon} + \dot{\delta}_{\mathcal{D}}\right)\overline{J}_1 - p_2\dot{\varepsilon}_2\overline{J}_2 + \overline{\mathcal{H}}_s\left(\overline{\mathbf{J}}; \theta\right), \tag{3.74}$$

$$\overline{\mathcal{H}}_1\left(\overline{\mathbf{J}}, \overline{\alpha}_1; \theta\right) = \mathcal{D}(\theta)\mathcal{F}_D\left(\overline{\mathbf{J}}\right) \cos\overline{\alpha}_1, \tag{3.75}$$

where the following notations

$$\overline{\mathcal{H}}_s\left(\overline{\mathbf{J}}; \theta\right) = \mathcal{H}_s\left(2p_1\overline{J}_1,\ \pm p_2\left(2\overline{J}_1 - C_\pm\right);\ \theta\right), \tag{3.76}$$

$$\mathcal{F}_D\left(\overline{\mathbf{J}}\right) = F_{\mathcal{D}}\left(2p_1\overline{J}_1,\ \pm p_2\left(2\overline{J}_1 - C_\pm\right)\right), \tag{3.77}$$

have been introduced. To simplify notations in what follows we skip the index 1 as well as the over-line in the action-angle variables $\overline{J}_1$ and $\overline{\alpha}_1$. Thus, we can write the resonant Hamiltonian as

$$\overline{\mathcal{H}}(J,\alpha;\theta) = \left(\overline{\varepsilon} + \dot{\delta}_{\mathcal{D}}\right) J + \overline{\mathcal{H}}_s(J;\theta) + \Delta\mathcal{D}(\theta)\mathcal{F}_D(J)\cos\alpha. \tag{3.78}$$

The main goal of our further exposition in this paragraph is to eliminate the explicit dependence on the angle variable in the resonant Hamiltonian (3.78). To do so we employ the canonical perturbation theory[4] and consider a canonical transformation defined by the generating function

$$S(\mathcal{J},\alpha;\theta) = \alpha\mathcal{J} + \Delta\mathcal{G}(\mathcal{J},\alpha;\theta). \tag{3.79}$$

Note that the generating function of the second kind given by equation (3.79) defines a nearly identity canonical transformation. The perturbing part $\mathcal{G}$ can be formally expanded in the small parameter Δ as

$$\mathcal{G}(\mathcal{J},\alpha;\theta) = \sum_{m=1}^{\infty} \Delta^{m-1}\mathcal{G}_m(\mathcal{J},\alpha;\theta). \tag{3.80}$$

Moreover, it is assumed that apart from being arbitrary the functions $\mathcal{G}_m(\mathcal{J},\alpha;\theta)$ are periodic in the angle variable α. The old (J,α) and the new $(\mathcal{J},a)$ action-angle variables are related according to the expressions

$$a = \frac{\partial S}{\partial \mathcal{J}} = \alpha + \Delta\frac{\partial \mathcal{G}}{\partial \mathcal{J}}, \qquad J = \frac{\partial S}{\partial \alpha} = \mathcal{J} + \Delta\frac{\partial \mathcal{G}}{\partial \alpha}. \tag{3.81}$$

With the equations (3.79) – (3.81) in hand we expand the new Hamiltonian in a perturbation series in powers of the small parameter Δ as follows:

$$\begin{aligned}
\mathcal{H}_r = {} & \left(\overline{\varepsilon} + \dot{\delta}_{\mathcal{D}}\right)\left(\mathcal{J} + \Delta\frac{\partial \mathcal{G}_1}{\partial \alpha} + \Delta^2\frac{\partial \mathcal{G}_2}{\partial \alpha} + \dots\right) + \mathcal{H}_{sr} + \Delta\mathcal{H}'_{sr}\frac{\partial \mathcal{G}_1}{\partial \alpha} \\
& + \Delta^2\left[\frac{\mathcal{H}''_{sr}}{2}\left(\frac{\partial \mathcal{G}_1}{\partial \alpha}\right)^2 + \mathcal{H}'_{sr}\frac{\partial \mathcal{G}_2}{\partial \alpha}\right] + \dots + \Delta\mathcal{D}(\theta)\left\{\mathcal{F}_{Dr} + \Delta\mathcal{F}'_{Dr}\frac{\partial \mathcal{G}_1}{\partial \alpha}\right. \\
& \left. + \Delta^2\left[\frac{\mathcal{F}''_{Dr}}{2}\left(\frac{\partial \mathcal{G}_1}{\partial \alpha}\right)^2 + \mathcal{F}'_{Dr}\frac{\partial \mathcal{G}_2}{\partial \alpha}\right] + \dots\right\}\cos\alpha + \Delta\frac{\partial \mathcal{G}_1}{\partial \theta} + \Delta^2\frac{\partial \mathcal{G}_2}{\partial \theta} + \dots,
\end{aligned} \tag{3.82}$$

[4]Various aspects of the canonical perturbation theory and its application to particular problems of nonlinear beam dynamics will be discussed in more detail in forthcoming chapters.

where

$$\mathcal{H}_{sr} = \overline{\mathcal{H}}_s(\mathcal{J};\theta), \qquad \mathcal{H}'_{sr} = \left[\frac{\partial \overline{\mathcal{H}}_s(J;\theta)}{\partial J}\right]_{J=\mathcal{J}}, \qquad \ldots, \tag{3.83}$$

$$\mathcal{F}_{Dr} = \mathcal{F}_D(\mathcal{J}), \qquad \mathcal{F}'_{Dr} = \left[\frac{\partial \mathcal{F}_D(J)}{\partial J}\right]_{J=\mathcal{J}}, \qquad \ldots, \tag{3.84}$$

Equating terms of same order in the expansion (3.82) yields the lowest order Hamiltonian

$$\mathcal{H}_{r0} = \left(\bar{\varepsilon} + \dot{\delta}_{\mathcal{D}}\right)\mathcal{J} + \mathcal{H}_{sr}. \tag{3.85}$$

Since the yet unknown functions $\mathcal{G}_m(\mathcal{J}, \alpha; \theta)$ are to this end arbitrary we can without loss of generality assume that the average part of $\mathcal{G}_1(\mathcal{J}, \alpha; \theta)$ over the angle variable α vanishes. Therefore, $\mathcal{H}_{r1} \equiv 0$, from which if follows

$$\left(\bar{\varepsilon} + \dot{\delta}_{\mathcal{D}} + \mathcal{H}'_{sr}\right)\frac{\partial \mathcal{G}_1}{\partial \alpha} + \frac{\partial \mathcal{G}_1}{\partial \theta} = -\mathcal{F}_{Dr}\mathcal{D}(\theta)\cos\alpha. \tag{3.86}$$

We represent the function $\mathcal{G}_1$ in the form

$$\mathcal{G}_1(\mathcal{J}, \alpha; \theta) = \mathcal{A}(\mathcal{J}; \theta)\cos\alpha + \mathcal{B}(\mathcal{J}; \theta)\sin\alpha, \tag{3.87}$$

where the new unknowns $\mathcal{A}(\mathcal{J}; \theta)$ and $\mathcal{B}(\mathcal{J}; \theta)$ satisfy the equations

$$\frac{\partial \mathcal{A}}{\partial \theta} + \left(\bar{\varepsilon} + \dot{\delta}_{\mathcal{D}} + \mathcal{H}'_{sr}\right)\mathcal{B} = -\mathcal{F}_{Dr}\mathcal{D}(\theta),$$

$$\frac{\partial \mathcal{B}}{\partial \theta} - \left(\bar{\varepsilon} + \dot{\delta}_{\mathcal{D}} + \mathcal{H}'_{sr}\right)\mathcal{A} = 0.$$

Defining

$$\mathcal{C}(\mathcal{J}; \theta) = \mathcal{A}(\mathcal{J}; \theta) + i\mathcal{B}(\mathcal{J}; \theta), \tag{3.88}$$

we can write the general solution of the system of equations for $\mathcal{A}$ and $\mathcal{B}$ in the form

$$\mathcal{C}(\mathcal{J}; \theta) = -\mathcal{F}_{Dr}(\mathcal{J})e^{-\gamma_r(\mathcal{J};\theta)}\int_0^\theta d\tau \mathcal{D}(\tau)e^{\gamma_r(\mathcal{J};\tau)}, \tag{3.89}$$

where

$$\gamma_r(\mathcal{J};\tau) = -i \int_0^\theta d\tau \Big[\overline{\varepsilon}(\tau) + \dot{\delta}_{\mathcal{D}}(\tau) + \mathcal{H}'_{sr}(\mathcal{J};\tau) \Big]. \tag{3.90}$$

Further determination of $\mathcal{G}_2$, $\mathcal{G}_3$, etc. can be accomplished in analogy with the above procedure used to find $\mathcal{G}_1$. Thus we can construct the new Hamiltonian $\mathcal{H}_r$, depending on the new action variable $\mathcal{J}$ only. The latter is clearly an invariant of the motion and can be determined up to any order in the perturbation parameter Δ according to equations (3.81).

Let us first consider the case of fast crossing of a nonlinear resonance. For simplicity we again confine ourselves to the case of a resonance without stabilization ($\mathcal{H}_s = 0$). Because the function

$$\omega(\theta) = \overline{\varepsilon}(\theta) + \dot{\delta}_{\mathcal{D}}(\theta)$$

under the integral in equation (3.90) is a fast varying function of θ, we can use the method of stationary phase [Nayfeh (1981)] to compute the integral in expression (3.89). In the non degenerate case $\dot{\omega}(\theta_0) \neq 0$ this gives

$$\mathcal{C}(\mathcal{J};\theta) = -\sqrt{\frac{2\pi}{|\dot{\omega}(\theta_0)|}}\, \mathcal{F}_{\mathcal{D}r}(\mathcal{J}) \mathcal{D}(\theta_0) \exp\left[i\lambda(\theta,\theta_0)\right], \tag{3.91}$$

where

$$\lambda(\theta,\theta_0) = \int_{\theta_0}^{\theta} d\tau \omega(\tau) \mp \frac{\pi}{4} = \varepsilon(\theta) + \delta_{\mathcal{D}}(\theta) - \varepsilon(\theta_0) - \delta_{\mathcal{D}}(\theta_0) \mp \frac{\pi}{4}. \tag{3.92}$$

In the expressions (3.91) and (3.92) θ_0 denotes the stationary point, that is the solution of the equation

$$\omega(\theta_0) = 0.$$

The "−" sign in equation (3.92) corresponds to the case when $\dot{\omega}(\theta_0) > 0$, while the "+" sign corresponds to the case of $\dot{\omega}(\theta_0) < 0$. In the degenerate case

$$\dot{\omega}(\theta_0) = \ddot{\omega}(\theta_0) = \cdots = \omega^{(n-1)}(\theta_0) = 0$$

however, we have

$$\mathcal{C}(\mathcal{J};\theta) = -\frac{\Gamma(1/n)}{n} \left[\frac{n!}{\left|\omega^{(n)}(\theta_0)\right|} \right]^{1/n} \mathcal{F}_{\mathcal{D}r}(\mathcal{J}) \mathcal{D}(\theta_0) \exp\left[i\lambda_1(\theta,\theta_0)\right], \tag{3.93}$$

where

$$\lambda_1(\theta, \theta_0) = \lambda(\theta, \theta_0) \pm \frac{\pi}{4}\left(1 - \frac{2}{n}\right), \tag{3.94}$$

and $\Gamma(z)$ is the well-known gamma-function [Abramowitz (1984)].

Let us now consider the case of a slow crossing of a nonlinear resonance. We assume a linear dependence of the resonance detuning ε and the amplitude of the driving term $\mathcal{D}$ on the azimuth θ

$$\overline{\varepsilon}(\theta) = \varepsilon_0 - 2\dot{\varepsilon}\theta, \qquad \mathcal{D}(\theta) = \mathcal{D}_0 + \dot{\mathcal{D}}\theta, \tag{3.95}$$

where

$$\varepsilon_0 = 2p_1\nu_{10} \pm 2p_2\nu_{20} - m, \qquad \dot{\varepsilon} = -(2p_1\dot{\nu}_1 \pm 2p_2\dot{\nu}_2) > 0. \tag{3.96}$$

By substituting the expressions (3.95) and (3.96) into equation (3.89), and taking into account (3.90) we obtain

$$\mathcal{C}(\mathcal{J}; \theta) = -\frac{\mathcal{F}_{Dr}(\mathcal{J})}{\sqrt{\dot{\varepsilon}}} e^{-i\lambda_2(\theta)}$$

$$\times \left\{ \sqrt{\frac{\pi}{2}} \left[\mathcal{D}_0 + \frac{\dot{\mathcal{D}}\left(\varepsilon_0 + \dot{\delta}_{\mathcal{D}}\right)}{2\dot{\varepsilon}} \right] \left[C\left(\sqrt{\frac{2}{\pi}}\kappa\right) + iS\left(\sqrt{\frac{2}{\pi}}\kappa\right) \right] - \frac{i\dot{\mathcal{D}}e^{i\kappa^2}}{2\sqrt{\dot{\varepsilon}}} \right\}_{\kappa=\kappa_0}^{\kappa=\kappa_1}, \tag{3.97}$$

where

$$\lambda_2(\theta) = \left[\sqrt{\dot{\varepsilon}}\theta - \frac{\varepsilon_0 + \dot{\delta}_{\mathcal{D}}}{2\sqrt{\dot{\varepsilon}}} \right]^2, \tag{3.98}$$

$$\kappa_0 = -\frac{\varepsilon_0 + \dot{\delta}_{\mathcal{D}}}{2\sqrt{\dot{\varepsilon}}}, \qquad \kappa_1 = \sqrt{\dot{\varepsilon}}\theta. \tag{3.99}$$

Here $C(z)$ and $S(z)$ are the *Fresnel*'s integrals [Abramowitz (1984)].

The first order generating function $\mathcal{G}_1(\mathcal{J}, \alpha; \theta)$ thus found, enables us to write the new Hamiltonian $\mathcal{H}_r$ from equation (3.82) up to second order in the perturbation parameter Δ as

$$\mathcal{H}_r = \left(\overline{\varepsilon} + \dot{\delta}_{\mathcal{D}}\right)\mathcal{J} + \mathcal{H}_{sr} + \frac{\Delta^2}{2}\left[\frac{1}{2}\mathcal{H}''_{sr}|\mathcal{C}|^2 + \mathcal{D}(\theta)\mathcal{F}'_{Dr}\mathcal{B}\right]. \tag{3.100}$$

Using the transformation equations (3.81) we can proceed further and determine the growth of the amplitude of betatron oscillations in the process of crossing of a nonlinear resonance. Since

$$J = \mathcal{J} - \Delta[\mathcal{A}(\mathcal{J}, \theta) \sin \alpha - \mathcal{B}(\mathcal{J}, \theta) \cos \alpha],$$

for the maximum amplitude growth we have

$$(\Delta J)_{max} = \Delta |\mathcal{C}|, \tag{3.101}$$

where $(\Delta J) = J - \mathcal{J}$. In the case of a non degenerate fast crossing of a nonlinear resonance the maximum increase of the amplitude of betatron oscillations is

$$(\Delta J)_{max} = \sqrt{\frac{2\pi}{|\dot{\omega}(\theta_0)|}} |\mathcal{D}(\theta_0)| \mathcal{F}_{Dr}. \tag{3.102}$$

In the case of slow resonance crossing for the maximum growth of the amplitude we obtain

$$(\Delta J)_{max} = \frac{\mathcal{F}_{Dr}}{\sqrt{\dot{\varepsilon}}} \sqrt{\mathrm{Re}^2 \mathcal{C}_{\{\}} + \mathrm{Im}^2 \mathcal{C}_{\{\}}}, \tag{3.103}$$

where $\mathrm{Re}\mathcal{C}_{\{\}}$ and $\mathrm{Im}\mathcal{C}_{\{\}}$ are the real and imaginary part of the expression in curly brackets of equation (3.97), respectively.

¿From the equations (3.102) and (3.103) it becomes evident that the increase of the amplitude of betatron oscillations when crossing a nonlinear resonance is inversely proportional to the square root of the crossing rate. This fact is not surprising since it is intuitively clear that the faster a dynamical system passes through a resonance, the less time is available to the resonance to exercise influence over the system and vice versa.

3.7 Periodic Crossing of a Nonlinear Resonance

In the present paragraph we will study the case of successive resonance crossings due to periodic perturbation (modulation) of the betatron tunes. In other words, instead of the unperturbed betatron tunes ν_i $(i = 1, 2)$ entering the Hamiltonian (3.14) we consider [Tzenov (1992a)]

$$\widetilde{\nu}_i = \nu_i - \nu_{iD} \sin(\nu_m \theta + \alpha_{m0}), \tag{3.104}$$

where ν_{iD} is the depth of modulation and ν_m is the modulation frequency.

The betatron tune can be modulated by various external factors, such as the power supply ripple, synchro-betatron coupling, via the residual uncompensated chromaticity, or by ground motion effects. In the case of tune modulation caused by synchrotron oscillations, we specify the depth and the tune of modulation as

$$\nu_{iD} = \frac{R}{2}\left\langle \frac{1}{\beta_i}\left(1+\alpha_i^2\right)\right\rangle_{2\pi}\sqrt{\frac{2J_{s0}}{\mathcal{B}_s}}, \qquad \nu_m = \nu_s, \tag{3.105}$$

where $(i = 1, 2)$. Here again, R is the mean radius of the ring, $\langle \dots \rangle_{2\pi}$ denotes averaging over one turn[5], α_i and β_i are the *Twiss* parameters defined in equations (2.39) – (2.41), and J_{s0} is the initial value of the longitudinal action. Furthermore, $\mathcal{B}_s$ is the longitudinal "beta"-function, and ν_s is the synchrotron tune. They are defined according to

$$\frac{1}{\mathcal{B}_s^2} = \frac{\Delta E_0}{E_s}\frac{k}{2\pi\beta_s^2 R^2 \mathcal{K}}\cos\Phi_s \qquad \nu_s^2 = \frac{\Delta E_0}{E_s}\frac{k\mathcal{K}}{2\pi\beta_s^2}\cos\Phi_s, \tag{3.106}$$

where Φ_s is the synchronous phase, and the other parameters have been already introduced in chapter 1. Let us also note that the case of modulation of the driving term phase

$$\widetilde{\delta}_{\mathcal{D}} = \delta_{\mathcal{D}} + \frac{2p_1\widetilde{\nu}}{\nu_m}\cos\left(\nu_m\theta + \alpha_{m0}\right), \tag{3.107}$$

can be reduced by a canonical transformation with a generating function

$$E_2\Big(\alpha_i, \widetilde{J}_i; \theta\Big) = \left[\alpha_1 + \frac{\widetilde{\nu}}{\nu_m}\cos\left(\nu_m\theta + \alpha_{m0}\right)\right]\widetilde{J}_1 + \alpha_2\widetilde{J}_2 \tag{3.108}$$

exactly to our initial problem with modulated betatron tune of the form, specified by equation (3.104) with $\nu_{2D} = 0$.

Repeating the steps from equation (3.17) to equation (3.23) we obtain the resonant Hamiltonian

$$\overline{\mathcal{H}}(J,\alpha;\theta) = \overline{\mathcal{H}}_0(J;\theta) + \Delta\overline{\mathcal{H}}_1(J,\alpha), \tag{3.109}$$

where

$$\overline{\mathcal{H}}_0(J;\theta) = \left[\varepsilon - \nu_{12}\sin\left(\nu_m\theta + \alpha_{m0}\right)\right]J + \mathcal{H}_s(2p_1J, \pm p_2(2J - C_\pm)), \tag{3.110}$$

$$\overline{\mathcal{H}}_1(J,\alpha) = \mathcal{D}\mathcal{F}_D(J)\cos\alpha, \tag{3.111}$$

[5]The procedure of averaging over one revolution is justified, because betatron oscillations are much faster than synchrotron ones.

$$\nu_{12} = 2p_1\nu_{1D} \pm 2p_2\nu_{2D}. \tag{3.112}$$

For simplicity of notations the index 1 and the over-line in the action-angle variables $\overline{J}_1$ and $\overline{\alpha}_1$ have been dropped again.

The detailed treatment of the periodic crossing of a nonlinear resonance by means of the canonical perturbation method will be postponed to a later chapter. We rather assume now that a non-stationary perturbation

$$V(J;\theta) = -\nu_{12} J \sin(\nu_m \theta + \alpha_{m0}) \tag{3.113}$$

acts on the isolated resonance system

$$H_{00}(J,\alpha) = \varepsilon J + \mathcal{H}_s(J) + \Delta\mathcal{D}\mathcal{F}_D(J)\cos\alpha. \tag{3.114}$$

Since the excursion of the action variable J is of the order of the resonance driving term (assumed to be small), we can expand the resonant Hamiltonian (3.114) about the stationary point J_r that solves the equation

$$\varepsilon + \mathcal{H}_s'(J_r) = 0$$

Then the Hamiltonian (3.114) can be written in the form governing the motion of a pendulum

$$\overline{H}_{00}(\overline{I},\alpha) = \frac{\overline{I}^2}{2} + \omega_0^2 \cos\alpha, \tag{3.115}$$

where

$$\overline{I} = (J - J_r)\mathcal{H}_s''(J_r), \qquad \overline{H}_{00} = H_{00}\mathcal{H}_s''(J_r), \tag{3.116}$$

$$\omega_0^2 = \Delta\mathcal{D}\mathcal{H}_s''(J_r)\mathcal{F}_D(J_r). \tag{3.117}$$

The perturbation (3.113) in the new scaled variables (3.116) now reads as

$$\overline{V}(\overline{I};\theta) = -\nu_{12}\overline{I}\sin(\nu_m\theta + \alpha_{m0}). \tag{3.118}$$

Before proceeding further we briefly review some of the basic properties and characteristics of the dynamics of the pendulum [Lichtenberg (1982)], [Sagdeev (1988)]. Without loss of generality we consider in what follows the quantity $\mathcal{H}_s''(J_r)$ to be positive. It is well-known [Lichtenberg (1982)], [Sagdeev (1988)] that $[\alpha_{st}, \overline{I}_{st}] = [(2n+1)\pi, 0]$ are stable fixed points, while

$\left[\alpha_{unst}, \overline{I}_{unst}\right] = [2n\pi, 0]$ are unstable fixed points for $n = 0, \pm 1, \pm 2, \ldots$. Let us define the parameter

$$\kappa^2 = \frac{1}{2}\left(1 + \frac{\overline{H}_{00}}{\omega_0^2}\right). \tag{3.119}$$

If $\kappa > 1$ $\left(\overline{H}_{00} > \omega_0^2\right)$, then $\overline{I}$ is always different from zero, resulting in unbounded motion in α (rotation). The case $\kappa < 1$ corresponds to bounded motion (libration), while for $\kappa = 1$ we have separatrix motion, in which the oscillation period becomes infinite. Figure 3.1 shows the phase space portrait for the Hamiltonian (3.115) for different values of κ.

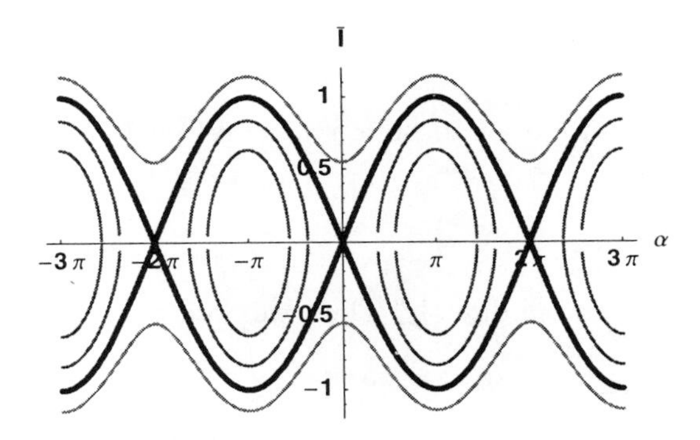

Fig. 3.1 Phase space portrait for the pendulum Hamiltonian. The thick curve represents the separatrix.

Using the definition of the action integral

$$\mathcal{J} = \frac{1}{2\pi} \oint p\mathrm{d}q, \tag{3.120}$$

where q and p are the canonical coordinate and momentum, and the integral is taken over one cycle of oscillation in time, we can write

$$\mathcal{J}\left(\overline{H}_{00}\right) = \frac{2}{\pi} \int\limits_0^{\alpha_{max}} \mathrm{d}\alpha \sqrt{2\left(\overline{H}_{00} + \omega_0^2 \cos\alpha\right)}. \tag{3.121}$$

Here the turning point $\alpha_{max} = \pi$ for rotation $(\kappa > 1)$, and $\cos\alpha_{max} =$

$-\overline{H}_{00}/\omega_0^2$ for libration ($\kappa < 1$). The action integral (3.121) can be expressed in terms of the complete elliptic integrals $\mathbf{K}(\kappa)$ and $\mathbf{E}(\kappa)$ of the first and second kind [Gradshteyn (2000)] as

$$\mathcal{J}\big(\overline{H}_{00}\big) = \frac{8\omega_0}{\pi}\begin{cases} \mathbf{E}(\kappa) - \big(1-\kappa^2\big)\mathbf{K}(\kappa) & (\kappa \le 1), \\ \kappa\mathbf{E}\big(\frac{1}{\kappa}\big) & (\kappa \ge 1). \end{cases} \tag{3.122}$$

From $\mathrm{d}\mathcal{J}/\mathrm{d}\overline{H}_{00} = 1/\omega$, we obtain the pendulum frequency

$$\omega\big(\overline{H}_{00}\big) = \frac{\pi\omega_0}{2}\begin{cases} 1/\mathbf{K}(\kappa) & (\kappa \le 1), \\ \kappa/\mathbf{K}\big(\frac{1}{\kappa}\big) & (\kappa \ge 1). \end{cases} \tag{3.123}$$

Using the asymptotics of the elliptic integrals [Gradshteyn (2000)] the oscillation frequency of the pendulum near the separatrix can be written as

$$\omega\big(\overline{H}_{00}\big) = \pi\omega_0\left[\ln\left(\frac{32\omega_0^2}{\big|\overline{H}_{00} - \omega_0^2\big|}\right)\right]^{-1}. \tag{3.124}$$

The separatrix trajectory can be found by using the separatrix condition $\kappa = 1$ $\big(\text{or } \overline{H}_{00} = \omega_0^2\big)$ to write

$$\overline{I} = \dot{\alpha} = \pm 2\omega_0 \sin\frac{\alpha}{2},$$

which can be integrated directly. Thus, we obtain

$$\alpha(\theta) = 4\arctan\big(e^{\pm\omega_0\theta}\big). \tag{3.125}$$

If the isolated resonance system (3.115) was not perturbed by the non-stationary perturbation (3.118), its energy $E = \overline{H}_{00}$ would have been constant. The change of the pendulum energy due to the perturbation (3.118) is

$$\begin{aligned} \dot{E} &= \big[E, \overline{V}\big] = \omega_0^2\nu_{12}\sin\alpha\sin\left(\nu_m\theta + \alpha_{m0}\right) \\ &= \frac{\omega_0^2\nu_{12}}{2}\left[\cos\left(\alpha - \nu_m\theta - \alpha_{m0}\right) - \cos\left(\alpha + \nu_m\theta + \alpha_{m0}\right)\right]. \end{aligned} \tag{3.126}$$

The energy gain per half a period of the pendulum oscillations is

$$\Delta E = \frac{\omega_0\nu_{12}}{2}\cos\alpha_{m0}\int\limits_{-\infty}^{\infty}\mathrm{d}\tau\left\{\cos\left[\widetilde{\alpha}(\tau) + \frac{\nu_m\tau}{\omega_0}\right] - \cos\left[\widetilde{\alpha}(\tau) - \frac{\nu_m\tau}{\omega_0}\right]\right\}, \tag{3.127}$$

where

$$\widetilde{\alpha}(\tau) = 4\arctan\left(e^{\tau}\right) - \pi, \tag{3.128}$$

and the fact that $\widetilde{\alpha}(\tau)$ is an odd function of τ has been taken into account. The integral on the right-hand-side of equation (3.127) is known as the *Melnikov-Arnold* integral [Lichtenberg (1982)], [Chirikov (1979)]

$$\mathcal{A}_m(\lambda) = \int\limits_{-\infty}^{\infty} d\tau \cos\left[\frac{m\widetilde{\alpha}(\tau)}{2} - \lambda\tau\right], \tag{3.129}$$

where $m = 2$ and $\lambda = \pm\nu_m/\omega_0$.

Because the cosine argument is antisymmetric the *Melnikov-Arnold* integral can be written as

$$\mathcal{A}_m(\lambda) = \int\limits_{-\infty}^{\infty} d\tau \exp\left\{i\left[\frac{m\widetilde{\alpha}(\tau)}{2} - \lambda\tau\right]\right\}.$$

Introducing the new variable $\xi = 2\arctan(e^\tau)$, we have

$$e^\tau = \tan\left(\frac{\xi}{2}\right) = \frac{i - ie^{i\xi}}{1 + e^{i\xi}}, \qquad e^{i\widetilde{\alpha}/2} = -ie^{i\xi} = \frac{1 + ie^\tau}{i + e^\tau}.$$

Therefore

$$\mathcal{A}_m(\lambda) = \int\limits_{-\infty}^{\infty} d\tau e^{-i\lambda\tau}\left(\frac{1 + ie^\tau}{i + e^\tau}\right)^m, \tag{3.130}$$

and for integer m the *Melnikov-Arnold* integral can be evaluated by means of the integrand residues. For $m > 0$ the integrand poles are purely imaginary and are located at the points

$$e^{\tau_0} = -i, \qquad \tau_0 = -i\pi\left(2n + \frac{1}{2}\right).$$

where n is any integer including zero. For positive λ we close the integration contour[6] in the lower half-plane of the complex τ. If $\lambda < 0$, we perform a formal change of the integration variable according to $\tau \to -\tau + i\pi$ and obtain

$$\mathcal{A}_m(-|\lambda|) = (-1)^m \mathcal{A}_m(|\lambda|)e^{-\pi|\lambda|}. \tag{3.131}$$

Integrating equation (3.130) by parts along the real τ-axis and neglecting the oscillating term [Chirikov (1979)] we arrive at the important recurrence

[6] For further details and clarification of this operation the reader is referred to [Chirikov (1979)].

relation:

$$\mathcal{A}_{m+1}(\lambda) = \frac{2\lambda}{m}\mathcal{A}_m(\lambda) - \mathcal{A}_{m-1}(\lambda). \tag{3.132}$$

Substituting a new variable $\zeta = \tau - \tau_0$ into equation (3.130) and evaluating the residues we obtain

$$\mathcal{A}_m(\lambda) = \frac{2\pi i e^{-\pi\lambda/2}}{i^m (m-1)!} \lim_{\zeta \to 0} \frac{\mathrm{d}^{m-1}}{\mathrm{d}\zeta^{m-1}} \left[e^{-i\lambda\zeta} \left(\zeta \frac{1+e^\zeta}{1-e^\zeta} \right)^m \right] \sum_{n=0}^{\infty} e^{-2\pi n\lambda}$$

$$= \frac{\pi}{i^{m-1}(m-1)!} \frac{e^{\pi\lambda/2}}{\sinh(\pi\lambda)} \lim_{\zeta \to 0} \frac{\mathrm{d}^{m-1}}{\mathrm{d}\zeta^{m-1}} \left[e^{-i\lambda\zeta} \left(\zeta \frac{1+e^\zeta}{1-e^\zeta} \right)^m \right], \tag{3.133}$$

where the well-known identity [Gradshteyn (2000)]

$$\frac{1}{\sinh(z)} = 2\sum_{k=0}^{\infty} e^{-(2k+1)z} \qquad (z > 0)$$

has been taken into account. For the first two *Melnikov-Arnold* integrals we have

$$\mathcal{A}_0(\lambda) = 0, \qquad \mathcal{A}_1(\lambda) = -\frac{2\pi e^{\pi\lambda/2}}{\sinh(\pi\lambda)}, \tag{3.134}$$

and by making use of the recurrence relation (3.132), we find

$$\mathcal{A}_2(\lambda) = -\frac{4\pi\lambda e^{\pi\lambda/2}}{\sinh(\pi\lambda)}, \qquad \mathcal{A}_3(\lambda) = (2\lambda^2 - 1)\mathcal{A}_1(\lambda), \tag{3.135}$$

and so on.

Recalling now the expression (3.127) for the energy gain of the pendulum due to the non-stationary perturbation, and taking into account equations (3.131) and (3.135), we obtain

$$\Delta E = \frac{2\pi\nu_m\nu_{12}}{\cosh(\pi\nu_m/2\omega_0)} \cos\alpha_{m0}. \tag{3.136}$$

The phase of the perturbation is changed after half a period of the pendulum oscillations by a value of

$$\Delta\varphi = \frac{\pi\nu_m}{\omega(\overline{E})} = \frac{\nu_m}{\omega_0} \ln\left(\frac{32\omega_0^2}{|\overline{E} - \omega_0^2|} \right). \tag{3.137}$$

Finally, we can write the separatrix mapping as

$$\overline{E} = E + \frac{2\pi\nu_m\nu_{12}}{\cosh\left(\pi\nu_m/2\omega_0\right)}\cos\varphi, \qquad \overline{\varphi} = \varphi + \frac{\nu_m}{\omega_0}\ln\left(\frac{32\omega_0^2}{\left|\overline{E}-\omega_0^2\right|}\right). \quad (3.138)$$

The quantity

$$K = \left|\frac{\delta\overline{\varphi}}{\delta\varphi} - 1\right| \quad (3.139)$$

determines the extension of an infinitesimally small phase interval. If

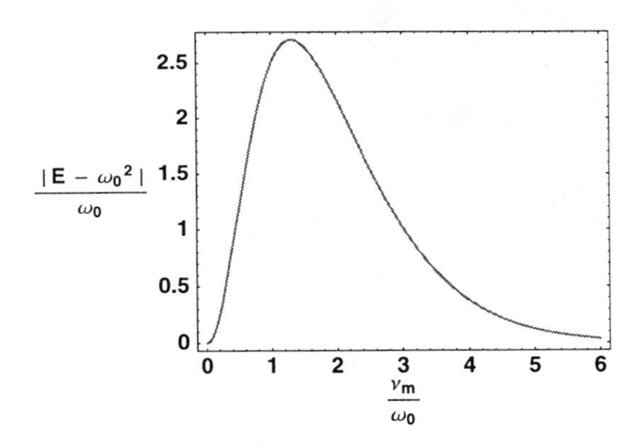

Fig. 3.2 Dependence of the width of the stochastic layer on the modulation tune.

$$K \geq 1, \quad (3.140)$$

one can expect a local instability to occur, and the relation (3.140) may serve as a good phenomenological criterion for the onset of stochasticity. For the separatrix mapping (3.138) the parameter K is

$$K = \frac{2\pi|\nu_{12}|}{|E-\omega_0^2|}\frac{\nu_m^2}{\omega_0}\frac{|\sin\varphi|}{\cosh\left(\pi\nu_m/2\omega_0\right)} \approx \frac{2\pi|\nu_{12}|}{|E-\omega_0^2|}\frac{\nu_m^2}{\omega_0\cosh\left(\pi\nu_m/2\omega_0\right)}. \quad (3.141)$$

The width of the stochastic layer is found in a straightforward manner from equation (3.140) to be

$$\left|E - \omega_0^2\right| \leq \frac{2\pi|\nu_{12}|}{\omega_0} \frac{\nu_m^2}{\cosh\left(\pi\nu_m/2\omega_0\right)}. \tag{3.142}$$

It is worth to mention that the proportionality of the width of the stochastic layer to the depth of the betatron tune modulation $|\nu_{12}|$ is physically justified and therefore not surprising. More interesting is however, its dependence on the modulation frequency ν_m, exhibiting a maximum at ν_{m0}, which is a root of the transcendental equation

$$\coth\left(\frac{\pi\nu_{m0}}{2\omega_0}\right) = \frac{1}{2}\frac{\pi\nu_{m0}}{2\omega_0}. \tag{3.143}$$

The relative width $\left|E - \omega_0^2\right|/\omega_0$ of the stochastic layer as a function of ν_m/ω_0 is shown in figure 3.2.

Chapter 4

Canonical Perturbation Theory

4.1 Introduction

In chapter 3, we came round to the opinion that the dynamics of particles in modern high-energy accelerators and storage rings is essentially nonlinear. Nonlinear dynamical systems are in general not integrable, that is the associated *Hamilton-Jacobi* equation does not possess exact solution. However, for systems that are close enough to integrable ones, one can attempt to obtain approximate solutions to a desired degree of accuracy, which are extremely useful for theoretical calculations in a number of cases.

A basic method to analyze the nonlinear effects of particle motion in accelerators is by perturbation from known, integrable solutions to a system that differs slightly from the original one. The measure of this difference is the small perturbation parameter ϵ, naturally used as an expansion parameter. As expected, the closer the system under consideration is to the model integrable system, the better the approximate solution thus obtained will describe the real processes going off.

In order the above perturbation scheme to work, one has to implicitly assume that the sought-for solution actually exists. This however, is not generally the case, and in this aspect the merit of the perturbation series solution is that it may still approximate the motion in some coarse-grained sense.

In paragraph 4.5, we will deal with the method of formal series, which constitutes a non perturbative technique for solving the *Hamilton-Jacobi* equation. The method of formal series represents a novel approach to the solution of the *Hamilton-Jacobi* equation, whose basic idea is to represent the solution of an arbitrary nonlinear equation as a ratio of two formal *Volterra* series in powers of the perturbation (coupling) parameter, rather than a conventional power series, widely used in classical perturbation theory. It turns out that this ratio is well behaved even for large values of the coupling parameter.

4.2 Classical Perturbation Theory

As we have seen (chapter 3) the nonlinear betatron motion of particles in cyclic accelerators and storage rings can be described by a Hamiltonian

$$H(\mathbf{J}, \boldsymbol{\alpha}; \theta) = H_0(\mathbf{J}) + \epsilon V(\mathbf{J}, \boldsymbol{\alpha}; \theta). \tag{4.1}$$

Here H has been written in terms of the n-dimensional action-angle variables $(\mathbf{J}, \boldsymbol{\alpha})$ of the unperturbed problem, and the unperturbed Hamiltonian $H_0(\mathbf{J})$ includes nonlinear terms depending on $\mathbf{J}$ only. Thus, the unperturbed tune depends in general on the amplitude of betatron oscillations. In the absence of the perturbation $V(\mathbf{J}, \boldsymbol{\alpha}; \theta)$ the action variables remain constant and the motion is accomplished on a $(n+1)$-dimensional torus in the extended phase space $(\mathbf{J}, \boldsymbol{\alpha}; \theta)$. The effect of the perturbation is expressed in distortion of this torus at the least, to its complete destruction, depending on the strength and the character of the perturbation.

Consider a canonical transformation from the old variables $(\mathbf{J}, \boldsymbol{\alpha})$ to the new variables $(\mathbf{J}_1, \boldsymbol{\alpha}_1)$ specified by the generating function

$$\mathcal{F}_2(\mathbf{J}_1, \boldsymbol{\alpha}; \theta) = \boldsymbol{\alpha} \cdot \mathbf{J}_1 + \epsilon \mathcal{G}(\mathbf{J}_1, \boldsymbol{\alpha}; \theta). \tag{4.2}$$

The new coordinates and the Hamiltonian can be written as

$$\mathbf{J} = \mathbf{J}_1 + \epsilon \nabla_\alpha \mathcal{G}, \tag{4.3}$$

$$\boldsymbol{\alpha}_1 = \boldsymbol{\alpha} + \epsilon \nabla_{J_1} \mathcal{G}, \tag{4.4}$$

$$H_1 = H + \epsilon \frac{\partial \mathcal{G}}{\partial \theta}. \tag{4.5}$$

Note that the lowest order term in the generating function $\mathcal{F}_2$ has been chosen to generate the identity transformation. The new Hamiltonian after substituting the transformed variables reads as

$$H_1 = H_0\left(\mathbf{J}_1 + \sum_{k=1}^{\infty} \epsilon^k \nabla_\alpha \mathcal{G}_k\right) + \epsilon V\left(\mathbf{J}_1 + \sum_{k=1}^{\infty} \epsilon^k \nabla_\alpha \mathcal{G}_k, \boldsymbol{\alpha}; \theta\right)$$

$$+\sum_{k=1}^{\infty}\epsilon^k\frac{\partial\mathcal{G}_k}{\partial\theta} = \sum_{m=0}^{\infty}\frac{1}{m!}\left(\sum_{k=1}^{\infty}\epsilon^k\nabla_\alpha\mathcal{G}_k\cdot\nabla_{J_1}\right)^m H_0(\mathbf{J}_1)$$

$$+\epsilon\sum_{m=0}^{\infty}\frac{1}{m!}\left(\sum_{k=1}^{\infty}\epsilon^k\nabla_\alpha\mathcal{G}_k\cdot\nabla_{J_1}\right)^m V(\mathbf{J}_1,\boldsymbol{\alpha};\theta) + \sum_{k=1}^{\infty}\epsilon^k\frac{\partial\mathcal{G}_k}{\partial\theta}. \tag{4.6}$$

where $\mathcal{G}$ has been formally expanded [see equation (3.80)] according to

$$\mathcal{G}(\mathbf{J}_1,\boldsymbol{\alpha};\theta) = \sum_{k=1}^{\infty}\epsilon^{k-1}\mathcal{G}_k(\mathbf{J}_1,\boldsymbol{\alpha};\theta).$$

It is important to point out that our substitution in equation (4.6) is not complete; the new Hamiltonian is a function of the old angles and the new actions (as is the generating function $\mathcal{G}$ itself). For the moment it will prove more convenient to work with the mixed variables.

Equation (4.6) yields in zero order

$$H_1^{(0)} = H_0(\mathbf{J}_1), \tag{4.7}$$

and in first order

$$H_1^{(1)} = \boldsymbol{\nu}(\mathbf{J}_1)\cdot\nabla_\alpha\mathcal{G}_1 + V(\mathbf{J}_1,\boldsymbol{\alpha};\theta) + \frac{\partial\mathcal{G}_1}{\partial\theta},$$

where

$$\boldsymbol{\nu}(\mathbf{J}) = \nabla_J H_0(\mathbf{J}), \tag{4.8}$$

We proceed in the same way as in chapter 3 and require that $H_1^{(1)}$ vanishes. Thus, if we are able to find a solution to the linearized *Hamilton-Jacobi* equation

$$\frac{\partial\mathcal{G}_1}{\partial\theta} + \boldsymbol{\nu}(\mathbf{J}_1)\cdot\nabla_\alpha\mathcal{G}_1 + V(\mathbf{J}_1,\boldsymbol{\alpha};\theta) = 0 \tag{4.9}$$

the new Hamiltonian will contain perturbative terms of at least second order in the perturbation parameter ϵ.

Since both V and $\mathcal{G}_1$ are periodic functions of $\boldsymbol{\alpha}$, they can be expanded in a *Fourier* series,

$$V(\mathbf{J}_1,\boldsymbol{\alpha};\theta) = \sum_{\mathbf{n}}\mathcal{V}_{\mathbf{n}}(\mathbf{J}_1;\theta)e^{i\mathbf{n}\cdot\boldsymbol{\alpha}},$$

$$\mathcal{G}_1(\mathbf{J}_1,\boldsymbol{\alpha};\theta) = \sum_{\mathbf{n}}\mathcal{G}_{\mathbf{n}}^{(1)}(\mathbf{J}_1;\theta)e^{i\mathbf{n}\cdot\boldsymbol{\alpha}}.$$

Then the linearized *Hamilton-Jacobi* equation (4.9) becomes

$$\left[\frac{\partial}{\partial\theta} + i\mathbf{n}\cdot\boldsymbol{\nu}(\mathbf{J}_1)\right]\mathcal{G}_{\mathbf{n}}^{(1)} = -\mathcal{V}_{\mathbf{n}}. \tag{4.10}$$

We are looking for the distortions of the invariant torus, that is we must select the periodic solutions to equation (4.9), provided V is periodic in θ and its average vanishes. The periodic in θ solution to the equation (4.10) can be written as

$$\mathcal{G}_{\mathbf{n}}^{(1)}(\mathbf{J}_1;\theta) = \int\limits_{\theta}^{\theta+2\pi} \mathrm{d}\theta' G_{\mathbf{n}}(\theta-\theta')\mathcal{V}_{\mathbf{n}}(\mathbf{J}_1;\theta'), \tag{4.11}$$

where $G_{\mathbf{n}}(\tau)$ is the *Green*'s function satisfying the equation

$$\left(\frac{\partial}{\partial\tau} + i\mathbf{n}\cdot\boldsymbol{\nu}\right)G_{\mathbf{n}}(\tau) = -\delta_p(\tau),$$

and $\delta_p(\tau)$ denotes the periodic δ-function. By taking into account the identity [Prudnikov (1986)]

$$\sum_{l=-\infty}^{\infty}\frac{e^{ilx}}{l+a} = \frac{\pi}{\sin\pi a}e^{i[\pi-\mathrm{mod}(x)]a},$$

where

$$\mathrm{mod}(x) = x - 2n\pi \qquad \text{for} \qquad 2n\pi < x < 2(n+1)\pi,$$

it can be easily checked that the *Green*'s function $G_{\mathbf{n}}$ is given by

$$G_{\mathbf{n}}(\tau) = \frac{i}{2\sin(\pi\mathbf{n}\cdot\boldsymbol{\nu})}e^{i\mathbf{n}\cdot\boldsymbol{\nu}[\pi-\mathrm{mod}(\tau)]},$$

so that equation (4.11) becomes

$$\mathcal{G}_{\mathbf{n}}^{(1)} = \frac{i}{2\sin(\pi\mathbf{n}\cdot\boldsymbol{\nu})}\int\limits_{\theta}^{\theta+2\pi}\mathrm{d}\theta'\mathcal{V}_{\mathbf{n}}(\mathbf{J}_1;\theta')e^{i\mathbf{n}\cdot\boldsymbol{\nu}(\theta'-\theta-\pi)}. \tag{4.12}$$

Finally, for $\mathcal{G}_1$ we obtain

$$\mathcal{G}_1(\mathbf{J}_1,\boldsymbol{\alpha};\theta) = \frac{i}{2}\sum_{\mathbf{n}}\frac{1}{\sin(\pi\mathbf{n}\cdot\boldsymbol{\nu})}\int\limits_{\theta}^{\theta+2\pi}\mathrm{d}\theta'\mathcal{V}_{\mathbf{n}}(\mathbf{J}_1;\theta')e^{i\mathbf{n}\cdot[\boldsymbol{\alpha}+\boldsymbol{\nu}(\theta'-\theta-\pi)]}. \tag{4.13}$$

Equation (4.6) yields the new Hamiltonian in second order

$$H_1^{(2)} = \frac{\partial \mathcal{G}_2}{\partial \theta} + \boldsymbol{\nu} \cdot \nabla_\alpha \mathcal{G}_2 + \nabla_\alpha \mathcal{G}_1 \cdot \nabla_{J_1} V + \frac{1}{2} \nabla_\alpha \mathcal{G}_1 \cdot \nabla_{J_1} \boldsymbol{\nu} \cdot \nabla_\alpha \mathcal{G}_1. \quad (4.14)$$

Extracting the part independent of the angle variables and the azimuthal (independent) variable θ from the last two terms in equation (4.14), we can set it to determine the new second order Hamiltonian. The remainder, together with the first two terms in equation (4.14) will enter an equation similar to equation (4.9) to solve for the second order generating function $\mathcal{G}_2$. Thus, the new second order Hamiltonian can be written as

$$H_1^{(2)}(\mathbf{J}_1) = \sum_{\mathbf{m},n} \left[\frac{|\mathcal{V}_{\mathbf{m}n}|^2}{(\mathbf{m} \cdot \boldsymbol{\nu} - n)^2} \mathbf{m} \cdot \nabla_{J_1} \boldsymbol{\nu} \cdot \mathbf{m} - \frac{\mathcal{V}_{\mathbf{m}n} \mathbf{m} \cdot \nabla_{J_1} \mathcal{V}_{\mathbf{m}n}^*}{\mathbf{m} \cdot \boldsymbol{\nu} - n} \right]. \quad (4.15)$$

In deriving equation (4.15) use has been made of the representation of the perturbation V in a double *Fourier* series

$$V(\mathbf{J}_1, \boldsymbol{\alpha}; \theta) = \sum_{\mathbf{m},n} \mathcal{V}_{\mathbf{m}n}(\mathbf{J}_1) e^{i(\mathbf{m} \cdot \boldsymbol{\alpha} - n\theta)}.$$

The new frequency is given by the expression

$$\widetilde{\boldsymbol{\nu}}(\mathbf{J}_1) = \boldsymbol{\nu}(\mathbf{J}_1) + \epsilon^2 \nabla_{J_1} H_1^{(2)}. \quad (4.16)$$

Proceeding further order by order, we will obtain a new Hamiltonian which depends only upon the new action variables. Therefore, these new action variables are approximate integrals of the motion, so that the dynamics according to equation (4.3) is restricted to the $(n+1)$-dimensional torus $\mathbf{J}(\mathbf{J}_1, \boldsymbol{\alpha}; \theta)$ in phase space.

The approach we have presented above would work fine if $\mathcal{G}$ were always small, however, we are at once faced with the problem of small denominators. For any $\mathbf{J}_1$ a harmonic number $\mathbf{m}$ can be found such that $\mathbf{m} \cdot \boldsymbol{\nu}(\mathbf{J}_1)$ is arbitrarily close to an integer. This phenomenon represents a serious physical and mathematical difficulty, and is called a nonlinear resonance (see chapter 3). Note that we came across the problem of small denominators because we have required periodic solutions to the equation for $\mathcal{G}$. It will become evident later that once the resonance condition

$$\mathbf{m} \cdot \boldsymbol{\nu}(\mathbf{J}_1) = n$$

is satisfied, there are no periodic solutions to equation (4.10), and the amplitude of the solution grows linearly in θ.

Near a resonance in the unperturbed Hamiltonian the standard methods of canonical perturbation theory cannot be applied at least insofar the particular resonance is concerned. We can still use the perturbation theory for the non resonant terms, but a special treatment of the resonant term(s) is needed. In chapter 3 such a treatment was performed in detail, based on the method of effective potential. Later in this chapter we will discuss the secular perturbation theory as applied to isolated nonlinear resonances.

4.3 Effect of Linear and Nonlinear Perturbations in One Dimension

In chapter 2 we have assumed perfect quadrupole magnets with an ideal design orbit that passes through the center of all quadrupoles. It is interesting and useful to study the effect of linear perturbations in the confining magnetic field, the source of which can be a variation in the length of quadrupoles, errors in the quadrupole power supply, horizontal closed orbit deviation in sextupoles, etc. Although in these cases the perturbed problem can be solved exactly, it is instructive to apply the classical perturbation theory in deriving analytic formulae which describe the effect of small perturbation.

The Hamiltonian in the case of a quadrupole gradient perturbation is [see equation (1.77)]

$$\widehat{H} = \frac{R}{2}\widehat{p}^2 + \frac{G(\theta)}{2R}\widehat{x}^2 + \frac{g(\theta)}{2R}\widehat{x}^2, \tag{4.17}$$

where the coefficient of quadrupole perturbation $g(\theta)$ is considered small. Passing over to action-angle variables, we write equation (4.17) in the form

$$H = \dot{\chi}J + \frac{g(\theta)\beta(\theta)J}{2R}(1 + \cos 2\alpha). \tag{4.18}$$

Before proceeding further it is necessary to rearrange terms in equation (4.18) by including the angle averaged part of the perturbation in H_0. Thus, we obtain

$$H(J, \alpha; \theta) = H_0(J) + \epsilon V(J, \alpha; \theta), \tag{4.19}$$

where[1]

$$H_0(J) = \widetilde{\nu} J, \qquad V(J, \alpha; \theta) = \frac{g(\theta)\beta(\theta)J}{2R} \cos 2[\alpha + \psi(\theta) - \widetilde{\nu}\theta]. \quad (4.20)$$

Note that $\psi(\theta)$ is the new phase advance including the shift due to the first order contribution of the perturbation

$$\psi(\theta) = \psi(0) + \chi(\theta) + \frac{1}{2R} \int_0^\theta \mathrm{d}\theta' g(\theta')\beta(\theta'), \quad (4.21)$$

and $\widetilde{\nu}$ is the perturbed betatron tune including the first order tune shift $\Delta\nu$

$$\widetilde{\nu} = \nu + \Delta\nu, \qquad \Delta\nu = \frac{1}{2\mathcal{C}} \int_0^{2\pi} \mathrm{d}\theta' g(\theta')\beta(\theta'), \quad (4.22)$$

where $\mathcal{C}$ is the circumference of the machine.

The first order distortions of the invariant curves are obtained by a simple substitution of the relevant quantities into the expression for the generating function in equation (4.13) and performing the inverse of the transformation specified in the footnote below. As as result we have

$$\mathcal{G}_1(J_1, \alpha; \theta) = \frac{-J_1}{4R \sin(2\pi\widetilde{\nu})} \int_\theta^{\theta+2\pi} \mathrm{d}\theta' g(\theta')\beta(\theta') \sin 2[\alpha + \psi(\theta') - \psi(\theta) - \pi\widetilde{\nu}], \quad (4.23)$$

The invariant curves are given by

$$J = J_1 + \frac{\partial \mathcal{G}}{\partial \alpha},$$

where J_1 is constant (to first order in the strength of the quadrupole perturbation g). From equation (4.23) we obtain explicitly

$$J = J_1 \left\{ 1 - \frac{1}{2R \sin(2\pi\widetilde{\nu})} \int_\theta^{\theta+2\pi} \mathrm{d}\theta' g(\theta')\beta(\theta') \cos 2[\alpha + \psi(\theta') - \psi(\theta) - \pi\widetilde{\nu}] \right\}. \quad (4.24)$$

[1]To obtain equation (4.20) use has been made of a trivial canonical transformation $\alpha \to \alpha + \psi(\theta) - \widetilde{\nu}\theta$, $J \to J$ specified by the generating function $\mathcal{S}\left(\widetilde{J}, \alpha; \theta\right) = [\alpha - \psi(\theta) + \widetilde{\nu}\theta]\widetilde{J}$.

The second term in the curly brackets represents the effect of the quadrupole gradient perturbation. Clearly, this effect is very sensitive to the distance from a parametric resonance of the form $2\widetilde{\nu} = \text{integer}$.

Recall that the dots in equation (1.69) represent terms proportional to the momentum deviation η and its higher powers. Let us examine the term linear in η which can be written as

$$V = -\frac{R\eta}{2}\widehat{p}^2. \tag{4.25}$$

This term introduces a linear dependence of the amplitude of betatron oscillations and the betatron tune on the deviation of the actual particle momentum from the momentum of the synchronous particle. Since η varies slowly in comparison with $\widehat{x}$ and $\widehat{p}$, it can be assumed constant. Introducing the action-angle variables we cast equation (4.25) in the form

$$V(J,\phi;\theta) = -\frac{R\eta\gamma(\theta)}{2}J - \frac{\eta\dot{\chi}(\theta)}{2}\left\{\left[\alpha^2(\theta)-1\right]\cos 2\phi + 2\alpha(\theta)\sin 2\phi\right\}. \tag{4.26}$$

Here a different notation ϕ for the angle variable is used to distinguish it from the *Twiss* parameter $\alpha(\theta)$, and $\gamma(\theta)$ is the third *Twiss* parameter defined in equation (2.42). The first term on the right-hand-side of equation (4.26) yields a *chromatic* shift in the betatron tune

$$\Delta\nu_c = -\frac{R\eta}{4\pi}\int_0^{2\pi} d\theta\gamma(\theta). \tag{4.27}$$

Integrating equation (2.40) and taking into account the periodicity of the *Twiss* parameters we have

$$\int_0^{2\pi} d\theta\left[\frac{G(\theta)\beta(\theta)}{R} - R\gamma(\theta)\right] = \alpha(2\pi) - \alpha(0) = 0.$$

Thus, we can write equation (4.27) in alternative form

$$\Delta\nu_c = \xi_N\eta, \tag{4.28}$$

where

$$\xi_N = -\frac{1}{2C}\int_0^{2\pi} d\theta G(\theta)\beta(\theta) \tag{4.29}$$

is the so-called *natural chromaticity*.

As an example of a nonlinear perturbation, we apply the perturbation theory to study the effect of a sextupole on the beam dynamics in one degree of freedom. After substituting the expression for $\widetilde{x}$ from equation (1.73) into the sextupole contribution (with $\widetilde{z} = 0$) on the right-hand-side of equation (1.70) we have four terms at our disposal. In what follows, we focus our attention on two of them; the term proportional to the second power of $\widehat{x}$ and the term proportional to the third power of $\widehat{x}$.

The quadratic in $\widehat{x}$ term with

$$g(\theta) = \frac{\lambda_0(\theta)D(\theta)\eta}{R} \tag{4.30}$$

is responsible for the so-called *residual chromatic effect* in cyclic accelerators and storage rings. Substituting $g(\theta)$ from equation (4.30) under the integral in equation (4.22), we obtain the residual chromatic tune shift

$$\Delta\nu_c = \xi_R \eta, \tag{4.31}$$

where

$$\xi_R = \frac{1}{2RC} \int_0^{2\pi} \mathrm{d}\theta' \lambda_0(\theta') D(\theta') \beta(\theta') \tag{4.32}$$

is the *residual chromaticity* due to sextupoles[2]. It provides an effective method to compensate the undesirable natural chromaticity by choosing the locations and the strengths of the sextupole correctors appropriately.

We now consider the non-chromatic part of the sextupole perturbation and write the Hamiltonian as

$$H = \frac{R}{2}\widehat{p}^2 + \frac{G(\theta)}{2R}\widehat{x}^2 + \frac{\lambda_0(\theta)}{6R^2}\widehat{x}^3. \tag{4.33}$$

Transforming to the action-angle variables, we obtain the new Hamiltonian

$$H(J, \alpha; \theta) = \dot{\chi} J + \epsilon V(J, \alpha; \theta), \tag{4.34}$$

where the perturbation is given explicitly by

$$V(J, \alpha; \theta) = \frac{\lambda_0(\theta)}{6\sqrt{2}R^2} [J\beta(\theta)]^{3/2} (\cos 3\alpha + 3\cos\alpha). \tag{4.35}$$

---

[2] It should be pointed out that all magnetic multipoles give rise to a residual chromaticity.

From equation (4.13) we obtain the generating function

$$\mathcal{G}_1(J_1, \alpha; \theta) = -\frac{J_1^{3/2}}{4\sqrt{2}R^2}$$

$$\times \left\{ \frac{1}{\sin \pi\nu} \int\limits_{\theta}^{\theta+2\pi} d\theta' \lambda_0(\theta') \beta^{3/2}(\theta') \sin\left[\alpha + \chi(\theta') - \chi(\theta) - \pi\nu\right] \right.$$

$$\left. + \frac{1}{3\sin 3\pi\nu} \int\limits_{\theta}^{\theta+2\pi} d\theta' \lambda_0(\theta') \beta^{3/2}(\theta') \sin 3[\alpha + \chi(\theta') - \chi(\theta) - \pi\nu] \right\}. \tag{4.36}$$

To find the nonlinear tune shift due to the sextupole perturbation, we have to calculate the average of the third and fourth term in equation (4.14). Since the fourth term is proportional to J_1^3 its average can be neglected as compared to the average of the third term, which is

$$\left\langle \frac{\partial \mathcal{G}_1}{\partial \alpha} \frac{\partial V}{\partial J_1} \right\rangle = -\frac{J_1^2}{64R^3\mathcal{C}} \int\limits_0^{2\pi} d\theta \lambda_0(\theta) \beta^{3/2}(\theta) \int\limits_{\theta}^{\theta+2\pi} d\theta' \lambda_0(\theta') \beta^{3/2}(\theta')$$

$$\times \left\{ \frac{3\cos\left[\chi(\theta') - \chi(\theta) - \pi\nu\right]}{\sin \pi\nu} + \frac{\cos 3[\chi(\theta') - \chi(\theta) - \pi\nu]}{\sin 3\pi\nu} \right\}. \tag{4.37}$$

The new Hamiltonian can be written as

$$H_1(J_1) = \nu J_1 + \left\langle \frac{\partial \mathcal{G}_1}{\partial \alpha} \frac{\partial V}{\partial J_1} \right\rangle,$$

and the new tune is

$$\nu_1(J_1) = \nu + \frac{\partial}{\partial J_1} \left\langle \frac{\partial \mathcal{G}_1}{\partial \alpha} \frac{\partial V}{\partial J_1} \right\rangle. \tag{4.38}$$

Note that if the actual distribution of sextupoles is known, the integral in equation (4.37) can be evaluated in principle. It is clear that the tune shift given by equation (4.38) varies linearly with J_1, which is similar to the first-order effect of an octupole perturbation. In the vicinity of the third order resonance 3ν = integer, we are facing again the already discussed problem of small denominators and the perturbation theory is not appropriate.

4.4 Secular Perturbation Theory

In paragraph 4.2 we discovered that near a resonance

$$\mathbf{m} \cdot \boldsymbol{\nu}(\mathbf{J}) = n \tag{4.39}$$

in the unperturbed Hamiltonian a small (resonant) denominator appears when the first order action invariant is being calculated, following the prescription of the classical perturbation theory. As mentioned, a necessity for special treatment of the resonant terms by removing the secularity they introduce arises. This can be achieved by a canonical transformation to a new frame of reference, rotating with the resonant frequency, which was already used in paragraph 3.2 to eliminate the resonant variables.

A sufficiently large perturbation gives rise to secondary resonances, which modify or may even destroy the first-order invariant, calculated after the resonant terms have been removed. The secularity introduced by secondary resonances can be eliminated by a canonical transformation, analogous to the one used for the removal of the secularity due to the original (primary) resonance. As a matter of fact, there is a remarkable similarity in the effect of primary and secondary resonances on the zero-order and the first-order invariant, respectively. This forms the basis of effective quantitative methods to describe the transition from regular to stochastic behavior, which will be discussed in subsequent paragraphs.

Consider again the Hamiltonian (4.1) in two degrees of freedom, and let for concreteness assume the unperturbed part $H_0(\mathbf{J})$ to be of the form

$$H_0(\mathbf{J}) = \boldsymbol{\nu}_0 \cdot \mathbf{J} + \widetilde{H}_0(\mathbf{J}). \tag{4.40}$$

We perform a canonical transformation specified by the generating function

$$\mathcal{F}_2\left(\boldsymbol{\alpha}, \widehat{\mathbf{J}}; \theta\right) = (m_1\alpha_1 + m_2\alpha_2 - n\theta)\widehat{J}_1 + \alpha_2\widehat{J}_2, \tag{4.41}$$

which defines the new canonical variables

$$J_1 = m_1\widehat{J}_1, \qquad J_2 = \widehat{J}_2 + m_2\widehat{J}_1, \tag{4.42}$$

$$\widehat{\alpha}_1 = m_1\alpha_1 + m_2\alpha_2 - n\theta, \qquad \widehat{\alpha}_2 = \alpha_2. \tag{4.43}$$

Applying the canonical transformation given by equations (4.41) – (4.43), we obtain the new Hamiltonian

$$\widehat{H}\left(\widehat{\mathbf{J}}, \widehat{\boldsymbol{\alpha}}; \theta\right) = \delta\widehat{J}_1 + \widehat{H}_0\left(\widehat{\mathbf{J}}\right) + \epsilon\sum_l \mathcal{V}_{l\mathbf{m}\, ln}\left(\widehat{\mathbf{J}}\right)e^{il\widehat{\alpha}_1}$$

$$+\epsilon \sum_{\mathbf{n}p}{}' \mathcal{V}_{\mathbf{n}p}\left(\widehat{\mathbf{J}}\right) \exp\left\{\frac{i}{m_1}[n_1\widehat{\alpha}_1 + (m_1 n_2 - m_2 n_1)\widehat{\alpha}_2 + (n_1 n - m_1 p)\theta]\right\}, \tag{4.44}$$

where

$$\delta = \mathbf{m} \cdot \boldsymbol{\nu}_0 - n \tag{4.45}$$

is the resonance detuning,

$$\widehat{H}_0\left(\widehat{\mathbf{J}}\right) = \nu_{02}\widehat{J}_2 + \widetilde{H}_0\left(m_1\widehat{J}_1, \widehat{J}_2 + m_2\widehat{J}_1\right), \tag{4.46}$$

$$\mathcal{V}_{\mathbf{n}p}\left(\widehat{\mathbf{J}}\right) = \mathcal{V}_{\mathbf{n}p}\left(m_1\widehat{J}_1, \widehat{J}_2 + m_2\widehat{J}_1\right), \tag{4.47}$$

and $\sum'$ in the last term of equation (4.44) denotes exclusion of the harmonics $\mathcal{V}_{k\mathbf{m}\,kn}$ from the sum. Now proceeding as in paragraph 4.2 we can average over the "fast" variable $\widehat{\alpha}_2$, which near the resonance varies much faster as compared to the resonant phase $\widehat{\alpha}_1$. We obtain the transformed Hamiltonian to first order

$$\overline{H}\left(\widehat{\mathbf{J}}, \widehat{\boldsymbol{\alpha}}\right) = \delta\widehat{J}_1 + \widehat{H}_0\left(\widehat{\mathbf{J}}\right) + \epsilon\sum_l \mathcal{V}_{l\mathbf{m}\,ln}\left(\widehat{\mathbf{J}}\right)e^{il\widehat{\alpha}_1}. \tag{4.48}$$

Since the new Hamiltonian (4.48) is independent of $\widehat{\alpha}_2$, similar to the result (3.25) we have the invariant

$$\widehat{J}_2 = J_2 - \frac{m_2}{m_1}J_1 = \text{const.} \tag{4.49}$$

Because $\widehat{J}_2$ is a constant, the motion governed by the Hamiltonian (4.48) is in fact the motion of a single degree of freedom, and therefore integrable. As a rule, the *Fourier* harmonics $\mathcal{V}_{l\mathbf{m}\,ln}\left(\widehat{\mathbf{J}}\right)$ decrease rapidly as l increases, and to a good approximation we can confine ourselves to the lowest ones with $l = \pm 1$:

$$\overline{H}\left(\widehat{\mathbf{J}}, \widehat{\boldsymbol{\alpha}}\right) = \delta\widehat{J}_1 + \widehat{H}_0\left(\widehat{\mathbf{J}}\right) - 2\epsilon\left|\mathcal{V}_{\mathbf{m}n}\left(\widehat{\mathbf{J}}\right)\right|\cos\widehat{\alpha}_1. \tag{4.50}$$

Without loss of generality, we have assumed that the argument of $\mathcal{V}_{\mathbf{m}n}\left(\widehat{\mathbf{J}}\right)$ is equal to π, which can always be accomplished by shifting the angle variable with a constant.

The stationary points in the phase space $\left(\widehat{J}_1, \widehat{\alpha}_1\right)$ where the trajectories are stationary are defined as solutions to the equations

$$\frac{\partial \overline{H}}{\partial \widehat{\alpha}_1} = 0, \qquad \frac{\partial \overline{H}}{\partial \widehat{J}_1} = 0, \tag{4.51}$$

which can be written explicitly as

$$\sin \widehat{\alpha}_1 = 0, \qquad \delta + \frac{\partial \widehat{H}_0}{\partial \widehat{J}_1} - 2\epsilon \frac{\partial |\mathcal{V}_{\mathbf{mn}}|}{\partial \widehat{J}_1} \cos \widehat{\alpha}_1 = 0. \tag{4.52}$$

From the first of equations (4.52) it follows that there are three stationary points located at $\widehat{\alpha}_{1s} = 0, \pm\pi$. The first one is stable, while the other two are unstable stationary points, provided that $\partial^2 \widehat{H}_0 / \partial \widehat{J}_{1s}^2 > 0$.

Since the excursion in $\widehat{J}_1$ is of the order of the perturbation we can expand the Hamiltonian (4.50) about the stationary point $\widehat{J}_{1s}$. Ignoring the constant term we can write the Hamiltonian describing the motion near resonance as

$$\Delta \overline{H}\left(\Delta \widehat{J}_1, \widehat{\alpha}_1\right) = \frac{G}{2}\left(\Delta \widehat{J}_1\right)^2 - F \cos \widehat{\alpha}_1, \tag{4.53}$$

where

$$G = \frac{\partial^2 \widehat{H}_0}{\partial \widehat{J}_{1s}^2} \tag{4.54}$$

is the *nonlinearity parameter*, and

$$F = 2\epsilon \left| \mathcal{V}_{\mathbf{mn}}\left(\widehat{\mathbf{J}}_s\right) \right|, \qquad \Delta \widehat{J}_1 = \widehat{J}_1 - \widehat{J}_{1s} \pm \frac{2\epsilon}{G} \frac{\partial |\mathcal{V}_{\mathbf{mn}}|}{\partial \widehat{J}_{1s}}. \tag{4.55}$$

The Hamiltonian (4.53) shows that the motion near every isolated resonance resembles that of a pendulum. This interesting result provides a universal description of the dynamics close to an arbitrary resonance, and sometimes $\Delta \overline{H}$ is referred to as the *standard Hamiltonian.*

The boundaries of the stable island around the stable stationary point are formed by separatrices, passing through the unstable stationary points. Since the Hamiltonian (4.50) is a constant on the separatrix, we have

$$\delta \widehat{J}_1 + \widehat{H}_0\left(\widehat{\mathbf{J}}\right) - 2\epsilon \left| \mathcal{V}_{\mathbf{mn}}\left(\widehat{\mathbf{J}}\right) \right| \cos \widehat{\alpha}_1 = \delta \widehat{J}_{1u} + \widehat{H}_0\left(\widehat{\mathbf{J}}_u\right) + 2\epsilon \left| \mathcal{V}_{\mathbf{mn}}\left(\widehat{\mathbf{J}}_u\right) \right|,$$

where $\widehat{J}_{1u}$ is the action at the unstable stationary point. Expanding $\widehat{J}_1$ about $\widehat{J}_{1u}$, we find the separatrix equation to be approximately

$$\left(\widehat{J}_1 - \widehat{J}_{1u}\right)^2 \simeq \frac{2F_u}{G_u}(1 + \cos\widehat{\alpha}_1), \tag{4.56}$$

so that the maximum separation or island width is

$$\Delta\widehat{J}_{1w} = \pm 2\sqrt{\frac{F_u}{G_u}}. \tag{4.57}$$

The last term in equation (4.44) gives rise to secondary resonances which, as we already mentioned, modify or destroy the invariant $\widehat{J}_2$, provided ϵ is not sufficiently small. To study this effect we have to express the Hamiltonian (4.48) in action-angle variables of the $\left(\widehat{J}_1, \widehat{\alpha}_1\right)$ motion and subsequently analyze the resonances between harmonics of the $\left(\widehat{J}_1, \widehat{\alpha}_1\right)$ phase oscillations and the unperturbed tune ν_2. Instead of casting (4.48) in action-angle variables exactly, we use the approximation (4.50). Application of the classical perturbation theory to the dynamics in the vicinity of the stable stationary point $\widehat{\alpha}_{1s} = 0$ yields

$$\overline{H}\left(I_1, \widehat{J}_2\right) = \widehat{H}_0\left(\widehat{J}_{1s}, \widehat{J}_2\right) + \omega_1 I_1 - \frac{\epsilon G}{16} I_1^2 + \ldots, \tag{4.58}$$

where the new action-angle variables (I_1, ϕ_1) are related to the old ones $\left(\widehat{J}_1, \widehat{\alpha}_1\right)$ via the expressions

$$\widehat{\alpha}_1 = \sqrt{\frac{2I_1}{\mathcal{R}}} \sin\phi_1, \qquad \widehat{J}_1 = \sqrt{2I_1\mathcal{R}} \cos\phi_1. \tag{4.59}$$

Here the quantity $\mathcal{R}$ and the phase oscillation frequency ω_1 depend on $\widehat{J}_2$ and are given by

$$\mathcal{R} = \sqrt{\frac{F}{G}}, \qquad \omega_1 = \sqrt{FG}.$$

The Hamiltonian (4.58) solves our problem entirely in the case when the averaging over $\widehat{\alpha}_2$ is valid. The basic characteristic of the motion is the existence of two invariants I_1 and $\widehat{J}_2$.

To take into account the contribution of a secondary resonance in the process of modification of the invariants I_1 and $\widehat{J}_2$ as prescribed by the Hamiltonian (4.58), we expand the last term in equation (4.44) about the stable singularity $\widehat{\alpha}_{1s} = 0$ and write it in the form

$$\epsilon \sum_{\mathbf{n}pl}' \mathcal{W}_{\mathbf{n}pl}\left(\widehat{J}_{1s}, I_1, \widehat{J}_2\right) \exp\left[il\phi_1 \mathrm{sgn}(n_1)\right]$$

$$\times \exp\left\{\frac{i}{m_1}\left[(m_1 n_2 - m_2 n_1)\widehat{\alpha}_2 + (n_1 n - m_1 p)\theta\right]\right\}, \tag{4.60}$$

where

$$\mathcal{W}_{\mathbf{n}pl}\left(\widehat{J}_{1s}, I_1, \widehat{J}_2\right) = \mathcal{V}_{\mathbf{n}p}\left(\widehat{J}_{1s}, \widehat{J}_2\right) \mathcal{J}_l\left(\frac{|n_1|}{m_1}\sqrt{\frac{2I_1}{\mathcal{R}}}\right). \tag{4.61}$$

To derive (4.60), the expression for $\widehat{\alpha}_1$ in equation (4.59) has been substituted accordingly, and use has been made of the identity

$$e^{iz \sin \phi} = \sum_{l=-\infty}^{\infty} \mathcal{J}_l(|z|) e^{il\phi \mathrm{sgn}(z)}, \tag{4.62}$$

where $\mathcal{J}_l(z)$ is the *Bessel* function of order l and $\mathrm{sgn}(z)$ is the well-known sign-function. It is now clear from (4.60) that higher order resonances between ϕ_1 and $\widehat{\alpha}_2$ are possible. We can use the same method as in the beginning of this paragraph to remove the secularity introduced by the secondary resonance. Assume a secondary resonance of the form

$$\mathbf{p} \cdot \boldsymbol{\omega}_0 = q + \delta_1, \tag{4.63}$$

where $\boldsymbol{\omega}_0 = (\omega_1, \nu_{02})$ and δ_1 is the resonance detuning. Next we perform a canonical transformation specified by the generating function

$$\mathcal{F}_2\left(\phi_1, \widehat{\alpha}_2, \widehat{\mathbf{I}}; \theta\right) = (p_1\phi_1 + p_2\widehat{\alpha}_2 - q\theta)\widehat{I}_1 + \widehat{\alpha}_2 \widehat{I}_2. \tag{4.64}$$

Averaging again over the fast phase $\widehat{\phi}_2 = \widehat{\alpha}_2$ and noting that the summation in (4.60) can be reduced to a summation over new summation indices j and k with

$$n_1 = km_1, \qquad n_2 = jp_2 + km_2, \qquad p = kn + jq, \qquad l = jp_1 \mathrm{sgn}(k), \tag{4.65}$$

we obtain the new Hamiltonian

$$\overline{\overline{H}}\left(\widehat{\mathbf{I}}, \widehat{\boldsymbol{\phi}}\right) = \delta_1 \widehat{I}_1 + \overline{H}_0\left(\widehat{\mathbf{I}}\right) + \epsilon \sum_j \mathcal{G}_{j\mathbf{p}\, jq}\left(\widehat{J}_{1s}, \widehat{\mathbf{I}}\right) e^{ij\widehat{\phi}_1}. \tag{4.66}$$

Here $\overline{H}_0\left(\widehat{\mathbf{I}}\right)$ is given by the expression

$$\overline{H}_0\left(\widehat{\mathbf{I}}\right) = \nu_{02}\widehat{I}_2 + \widetilde{H}_0\left(m_1\widehat{J}_{1s}, m_2\widehat{J}_{1s} + \widehat{I}_2 + p_2\widehat{I}_1\right), \tag{4.67}$$

while the *Fourier* coefficient of the j-th harmonic in $\widehat{\phi}_1$ can be written as

$$\mathcal{G}_{j\mathbf{p}\,jq} = \sum_k \mathcal{W}_{km_1\,jp_2+km_2\,kn+jp\,jp_1\mathrm{sgn}(k)}. \tag{4.68}$$

Note that the new Hamiltonian (4.66) does not depend on the angle variable $\widehat{\phi}_2$. This immediately implies that

$$\widehat{I}_2 = \widehat{J}_2 - \frac{p_2}{p_1}I_1 = \text{const} \tag{4.69}$$

is the adiabatic invariant for the island oscillation. The latter reduces the dynamics as described by (4.66) to the dynamics of a single degree of freedom, so that the $\left(\widehat{I}_1, \widehat{\phi}_1\right)$ motion is integrable.

4.5 The Method of Formal Series

In the preceding paragraphs we showed that the objective of the classical perturbation theory is to solve the *Hamilton-Jacobi* equation, and thus transform the original system to such that is trivial or easy to integrate. We saw that the direct perturbation expansion of the solution led us to the problem of small denominators. In this sense the perturbation solution is an asymptotic solution, signifying that it is well-serving for a relatively short period of time, inversely proportional to the strength of the perturbation.

The method of *formal series* of *Dubois-Violette* suggests a new approach to the solution of the *Hamilton-Jacobi* equation. The basic idea of *Dubois-Violette*'s method is to represent the solution of a generic nonlinear equation as a ratio of two formal *Volterra* series in powers of the perturbation parameter, rather than a conventional power series, given by perturbation theory. It turns out that the ratio mentioned above is well behaved even for large values of the coupling parameter. This latter fact should not be so surprising if one considers the solution to a nonlinear equation by *Dubois-Violette*'s method as a nonlinear generalization of the solution to a simple linear system by the ratio of two determinants, according to *Kramer*'s rule. Moreover, the denominator in the formal series solution is uniquely defined, which is not the case for *Pade* approximations and other non-perturbative tools, where one or more coefficients in the series expansion of the denominator need to be fixed taking into account initial conditions. Although the solution to the original problem obtained by the method of *Dubois-Violette* is given in a symbolic form, it is in principle possible to compute the numerator as well as the denominator to every desired order in the perturbation parameter.

Mathematical details concerning the method of formal series have been considered in the most precise and exhaustive way in the article [Dubois-Violette (1970a)]. Further the approach has been applied to study the motion of the anharmonic oscillator in classical mechanics [Dubois-Violette (1970b)] and to construct a functional formulation of quantum field theory [Dubois-Violette (1969)]. In the present paragraph we apply the method of formal series to study nonlinear beam dynamics in cyclic accelerators [Tzenov (1992c)], [Tzenov (1995)], governed by a Hamiltonian of the most general form.

Following *Dubois-Violette* and *Tzenov* we start with a brief review of some basic facts from the theory of formal series. Let us denote by $\mathcal{E}^{\mathbf{R}}$ the space of all real valued functions $\varphi(x)$ varying in the interval (a, b). We will restrict ourselves by regarding $\mathcal{E}^{\mathbf{R}}$ as a linear functional space. Let $\Phi_n(x_1, x_2, \dots, x_n)$ for $n = 0, 1, 2, \dots$ be given functions of the arguments x_i $(i = 1, 2, \dots, n)$ which vary also in the interval (a, b). Without loss of generality the functions $\Phi_n(x_1, x_2, \dots, x_n)$ can be taken to be symmetric in their arguments. Further we define the functional

$$P_n[\varphi] = \int\limits_a^b \mathrm{d}x_1 \int\limits_a^b \mathrm{d}x_2 \dots \int\limits_a^b \mathrm{d}x_n \Phi_n(x_1, x_2, \dots, x_n)\varphi(x_1)\varphi(x_2)\dots\varphi(x_n) \tag{4.70}$$

with kernels Φ_n, such that the integral (4.70) exists for $\forall\varphi(x) \in \mathcal{E}^{\mathbf{R}}$. The series

$$P[\varphi] = \sum_{n=0}^{\infty} P_n[\varphi] \tag{4.71}$$

is called *formal Volterra series.* Clearly, every term and therefore the whole series (4.71) is a functional of $\varphi(x) \in \mathcal{E}^{\mathbf{R}}$. Convergence of (4.71) is not necessarily supposed. Thus, $P[\varphi]$ is a symbol – an infinite series of numbers (asymptotic series) for which only the partial sums are well-defined. One can also introduce

$$P[t,\varphi] = \sum_{n=0}^{\infty} t^n P_n[\varphi], \tag{4.72}$$

that is an infinitely differentiable function with respect to t with

$$P[t,\varphi] - \sum_{m=0}^{n} t^m P_m[\varphi] = O(t^n). \tag{4.73}$$

It is possible to define a formal *Volterra* series with a result being a function, rather than a simple number

$$\Psi[x,\varphi] = \sum_{n=0}^{\infty} \Psi_n[x,\varphi], \tag{4.74}$$

where

$$\Psi_n[x,\varphi] = \int_a^b \mathrm{d}x_1 \int_a^b \mathrm{d}x_2 \ldots \int_a^b \mathrm{d}x_n \psi_n(x;x_1,x_2,\ldots,x_n)\varphi(x_1)\varphi(x_2)\ldots\varphi(x_n). \tag{4.75}$$

Note that the kernels $\psi_n(x;x_1,x_2,\ldots,x_n)$ have one more argument x not involved in the integration [compare with equation (4.70)].

One can perform all the algebraic manipulations with formal series depending on the same function $\varphi(x)$. It is straightforward to introduce the sum, the product and the ratio of two formal series with a result being a formal series as well [Dubois-Violette (1970a)]. The series

$$\varphi[x,\Psi] = \sum_{n=0}^{\infty} \varphi_n[x,\Psi], \tag{4.76}$$

where

$$\varphi_n[x,\Psi] = \int_a^b \mathrm{d}x_1 \int_a^b \mathrm{d}x_2 \ldots \int_a^b \mathrm{d}x_n \Phi_n(x;x_1,x_2,\ldots,x_n)\Psi(x_1)\Psi(x_2)\ldots\Psi(x_n) \tag{4.77}$$

is said to be the *inverse* of the series (4.74), if being substituted into equation (4.74) satisfies it identically.

A number of equations in physics and accelerator theory belong to the class of nonlinear integral equations of *Volterra* or *Fredholm* [Whittaker

(1963)]

$$u(x) + \epsilon G[x, u] = v(x), \tag{4.78}$$

or can be transformed to such equations. Here x is a real independent variable ($a < x < b$), $v(x) \in \mathcal{E}^{\mathbf{R}}$ is a given function of x and $u(x)$ is the unknown function, supposed to belong to the linear functional space $\mathcal{E}^{\mathbf{R}}$. Further $G[x, u]$ is an operator, generally nonlinear, acting on $u(x)$ and transforming it to another function also belonging to $\mathcal{E}^{\mathbf{R}}$. We assume also that the operator $G[x, u]$ has a formal series expansion in $u(x)$ of the type (4.74) and ϵ is a parameter not necessarily small.

In order to demonstrate how the method of *Dubois-Violette* as applied to the problem of solving nonlinear functional equations works, let us first consider the "discrete" version of equation (4.78). This means that we must find the solution to the following system of nonlinear algebraic equations

$$\mathbf{x} + \epsilon \mathbf{g}(\mathbf{x}) = \boldsymbol{\xi}, \tag{4.79}$$

where $\boldsymbol{\xi} = (\xi_1, \xi_2, \ldots, \xi_N)$ is a N-dimensional vector playing the role of N independent variables, $\mathbf{x} = (x_1, x_2, \ldots, x_N)$ is the unknown vector-variable and $\mathbf{g}(\mathbf{x}) = (g_1(\mathbf{x}), g_2(\mathbf{x}), \ldots, g_N(\mathbf{x}))$ is a given vector-function of the unknown variables $x_1, x_2, \ldots, x_N$. It is evident that the solution to the equation (4.79) will be represented by some vector-function of the independent variables (and of the parameter ϵ as well). Therefore

$$\mathbf{x} = \mathbf{f}(\boldsymbol{\xi}), \qquad \text{so that} \qquad \boldsymbol{\xi} = \mathbf{f}^{-1}(\mathbf{x}).$$

Thus, equation (4.79) become

$$\mathbf{x} + \epsilon \mathbf{g}(\mathbf{x}) = \mathbf{f}^{-1}(\mathbf{x}), \qquad \text{or} \qquad \mathbf{x} = \mathbf{f}(\mathbf{x} + \epsilon \mathbf{g}(\mathbf{x})). \tag{4.80}$$

If we formally write $\boldsymbol{\xi}$ instead of $\mathbf{x}$ in equation (4.80), we have

$$\boldsymbol{\xi} = \mathbf{f}(\boldsymbol{\xi} + \epsilon \mathbf{g}(\boldsymbol{\xi})). \tag{4.81}$$

This means that the vector-function $\mathbf{f}(\boldsymbol{\xi})$ we are seeking is the inverse of the translation operator by a vector-argument, that is precisely the given vector function $\mathbf{g}(\boldsymbol{\xi})$. The equation (4.81) can be written in a symbolic form as follows

$$\widehat{\mathbf{\Gamma}}_{inv} \exp\left[\epsilon \mathbf{g}(\boldsymbol{\xi}) \cdot \nabla_\xi\right] \bullet \mathbf{f}(\boldsymbol{\xi}) = \boldsymbol{\xi}. \tag{4.82}$$

The symbol $\widehat{\mathbf{\Gamma}}_{inv}$ implies that after the operator $\exp[\ldots]$ on the left-hand-side of equation (4.82) has been developed in a power series, the gradient operators ∇_ξ should be shifted to the right of the multipliers $\mathbf{g}(\boldsymbol{\xi})$, so that their action concerns $\mathbf{f}(\boldsymbol{\xi})$ only. The notation "$\bullet\mathbf{f}$" means that $\widehat{\mathbf{\Gamma}}_{inv}\exp[\dots]$ is an operator, acting on a given function $\mathbf{f}$. Let us now introduce the operator

$$\widehat{\mathbf{\Gamma}}\exp\left[-\epsilon\nabla_\xi\cdot\mathbf{g}(\boldsymbol{\xi})\right]\bullet \tag{4.83}$$

which is inverse in some sense to the operator defined in equation (4.82). The symbol $\widehat{\mathbf{\Gamma}}$ implies differentiation acting on the function $\mathbf{f}(\boldsymbol{\xi})$ (which should be added), as well as on the function $\mathbf{g}(\boldsymbol{\xi})$. All the gradient operators present in the series expansion of (4.83) must be shifted to the left, so that their action spreads on the function $\mathbf{g}(\boldsymbol{\xi})$ as well. One can check in a straightforward manner the validity of the identity

$$\widehat{\mathbf{\Gamma}}\exp\left[-\epsilon\nabla_\xi\cdot\mathbf{g}(\boldsymbol{\xi})\right]\bullet\widehat{\mathbf{\Gamma}}_{inv}\exp\left[\epsilon\mathbf{g}(\boldsymbol{\xi})\cdot\nabla_\xi\right]\bullet\mathbf{f}(\boldsymbol{\xi})$$

$$=\mathbf{f}(\boldsymbol{\xi})\widehat{\mathbf{\Gamma}}\exp\left[-\epsilon\nabla_\xi\cdot\mathbf{g}(\boldsymbol{\xi})\right]\bullet\mathbf{1}. \tag{4.84}$$

Applying the operator (4.83) on both sides of equation (4.82) and solving for $\mathbf{f}(\boldsymbol{\xi})$ with due account of the identity (4.84), we finally arrive at

$$\mathbf{x}=\mathbf{f}(\boldsymbol{\xi})=\frac{\widehat{\mathbf{\Gamma}}\exp\left[-\epsilon\nabla_\xi\cdot\mathbf{g}(\boldsymbol{\xi})\right]\bullet\boldsymbol{\xi}}{\widehat{\mathbf{\Gamma}}\exp\left[-\epsilon\nabla_\xi\cdot\mathbf{g}(\boldsymbol{\xi})\right]\bullet\mathbf{1}}. \tag{4.85}$$

It is worthwhile to note that the solution (4.85) to the equation (4.79) considered as a function of ϵ *is not* a conventional power series, but a ratio of two series. In fact, one arrives at a similar result regarding the solution to the simple linear system

$$\mathbf{x}+\epsilon\widehat{\mathcal{A}}\cdot\mathbf{x}=\boldsymbol{\xi},$$

where $\widehat{\mathcal{A}}$ is a $N\times N$ matrix. It is well-known that according to *Kramer*'s rule its solution is expressed as a ratio of two polynomials of order $N-1$ and N respectively, rather than a single polynomial in ϵ. Therefore, one can expect that (4.85) will be well-behaved even for large values of the perturbation parameter ϵ.

The generalization of the expression (4.85), suited to represent the solution to equation (4.78) is almost straightforward. One ought to replace the discrete index with a new continuous variable $x\in(a,b)$, that is the components x_i and ξ_i $(i=1,2,\dots,N)$ become now the functions $u(x)$ and $v(x)$ respectively, while the operator $G[x,u]$ stands for the vector-function

$\mathbf{g}(\mathbf{x})$. The differential operator ∇_ξ should be replaced by the functional derivative $\delta/\delta v(x)$, so that the expression (4.85) is transformed to

$$u(x) = F[x, v] = \frac{\widehat{\mathbf{\Gamma}} \exp\left\{-\epsilon \int \mathrm{d}y \frac{\delta}{\delta v(y)} G[y, v]\right\} \bullet v(x)}{\widehat{\mathbf{\Gamma}} \exp\left\{-\epsilon \int \mathrm{d}y \frac{\delta}{\delta v(y)} G[y, v]\right\} \bullet 1}. \tag{4.86}$$

The formula (4.86) can be obtained directly [Dubois-Violette (1970a)] by repeating the steps from equation (4.79) to equation (4.85), utilizing the *Taylor* expansion for functionals

$$F[y(x) + \varphi(x)] = \sum_{n=0}^{\infty} \frac{1}{n!} \left[y(x) \frac{\delta}{\delta \varphi(x)}\right]^n F[\varphi] \tag{4.87}$$

proposed by *Volterra*, where

$$\left[y(x) \frac{\delta}{\delta \varphi(x)}\right]^n$$

$$= \int \mathrm{d}x_1 \int \mathrm{d}x_2 \ldots \int \mathrm{d}x_n y(x_1) y(x_2) \ldots y(x_n) \frac{\delta^n}{\delta \varphi(x_1) \ldots \delta \varphi(x_n)}. \tag{4.88}$$

Since the solution (4.86) to the equation (4.78) is given in a symbolic form, it is necessary to clarify the way it should be applied. The first step is to develop the exponential $\exp\{\ldots\}$ in the numerator as well as in the denominator in a power series in ϵ (and certainly, in the functional G). The expression

$$\left\{\int \mathrm{d}y \frac{\delta}{\delta v(y)} G[y, v]\right\}^n$$

entering the coefficient of the n-th order term in the perturbation parameter ϵ has to be represented in the form of a n-fold integral, which means that we must introduce different integration variables, and according to the definition of the $\widehat{\mathbf{\Gamma}}$-symbol all functional derivatives should be shifted to the left of $G[y_i, v]$, namely

$$\widehat{\mathbf{\Gamma}} \left\{\int \mathrm{d}y \frac{\delta}{\delta v(y)} G[y, v(y)]\right\}^n$$

$$= \int \mathrm{d}y_1 \ldots \int \mathrm{d}y_n \frac{\delta^n G[y_1, v(y_1)] \ldots G[y_n, v(y_n)]}{\delta v(y_1) \ldots \delta v(y_n)}. \tag{4.89}$$

It is worth to note that the argument y in the operator $G[y, v(y)]$ means that the argument of the new function obtained as a result of the action of G on $v(\dots)$ must be y, rather than the function v enters G with the same argument y as in the functional derivative $\delta/\delta v(y)$. The next step is to substitute the representation (4.74) and (4.75) for each operator $G[y_i, v(y_i)]$ encountered on the right-hand-side of equation (4.89) and then to plug the symbols $\delta/\delta v(y_i)$ under the integrals coming from the representation (4.74) and (4.75). The function $v(x)$ in the numerator of (4.86) on which the entire operator acts must be placed under the integral as well. Further the products of integrals thus obtained have to be represented as a multiple integral of multiplicity equal to the sum of integral multiplicities entering the corresponding product. As a consequence we obtain two expressions in which all the arguments, except for the argument x are integration variables, that in principle should not be present in the final result. As far as the functional differentiations are concerned, we have to apply $\delta/\delta v(y_i)$ on a product of v-functions with different arguments, that is $v(z_1)v(z_2)\dots v(z_m)$. This gives

$$\frac{\delta}{\delta v(y_i)} v(z_1)v(z_2)\dots v(z_m)$$

$$= \sum_{k=1}^{m} v(z_1)\dots v(z_{k-1})\delta(z_k - y_i)v(z_{k+1})\dots v(z_m). \tag{4.90}$$

After the performance of all manipulations described above, we end up with integrals of known functions, which are products of v-functions and coefficients from the formal series representation of $G[y, v]$. In principle it should be possible to handle these integrals.

The problem of finding $u(x)$ has been reduced to computing the ratio of two series with terms that are combinations of integrals with increasing multiplicity. The form (4.86) representing the unknown function $u(x)$ is quite similar to the solution of linear *Fredholm* integral equations by the *N/D method* [Whittaker (1963)].

Let us now apply the method of formal series to the analysis of nonlinear beam dynamics. The Hamiltonian (4.1) describing the one-dimensional nonlinear betatron motion of particles in cyclic accelerators and storage rings can be written as

$$H(I, \alpha; \theta) = \dot{\chi}(\theta)I + V(I, \alpha; \theta), \tag{4.91}$$

where I denotes the action variable (unperturbed adiabatic invariant). Furthermore, the perturbation $V(I,\alpha;\theta)$ is periodic in the angle

$$V(I,\alpha;\theta) = \sum_{m=-\infty}^{\infty} V_m(I;\theta)e^{im\alpha}, \tag{4.92}$$

where $V_0(I;\theta) \neq 0$ in general. Following again the standard procedure described in paragraph 4.2, we seek a canonical transformation given by the generating function

$$S(J,\alpha;\theta) = \alpha J + G(J,\alpha;\theta), \tag{4.93}$$

such that the new action J is an exact invariant. The equations defining the new canonical variables (J,a) and the new Hamiltonian $\overline{H}(J;\theta)$ are

$$a = \alpha + G_J(J,\alpha;\theta), \qquad I = J + G_\alpha(J,\alpha;\theta), \tag{4.94}$$

$$\overline{H}(J;\theta) = G_\theta + \dot{\chi}(J + G_\alpha) + V(J + G_\alpha,\alpha;\theta), \tag{4.95}$$

where the subscripts indicate partial differentiation with respect to the variables indicated. We wish that the new Hamiltonian depends on the new action only, so that

$$\overline{H}(J;\theta) = \dot{\chi}J + \widehat{\mathcal{P}}V(J + G_\alpha(J,\alpha;\theta),\alpha;\theta), \tag{4.96}$$

where the projection operator

$$\widehat{\mathcal{P}}\ldots = \frac{1}{2\pi}\int_0^{2\pi} \mathrm{d}\alpha\ldots \tag{4.97}$$

implies averaging over the angle variable. Therefore the *Hamilton-Jacobi* equation can be written as

$$G_\theta + \dot{\chi}(\theta)G_\alpha + \left(1 - \widehat{\mathcal{P}}\right)V(J + G_\alpha,\alpha;\theta) = 0. \tag{4.98}$$

Its formal solution is given by the expression

$$G(J,\alpha;\theta) + \left(1 - \widehat{\mathcal{P}}\right)\int_0^\theta \mathrm{d}\tau_1 \widehat{\mathbf{T}}_\alpha(\chi(\tau_1) - \chi(\theta))V(J + G_\alpha(J,\alpha;\tau_1),\alpha;\tau_1)$$

$$= G^{(0)}(J,\alpha;\theta), \tag{4.99}$$

where

$$\widehat{\mathbf{T}}_\alpha(\mu) = \exp\left(\mu \frac{\partial}{\partial \alpha}\right) \tag{4.100}$$

is the translation operator, and the function $G^{(0)}(J, \alpha; \theta)$ satisfies the homogeneous equation

$$G_\theta^{(0)} + \dot{\chi}(\theta) G_\alpha^{(0)} = 0.$$

Differentiating both sides of equation (4.99) with respect to the angle α we transform it to a nonlinear integral equation of *Volterra* of the second kind

$$G_\alpha(J, \alpha; \theta) + \int_0^\theta \mathrm{d}\tau_1 \widehat{\mathbf{T}}_\alpha(\chi(\tau_1) - \chi(\theta)) \frac{\partial}{\partial \alpha} V(J + G_\alpha(J, \alpha; \tau_1), \alpha; \tau_1)$$

$$= f(J, \alpha; \theta), \tag{4.101}$$

where

$$f(J, \alpha; \theta) = G_\alpha^{(0)}(J, \alpha; \theta). \tag{4.102}$$

We are ready now to apply (with a slight generalization) the result (4.86) to equation (4.101), and its solution for times $\theta \in (0, 2k\pi)$ is found to be

$$G_\alpha(\alpha; \theta) = \frac{\widehat{\mathbf{\Gamma}} \exp\left\{ - \int\limits_0^{2k\pi} \mathrm{d}\sigma \frac{\delta}{\delta f(\sigma)} \mathcal{G}[\sigma, f(\sigma)] \right\} \bullet f(\theta)}{\widehat{\mathbf{\Gamma}} \exp\left\{ - \int\limits_0^{2k\pi} \mathrm{d}\sigma \frac{\delta}{\delta f(\sigma)} \mathcal{G}[\sigma, f(\sigma)] \right\} \bullet 1}, \tag{4.103}$$

where

$$\mathcal{G}[\sigma, f(\sigma)] = \int_0^\sigma \mathrm{d}\tau_1 \widehat{\mathbf{T}}_\alpha(\chi(\tau_1) - \chi(\sigma)) \frac{\partial}{\partial \alpha} V(J + f(\tau_1), \alpha; \tau_1). \tag{4.104}$$

Expanding the $\widehat{\mathbf{\Gamma}}$-exponent in the numerator as well as in the denominator of the right-hand-side of equation (4.103) in a formal *Volterra* series, we represent the solution as a ratio of two series in the form

$$G_\alpha(J, \alpha; \theta) = \frac{\sum\limits_{n=0}^{\infty} P_n(J, \alpha; \theta)}{\sum\limits_{n=0}^{\infty} Q_n(J, \alpha)}. \tag{4.105}$$

In order to proceed further, we note that the initial condition $G^{(0)}(J, \alpha; \theta) \equiv 0$ automatically implies that $f(J, \alpha; \theta)$ vanishes. Taking into account this fact for the first few terms we obtain

$$P_0(J, \alpha; \theta) = 0, \qquad Q_0(J, \alpha) = 1, \tag{4.106}$$

$$P_1(J, \alpha; \theta) = -\int_0^{\theta} \mathrm{d}\tau V_\alpha(J, \alpha + \chi(\tau) - \chi(\theta); \tau), \tag{4.107}$$

$$Q_1(J, \alpha) = -\int_0^{2k\pi} \mathrm{d}\tau V_{\alpha J}(J, \alpha; \tau), \tag{4.108}$$

$$P_2(J, \alpha; \theta) = P_1(J, \alpha; \theta) Q_1(J, \alpha)$$

$$-\int_0^{\theta} \mathrm{d}\tau P_1(J, \alpha; \tau) V_{\alpha J}(J, \alpha + \chi(\tau) - \chi(\theta); \tau), \tag{4.109}$$

$$Q_2(J, \alpha) = \frac{1}{2} Q_1^2(J, \alpha) - \int_0^{2k\pi} \mathrm{d}\sigma P_1(J, \alpha; \sigma) V_{\alpha J J}(J, \alpha; \sigma). \tag{4.110}$$

Once the generator G_α has been calculated, we can immediately reproduce the phase portrait of the system scanned at subsequent times. This can be done by utilizing equation (4.94) for different values of the action invariant J. The nonlinear tune shift can be calculated according to the expression

$$\Delta\nu(J) = \frac{1}{(2\pi)^2} \int_0^{2\pi} \mathrm{d}\theta \int_0^{2\pi} \mathrm{d}\alpha [1 + G_{\alpha J}(J, \alpha; \theta)] V_J(J + G_\alpha(J, \alpha; \theta), \alpha; \theta). \tag{4.111}$$

An important comment is in order at this point. Note that if we re-expand the denominator of the fraction on the right-hand-side of equation (4.105) and perform the multiplication, we will end up with an expression that can be obtained by a direct perturbation expansion of the solution to equation (4.101).

It is interesting and important to study the possibility of extending the method of *Dubois-Violette* to non autonomous systems with more than one degree of freedom. For that purpose it is necessary to define the functional version of the gradient (divergence) operator in the N-dimensional vector space. More details can be found in [Dubois-Violette (1970a)] and [Tzenov (1995)].

4.6 Renormalization Transformation for Two Resonances

In paragraph 4.4 we discussed the resemblance in the effect of primary and secondary resonances on the zero-order and first-order invariants. Based on this similarity *Escande* and *Doveil* developed a renormalization transformation method [Escande (1981)] for determining the destruction of an invariant curve, lying between two primary resonances.

The idea of the method consists in performing a sequence of transformations such that the resulting new Hamiltonians preserve the form of the original one, and describe successively higher-order resonances. At each step on finer and finer scales the two most important resonances bounding the invariant curve we are concerned with are analyzed, and a renormalization transformation in the resonance parameter space is being built. The invariant curve exists, provided the amplitudes of the resonances tend to zero.

To demonstrate the renormalization transformation method we consider the two resonance model [compare with equation (3.13)] in one dimension

$$H(J,\alpha;\theta) = \nu J + \frac{\alpha_N}{2} J^2 - \mathcal{D}_1 (2J)^{p_1} \cos(2p_1\alpha - m_1\theta)$$

$$-\mathcal{D}_2 (2J)^{p_2} \cos(2p_2\alpha - m_2\theta). \tag{4.112}$$

Note that the minus sign of the resonance terms in equation (4.112) can be always achieved by a proper shift of the angle variable and by redefinition of the initial azimuthal position. We consider now the existence of an invariant curve J_0 with inverse rotation number $Q(J_0)$, such that

$$\frac{2p_2}{m_2} < Q(J_0) < \frac{2p_1}{m_1}, \tag{4.113}$$

where by definition the inverse rotation number of each resonance described by the Hamiltonian (4.112) is given by

$$Q\left(J_r^{(1,2)}\right) = \frac{1}{\nu + \alpha_N J_r^{(1,2)}} = \frac{2p_{1,2}}{m_{1,2}} \tag{4.114}$$

at the resonance value of the action $J_r^{(1,2)}$. The canonical transformation defined by the generating function

$$\mathcal{F}_2(J_1, \alpha; \theta) = (2p_1\alpha - m_1\theta)J_1,$$

and by the new action-angle variables

$$\alpha_1 = 2p_1\alpha - m_1\theta, \qquad J = 2p_1J_1, \tag{4.115}$$

yields the new Hamiltonian

$$H_1(J_1, \alpha_1; \theta) = \delta_1 J_1 + 2p_1^2\alpha_N J_1^2 - \mathcal{D}_1(4p_1J_1)^{p_1}\cos\alpha_1$$

$$-\mathcal{D}_2(4p_1J_1)^{p_2}\cos(k\alpha_1 - \Delta\theta), \tag{4.116}$$

where

$$\delta_1 = 2p_1\nu - m_1, \qquad k = \frac{p_2}{p_1}, \qquad \Delta = \frac{p_1m_2 - p_2m_1}{p_1}. \tag{4.117}$$

In order to cast the Hamiltonian (4.116) in a simpler form originally used by *Escande* and *Doveil* in their analysis, it is necessary to perform two additional scaling transformations. The first one involves a change to a new independent variable, while the second concerns appropriate scaling of the action (momentum) variable. Although the second transformation does not preserve the *Poisson* bracket, it can be easily checked that if the transformed action differs from the original one by a constant multiplier, the form of *Hamilton*'s equations of motion is preserved, provided the new Hamiltonian differs from the original one by the same multiplier. In the first case the Hamiltonian should be multiplied by a factor reciprocal to the scaling factor of the independent variable. With all the above in hand we obtain

$$\mathcal{H}(p, x; t) = \frac{p^2}{2} - \epsilon_1\cos x - \epsilon_2\cos(kx - t), \tag{4.118}$$

where

$$\mathcal{H} = \frac{4p_1^2\alpha_N}{\Delta^2}H_1, \qquad p = \frac{4p_1^2\alpha_N}{\Delta}\left(J_1 + J_1^{(0)}\right), \tag{4.119}$$

$$J_1^{(0)} = \frac{\delta_1}{4p_1^2\alpha_N}, \qquad x = \alpha_1, \qquad t = \Delta\theta, \tag{4.120}$$

and

$$\epsilon_1 = \frac{4p_1^2\alpha_N}{\Delta^2}\mathcal{D}_1(2J_0)^{p_1}, \qquad \epsilon_2 = \frac{4p_1^2\alpha_N}{\Delta^2}\mathcal{D}_2(2J_0)^{p_2}. \tag{4.121}$$

The Hamiltonian (4.118) describes the motion of a pendulum, perturbed by a wave with wave number k and unit frequency. In the course of successive renormalization its form will be preserved yielding coupled *map equations* for the parameters k, ϵ_1 and ϵ_2. Instead of following [Escande (1981)] and representing (4.118) in terms of the exact pendulum action-angle variables, we will proceed in a way similar to that of paragraph 4.2. We consider a canonical transformation close to the identity generated by

$$\mathcal{F}_2(\overline{p}, x; t) = x\overline{p} + \mathcal{S}(\overline{p}, x; t),$$

and write the new Hamiltonian as

$$\overline{\mathcal{H}} = \frac{\partial \mathcal{S}}{\partial t} + \frac{1}{2}\left[\overline{p}^2 + 2\overline{p}\frac{\partial \mathcal{S}}{\partial x} + \left(\frac{\partial \mathcal{S}}{\partial x}\right)^2\right] - \epsilon_1 \cos x - \epsilon_2 \cos(kx - t). \tag{4.122}$$

We choose the function $\mathcal{S}(\overline{p}, x; t)$ so as to eliminate the third term on the right-hand-side of equation (4.122). To first order this yields

$$\frac{\partial \mathcal{S}}{\partial t} + \overline{p}\frac{\partial \mathcal{S}}{\partial x} = \epsilon_1 \cos x,$$

and therefore

$$\mathcal{S}(\overline{p}, x; t) = \frac{\epsilon_1}{\overline{p}} \sin x.$$

Taking into account the first-order approximation

$$x = \overline{x} + \frac{\epsilon_1}{\overline{p}^2} \sin \overline{x},$$

we obtain

$$\overline{\mathcal{H}}(\overline{p}, \overline{x}; t) = \frac{\overline{p}^2}{2} + \frac{\epsilon_1^2}{4\overline{p}^2} - \epsilon_2 \sum_{n=-\infty}^{\infty} \mathcal{J}_n\left(\frac{k\epsilon_1}{\overline{p}^2}\right) \cos[(k+n)\overline{x} - t], \tag{4.123}$$

where averaging over x has been performed and use of the identity (4.62) has been made. The Hamiltonian (4.123) is similar to (4.66) showing secondary resonances between harmonics of the rotation frequency and the driving unit frequency

$$\overline{p}_n - \frac{\epsilon_1^2}{2\overline{p}_n^3} = \frac{1}{k+n}. \tag{4.124}$$

Without loss of generality we can assume that the invariant curve lies between secondary resonances n and $n+1$. The next step is to average the Hamiltonian (4.123) over all secondary resonances except n and $n+1$. As a result we obtain again a two resonance model Hamiltonian of the form

$$\overline{\mathcal{H}}(\overline{p},\overline{x};t) = \frac{\overline{p}^2}{2} + \frac{\epsilon_1^2}{4\overline{p}^2} - \epsilon_2 \mathcal{J}_n\left(\frac{k\epsilon_1}{\overline{p}^2}\right)\cos\left[(k+n)\overline{x} - t\right]$$

$$-\epsilon_2 \mathcal{J}_{n+1}\left(\frac{k\epsilon_1}{\overline{p}^2}\right)\cos\left[(k+n+1)\overline{x} - t\right]. \tag{4.125}$$

Now that we have the new Hamiltonian in the desired form, it is easy to convert it to (4.118) by repeating the steps which lead us to the Hamiltonian (4.118) starting from (4.112). Depending on whether a certain resonance is closer to

$$Q = \left(\overline{p}_{00} - \frac{\epsilon_1^2}{2\overline{p}_{00}^3}\right)^{-1}, \qquad \overline{p}_{00} = \frac{1}{\Delta}(2p_1\alpha_N J_0 + \delta_1),$$

we choose it as the ϵ_1-resonance, with the remaining secondary resonance as the ϵ_2-resonance. Thus we obtain

$$\mathcal{H}'(p',x';t') = \frac{p'^2}{2} - \epsilon_1' \cos x' - \epsilon_2' \cos(k'x' - t'), \tag{4.126}$$

where

$$\epsilon_1' = (k+n+\lambda)P_{n+\lambda}\mathcal{J}_{n+\lambda}\left(\frac{k\epsilon_1}{\overline{p}_{00}^2}\right)\epsilon_2, \tag{4.127}$$

$$\epsilon_2' = (k+n+\lambda)P_{n+\lambda}\mathcal{J}_{n+1-\lambda}\left(\frac{k\epsilon_1}{\overline{p}_{00}^2}\right)\epsilon_2, \tag{4.128}$$

$$k' = \frac{k+n+1-\lambda}{k+n+\lambda}, \tag{4.129}$$

and

$$P_n = (k+n)^2\left[4(k+n) - \frac{3}{\overline{p}_n}\right]. \tag{4.130}$$

Here λ is equal to zero or unity depending on whether the n or $n+1$ resonance is closer to Q.

Equations (4.127) – (4.129) define a renormalization transformation which yields a self-similar problem. It is instructive to study the third transformation (4.129) separately since it is decoupled from the first two. To calculate the fixed point of the map, we set

$$k_n^\lambda = \frac{k_n^\lambda + n + 1 - \lambda}{k_n^\lambda + n + \lambda}, \tag{4.131}$$

which yields

$$k_{cn}^0 = \frac{1-n}{2} + \frac{1}{2}\sqrt{n^2 + 2n + 5}, \qquad k_{cn}^1 = -\frac{n}{2} + \frac{1}{2}\sqrt{n^2 + 4n}. \tag{4.132}$$

Let us now analyze the resonance strengths transformation, and suppose that we have iterated the map (4.129) to its fixed point. It is clear that the map defined by equations (4.127) and (4.128) has an attracting fixed point at $\epsilon_1 = \epsilon_2 = 0$. This means that for some initial ϵ_1 and ϵ_2 which are sufficiently small, the strengths of the secondary resonances surrounding the invariant curve in question tend to zero. Therefore, the invariant curve we are looking at does exist.

This however is not true for arbitrary ϵ_1 and ϵ_2. To see that the map defined by equations (4.127) and (4.128) has a critical fixed point, we assume that the argument of the Bessel functions is a small quantity and restrict ourselves to the first term in the series expansion of $\mathcal{J}_n$. We obtain the simplified map

$$\epsilon_1' = \frac{\left(k_{cn}^\lambda + n + \lambda\right)^4}{(n+\lambda)!} \left(\frac{k_{cn}^\lambda}{2\overline{p}_{00}^2}\right)^{n+\lambda} \epsilon_1^{n+\lambda} \epsilon_2, \tag{4.133}$$

$$\epsilon_2' = \frac{\left(k_{cn}^\lambda + n + \lambda\right)^4}{(n+1-\lambda)!} \left(\frac{k_{cn}^\lambda}{2\overline{p}_{00}^2}\right)^{n+1-\lambda} \epsilon_1^{n+1-\lambda} \epsilon_2, \tag{4.134}$$

which yields the critical fixed point

$$\epsilon_{1c} = \frac{2\overline{p}_{00}^2}{k_{cn}^\lambda} \left[\frac{(n+1-\lambda)!}{\left(k_{cn}^\lambda + n + \lambda\right)^4}\right]^{\frac{1}{n+1-\lambda}}, \tag{4.135}$$

$$\epsilon_{2c} = \frac{2\overline{p}_{00}^2}{k_{cn}^\lambda} (n+1)^{2\lambda-1} \left[\frac{(n+1-\lambda)!}{\left(k_{cn}^\lambda + n + \lambda\right)^4}\right]^{\frac{2(1-\lambda)}{n+1-\lambda}}. \tag{4.136}$$

To make the renormalization transformation (4.133) and (4.134) more feasible, we scale ϵ_1 and ϵ_2 according to

$$\xi = \frac{\epsilon_1}{\epsilon_{1c}}, \qquad \eta = \frac{\epsilon_2}{\epsilon_{2c}}, \tag{4.137}$$

and write the normalized renormalization map as

$$\xi' = \xi^{n+\lambda}\eta, \qquad \eta' = \xi^{n+1-\lambda}\eta. \tag{4.138}$$

Certain points in the (ξ, η) parameter space when iterated will approach the central fixed point, thus forming the basin of attraction, while others will drift away. It is interesting and useful to calculate the boundary curve, confining the basin of attraction for the central fixed point. Assuming a form [Ruth (1986)] for the boundary curve

$$\eta = \xi^{\sigma} \tag{4.139}$$

and substituting into equation (4.138), we find

$$\sigma^2 + (n - 1 + \lambda)\sigma - 1 - n + \lambda = 0. \tag{4.140}$$

Equation (4.140) has two roots

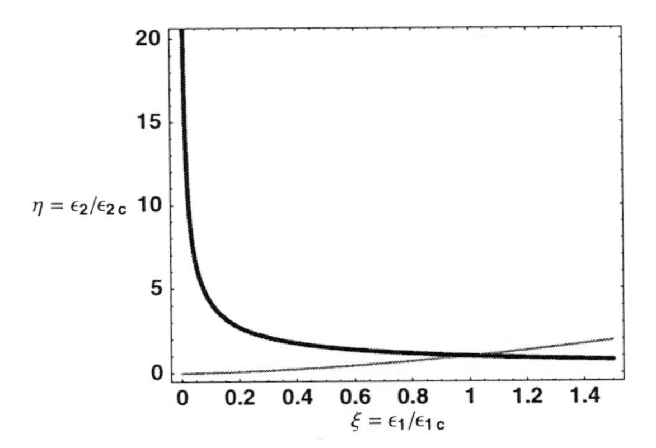

Fig. 4.1 Existence of an invariant curve in the (ξ, η) parameter space. The thick curve represents $\eta = \xi^{\sigma_2^\lambda}$, while the one originating from the central fixed point $\xi = \eta = 0$ represents $\eta = \xi^{\sigma_1^\lambda}$.

$$\sigma_{1,2}^{\lambda} = \frac{1 - n - \lambda}{2} \pm \frac{1}{2}\sqrt{n^2 + 2(1 + \lambda)n + \lambda^2 - 6\lambda + 5}, \tag{4.141}$$

defining the stable and unstable curves which are given by

$$\eta = \xi^{\sigma_2^\lambda}, \qquad \eta = \xi^{\sigma_1^\lambda}, \tag{4.142}$$

respectively.

In figure 4.1 the portrait in the (ξ, η) parameter space is shown. The points in the area below the curve $\eta = \xi^{\sigma_2^\lambda}$, after being iterated by the normalized renormalization map (4.138), tend to the fixed point $\xi = \eta = 0$. These points form the basin of attraction of the central fixed point, thus ensuring the existence of the sought-for invariant curve. Iteration of initial ξ and η lying above the stable curve $\eta = \xi^{\sigma_2^\lambda}$ leads to rapid growth. This region corresponds to the nonexistence of the invariant curve, and finally the stable curve itself corresponds to the transition case.

Chapter 5

Special Methods in Accelerator Theory

5.1 Introduction

There has been a long history in the development of asymptotic methods to study Hamiltonian flows by means of singular perturbation techniques such as the averaging method [Bogoliubov (1961)] and the method of multiple time-scales [Bender (1978)], [Kevorkian (1996)], [Murdock (1991)], [Nayfeh (1981)]. Recently a novel method based on the perturbative *renormalization group (RG) theory* has been developed [Chen (1996)] and has been successfully applied to ordinary differential equations and to some partial differential equations. An alternative formulation of the RG method based on the classical theory of envelopes has also been developed [Kunihiro (1995a)], [Kunihiro (1995b)].

The essence of the RG method is to remove secular or divergent terms from a naive perturbation series by renormalizing integration constants which usually appear in the lowest order of the expansion. As a result of the renormalization transformation a *renormalization group (RG) equation* governing the slow motion and/or the large-scale behavior of these constants (amplitudes) can be obtained. The RG method is extremely convenient for practical use because in the process of reduction of the asymptotic equation it is not necessary to introduce slow or large scales *a priori*, in contrast to other methods such as the reductive perturbation method [Mori (1998)] and the method of multiple scales. All needed to be done is to extract the structure of secular terms from the naive perturbation series, followed by application of the renormalization transformation.

Actually, the basic result of the RG method may be understood in the context of the construction of an approximate but *global* solution to a nonlinear problem. Starting from solutions of a local nature obtained within the framework of the naive perturbation theory, the RG equation is derived to improve the global behavior of the local perturbative solutions.

5.2 Renormalization Group Method

To demonstrate the basic features of the RG method, we consider the simple case of one-dimensional single-particle dynamics in a direction (say the horizontal one), transversal to the particle's design orbit. From equations (2.38), (3.1) and (3.2) it follows that the Hamiltonian can be written as

$$H(X, P; \theta) = \frac{\dot{\chi}(\theta)}{2}(P^2 + X^2) + \sum_{k=3}^{\infty} b_k(\theta) X^k, \tag{5.1}$$

where the coefficients $b_k(\theta)$ are functions of θ, and are related to $a_{k000}^{(I)}(\theta)$ in an obvious way. From the *Hamilton*'s equations of motion we obtain a single equation for X, namely

$$\frac{\mathrm{d}^2 X}{\mathrm{d}\chi^2} + X = \epsilon a_3(\theta) X^3 + \epsilon^2 a_5(\theta) X^5 + \epsilon^3 a_7(\theta) X^7 + \dots, \tag{5.2}$$

where

$$a_k(\theta) = -\frac{(k+1)\beta(\theta)}{R} b_{k+1}(\theta).$$

Here ϵ is again a formal small parameter measuring the magnitude of the multipole perturbation. Moreover, we have assumed that the multipole strengths decrease with order, and only even multipoles have been retained. The reason for that will become clear later [see expression (5.32)], for the RG technique provides a simple and effective method to find a global invariant of the motion, and to calculate the nonlinear tuneshift. In a similar manner the effect of odd multipoles can be treated. To simplify matters even further, in what follows we consider a smooth focusing approximation and regard the coefficients a_k as constants.

The most straightforward approach to the solution of equation (5.2) is to try a naive perturbation expansion of the form

$$X(\chi) = \mathcal{E} e^{-i\chi} + c.c. + \epsilon X_1(\chi) + \epsilon^2 X_2(\chi) + \dots, \tag{5.3}$$

where $\mathcal{E}$ is an arbitrary (*complex*) constant and *c.c.* stands for the complex conjugate counterpart. The amplitude and the phase of $\mathcal{E}$ should be defined in accordance with appropriate initial conditions at some initial "time" χ_0. For the first-order contribution $X_1(\chi)$ we obtain the equation

$$\left(\frac{\mathrm{d}^2}{\mathrm{d}\chi^2}+1\right)X_1 = a_3\left(3|\mathcal{E}|^2\mathcal{E}e^{-i\chi}+\mathcal{E}^3e^{-3i\chi}+c.c.\right). \tag{5.4}$$

The solution to equation (5.4) is given by

$$X_1(\chi) = a_3\left[3|\mathcal{E}|^2\mathcal{E}\mathcal{P}_1(\chi;\chi_0)e^{-i\chi}-\frac{\mathcal{E}^3}{8}e^{-3i\chi}+c.c.\right], \tag{5.5}$$

where $\mathcal{P}_1(\chi;\chi_0)$ is a polynomial (secular) solution to

$$\left(\frac{\mathrm{d}^2}{\mathrm{d}\chi^2}-2i\frac{\mathrm{d}}{\mathrm{d}\chi}\right)\mathcal{P}_1 = \widehat{\mathcal{L}}_1(\chi)\mathcal{P}_1 = 1 \tag{5.6}$$

with the initial condition $\mathcal{P}_1(\chi_0;\chi_0) = 0$ to be imposed. The next order $X_2(\chi)$ can be obtained from the second-order perturbation equation

$$\begin{aligned}\left(\frac{\mathrm{d}^2}{\mathrm{d}\chi^2}+1\right)X_2 &= \left[3a_3^2\left(3\mathcal{P}_1-\frac{1}{8}\right)+10a_5\right]|\mathcal{E}|^4\mathcal{E}e^{-i\chi}\\ &+\left[3a_3^2\left(3\mathcal{P}_1-\frac{1}{4}\right)+5a_5\right]|\mathcal{E}|^2\mathcal{E}^3e^{-3i\chi}\\ &+\left(a_5-\frac{3a_3^2}{8}\right)\mathcal{E}^5e^{-5i\chi}+c.c..\end{aligned} \tag{5.7}$$

From this, we find

$$\begin{aligned}X_2(\chi) &= \left\{3a_3^2\left[3\mathcal{P}_2(\chi;\chi_0)-\frac{\mathcal{P}_1(\chi;\chi_0)}{8}\right]+10a_5\mathcal{P}_1(\chi;\chi_0)\right\}|\mathcal{E}|^4\mathcal{E}e^{-i\chi}\\ &+\left[9a_3^2\mathcal{Q}_1(\chi;\chi_0)-\frac{1}{8}\left(5a_5-\frac{3a_3^2}{4}\right)\right]|\mathcal{E}|^2\mathcal{E}^3e^{-3i\chi}\\ &-\frac{1}{24}\left(a_5-\frac{3a_3^2}{8}\right)\mathcal{E}^5e^{-5i\chi}+c.c.,\end{aligned} \tag{5.8}$$

where

$$\left(\frac{\mathrm{d}^2}{\mathrm{d}\chi^2}-6i\frac{\mathrm{d}}{\mathrm{d}\chi}-8\right)\mathcal{Q}_1 = \widehat{\mathcal{L}}_2(\chi)\mathcal{Q}_1 = \mathcal{P}_1, \qquad \widehat{\mathcal{L}}_1(\chi)\mathcal{P}_2 = \mathcal{P}_1. \tag{5.9}$$

The explicit form of the secular terms $\mathcal{P}_1(\chi;\chi_0)$, $\mathcal{Q}_1(\chi;\chi_0)$, $\mathcal{P}_2(\chi;\chi_0)$, etc. is as follows:

$$\mathcal{P}_1(\chi;\chi_0) = \frac{i(\chi-\chi_0)}{2}, \qquad \mathcal{Q}_1(\chi;\chi_0) = -\frac{i(\chi-\chi_0)}{16} - \frac{3}{64}, \tag{5.10}$$

$$\mathcal{P}_2(\chi;\chi_0) = -\frac{(\chi-\chi_0)^2}{8} + \frac{i(\chi-\chi_0)}{8}. \tag{5.11}$$

Analyzing the first-order and the second-order results of equations (5.5) and (5.8), at once we detect the secular nature of the perturbation. Terms proportional to $\mathcal{P}_1$, $\mathcal{P}_2$ and $\mathcal{Q}_1$ become eventually much larger than the rest of the terms, thus limiting the uniform reliability of the solution and making it valid only locally in a small vicinity around χ_0. The RG approach suggests that we may still use the perturbative solution, if we could remove the terms divergent in the long-term limit. To do so, let us collect all the terms to second order which multiply the fundamental harmonic $e^{-i\chi}$. We have

$$X(\chi) = \left\{1 + 3\epsilon a_3|\mathcal{E}|^2\mathcal{P}_1 + \epsilon^2\left[3a_3^2\left(3\mathcal{P}_2 - \frac{\mathcal{P}_1}{8}\right) + 10a_5\mathcal{P}_1\right]|\mathcal{E}|^4\right\}\mathcal{E}e^{-i\chi}$$

$$+ c.c. + \text{higher harmonics}. \tag{5.12}$$

Introducing the intermediate "time" τ close to χ, we split $\chi-\chi_0$ as $\chi-\chi_0 = \chi-\tau+\tau-\chi_0$ and try to remove the last term $\tau-\chi_0$. This can be done by rewriting equation (5.12) as

$$X(\chi) = \left\{1 + 3\epsilon a_3|\mathcal{E}_R|^2\mathcal{P}_1(\chi;\tau) + \epsilon^2\left[3a_3^2\left(3\mathcal{P}_2(\chi;\tau) - \frac{\mathcal{P}_1(\chi;\tau)}{8}\right)\right.\right.$$

$$\left.\left. + 10a_5\mathcal{P}_1(\chi;\tau)\right]|\mathcal{E}_R|^4\right\}\mathcal{E}_R e^{-i\chi} + c.c. + \text{higher harmonics}, \tag{5.13}$$

where

$$\mathcal{E}_R(\tau;\chi_0) = \mathcal{E}e^{i\Theta(\tau;\chi_0)} \tag{5.14}$$

with

$$\Theta(\tau;\chi_0) = \epsilon\left[\frac{3a_3}{2} + 5\epsilon\left(\frac{3a_3^2}{16} + a_5\right)|\mathcal{E}|^2\right]|\mathcal{E}|^2(\tau-\chi_0) + \ldots. \tag{5.15}$$

The basic idea lying behind the operations performed above is that the initial condition was imposed at an arbitrary initial time χ_0 in the past, so that $\mathcal{E}$ is actually not known. Therefore, we can use it as a free adjustable parameter to "repair" the perturbation solution for suitable use and fit it subsequently to the current value of X at time χ. Since τ does not appear in the original problem X should not depend on τ. Therefore, its derivative with respect to τ equals zero, namely

$$\frac{\partial X}{\partial \tau} = 0.$$

In explicit form this reads

$$\frac{\mathrm{d}\mathcal{E}_R}{\mathrm{d}\tau} = \frac{3i\epsilon a_3}{2}|\mathcal{E}_R|^2\mathcal{E}_R + 5i\epsilon^2\left(\frac{3a_3^2}{16} + a_5\right)|\mathcal{E}_R|^4\mathcal{E}_R, \tag{5.16}$$

where we have discarded higher harmonic terms. The term proportional to $\chi - \tau$ coming from the first derivative of $\mathcal{P}_2(\chi;\tau)$ with respect to τ has also been dropped for reasons explained below. The renormalized solution (5.13) simplifies considerably if we set $\tau = \chi$. This yields

$$X(\chi) = \mathcal{E}_R(\chi)e^{-i\chi} + c.c. + \text{higher harmonics}, \tag{5.17}$$

where $\mathcal{E}_R(\chi)$ is a solution to the equation

$$\frac{\mathrm{d}\mathcal{E}_R}{\mathrm{d}\chi} = \frac{3i\epsilon a_3}{2}|\mathcal{E}_R|^2\mathcal{E}_R + 5i\epsilon^2\left(\frac{3a_3^2}{16} + a_5\right)|\mathcal{E}_R|^4\mathcal{E}_R, \tag{5.18}$$

The reason for dropping the term proportional to $\chi - \tau$ in equation (5.16) becomes now clear, because it disappears after setting $\tau = \chi$.

Equation (5.18) is the sought-for *RG equation* governing the global behavior of the solution. It can be obtained in a different manner [Goto (1999)] based on the use of the simple structure of the translational *(Lie)* group symmetry. Because of its simplicity, efficiency and elegance we will reproduce the *Lie-group* approach to the perturbative RG method here. Following *Goto et al.*, we define the renormalized amplitude $\widetilde{\mathcal{E}}(\chi)$ of the fundamental harmonic as

$$\widetilde{\mathcal{E}}(\chi) = \{1 + 3\epsilon a_3|\mathcal{E}|^2\mathcal{P}_1(\chi)$$

$$+\epsilon^2\left[3a_3^2\left(3\mathcal{P}_2(\chi) - \frac{\mathcal{P}_1(\chi)}{8}\right) + 10a_5\mathcal{P}_1(\chi)\right]|\mathcal{E}|^4\Big\}\mathcal{E} + O(\epsilon^3), \tag{5.19}$$

while the coefficients of the third harmonic $e^{-3i\chi}$ terms are summed up to form the amplitude $\epsilon\widetilde{\mathcal{E}}_3(\chi)$, given by

$$\epsilon\widetilde{\mathcal{E}}_3(\chi) = -\frac{\epsilon}{8}\left[a_3 + \epsilon\left(9a_3^2\mathcal{P}_1(\chi) + \frac{21a_3^2}{8} + 5a_5\right)|\mathcal{E}|^2 + O\left(\epsilon^3\right)\right]\mathcal{E}^3. \quad (5.20)$$

Here we have set $\chi_0 = 0$ and have introduced the notation $\mathcal{P}_i(\chi) = \mathcal{P}_i(\chi; 0)$ for $i = 1, 2$. Solving iteratively equation (5.19) for $\mathcal{E}$ in terms of $\widetilde{\mathcal{E}}(\chi)$, we obtain

$$\mathcal{E} = \widetilde{\mathcal{E}}(\chi) - 3\epsilon a_3\left|\widetilde{\mathcal{E}}(\chi)\right|^2\widetilde{\mathcal{E}}(\chi)\mathcal{P}_1(\chi) + O\left(\epsilon^2\right).$$

Substituting this into equation (5.20), we can eliminate the secular terms in the amplitude of the third harmonic order by order. The result is

$$\widetilde{\mathcal{E}}_3(\chi) = -\frac{1}{8}\left[a_3 + \epsilon\left(\frac{21a_3^2}{8} + 5a_5\right)\left|\widetilde{\mathcal{E}}(\chi)\right|^2 + O\left(\epsilon^2\right)\right]\widetilde{\mathcal{E}}^3(\chi). \quad (5.21)$$

Once the amplitude of the fundamental harmonic has been renormalized by the renormalization transformation (5.19), the process of elimination of secular terms in the higher harmonics is quite general. This is because the sources of secularity in the higher harmonics are precisely the resonant secular terms in the fundamental harmonic. In this sense the secular terms in the higher harmonics can be regarded as *non-resonant secular terms*, which automatically disappear after the renormalization of the amplitude of the fundamental harmonic.

Since the particle's dynamics as governed by the equation of motion (5.2) is *translationally invariant* we can perform a formal transformation $\chi \longrightarrow \chi + \Theta$ and write

$$\widetilde{\mathcal{E}}(\chi + \Theta) = \left\{1 + 3\epsilon a_3\left|\widetilde{\mathcal{E}}(\chi)\right|^2\mathcal{P}_1(\Theta)\right.$$

$$\left. +\epsilon^2\left[3a_3^2\left(3\mathcal{P}_2(\Theta) - \frac{\mathcal{P}_1(\Theta)}{8}\right) + 10a_5\mathcal{P}_1(\Theta)\right]\left|\widetilde{\mathcal{E}}(\chi)\right|^4\right\}\widetilde{\mathcal{E}}(\chi) + O\left(\epsilon^3\right). \quad (5.22)$$

Equation (5.22) is the explicit representation of the *Lie* group, reflecting the translational invariance embedded in the renormalization transformation (5.19). Clearly, the result (5.22) can be extended to arbitrary order, although calculations become rather tedious. Next, we observe that secular terms encountered in the coefficient of the fundamental harmonic $e^{-i\chi}$ in X_n are polynomials of degree n in the independent time variable χ, and proportional to $|\mathcal{E}|^{2n}\mathcal{E}$. Thus, we can write equation (5.22) in an alternative form as

$$\widetilde{\mathcal{E}}(\chi+\Theta) = \widetilde{\mathcal{E}}(\chi) + \sum_{n=1}^{\infty} \epsilon^n \sum_{k=1}^{n} g_{n,k}\Theta^k \left|\widetilde{\mathcal{E}}(\chi)\right|^{2n} \widetilde{\mathcal{E}}(\chi)$$

$$= \widetilde{\mathcal{E}}(\chi) + \epsilon g_{1,1}\Theta\left|\widetilde{\mathcal{E}}(\chi)\right|^2 \widetilde{\mathcal{E}}(\chi) + \epsilon^2\left(\Theta^2 g_{2,2} + \Theta g_{2,1}\right)\left|\widetilde{\mathcal{E}}(\chi)\right|^4 \widetilde{\mathcal{E}}(\chi) + \cdots, \quad (5.23)$$

where $g_{n,k}$ are constant coefficients given by

$$g_{1,1} = \frac{3ia_3}{2}, \qquad g_{2,1} = i\left(\frac{15a_3^2}{16} + 5a_5\right), \quad (5.24)$$

$$g_{3,1} = i\left(\frac{123a_3^3}{128} + 5a_3a_5 + \frac{37a_7}{2}\right), \quad (5.25)$$

$$g_{2,2} = -\frac{9a_3^2}{8}, \qquad g_{3,2} = -a_3\left(\frac{45a_3^2}{32} + \frac{15a_5}{2}\right), \quad (5.26)$$

$$g_{3,3} = -\frac{9ia_3^3}{16}, \qquad \text{etc.} \quad (5.27)$$

On the other hand the left-hand-side of equation (5.23) can be written as

$$\widetilde{\mathcal{E}}(\chi+\Theta) = \exp\left(\Theta\frac{\partial}{\partial\chi}\right)\widetilde{\mathcal{E}}(\chi) = \left(1 + \Theta\frac{\partial}{\partial\chi} + \frac{\Theta^2}{2}\frac{\partial^2}{\partial\chi^2} + \cdots\right)\widetilde{\mathcal{E}}(\chi). \quad (5.28)$$

Equating coefficients of equal powers in Θ, we obtain the set of equations

$$\frac{1}{n!}\frac{\mathrm{d}^n\widetilde{\mathcal{E}}}{\mathrm{d}\chi^n} = \epsilon^n \sum_{k=0}^{\infty} \epsilon^k g_{n+k,n}\left|\widetilde{\mathcal{E}}(\chi)\right|^{2(n+k)} \widetilde{\mathcal{E}}(\chi). \quad (5.29)$$

In particular, for $n = 1$ the above equation (5.29) gives an asymptotic representation for a *generator* of the translational group

$$\frac{\mathrm{d}\widetilde{\mathcal{E}}}{\mathrm{d}\chi} = \sum_{k=0}^{\infty} \epsilon^{1+k} g_{1+k,1}\left|\widetilde{\mathcal{E}}(\chi)\right|^{2(1+k)} \widetilde{\mathcal{E}}(\chi)$$

$$= \frac{3i\epsilon a_3}{2}\left|\widetilde{\mathcal{E}}(\chi)\right|^2 \widetilde{\mathcal{E}}(\chi) + 5i\epsilon^2\left(\frac{3a_3^2}{16} + a_5\right)\left|\widetilde{\mathcal{E}}(\chi)\right|^4 \widetilde{\mathcal{E}}(\chi) + \cdots. \quad (5.30)$$

This is precisely the RG equation (5.18) obtained by means of the standard renormalization procedure. In this sense the RG equation can be regarded as a generator of the translational group.

Since all coefficients on the right-hand-side of the RG equation (5.30) are purely imaginary, it is easy to see that

$$\left|\widetilde{\mathcal{E}}(\chi)\right|^2 = 2\mathcal{J} = \text{const} \tag{5.31}$$

is a constant of the motion. Solving the RG equation (5.30), for the deviation from the particle's design orbit we obtain

$$X(\chi) = \sqrt{2\mathcal{J}} e^{-i[1+\nu(\mathcal{J})]\chi + i\varphi}$$

$$-\frac{\epsilon}{8}(2\mathcal{J})^{3/2}\left[a_3 + \epsilon\left(\frac{21a_3^2}{8} + 5a_5\right)2\mathcal{J} + \cdots\right]e^{-3i[1+\nu(\mathcal{J})]\chi + 3i\varphi}$$

$$+c.c. + \text{higher harmonics}, \tag{5.32}$$

where

$$\nu(\mathcal{J}) = i\sum_{k=0}^{\infty} \epsilon^{1+k} g_{1+k,1}(2\mathcal{J})^{1+k} \tag{5.33}$$

is the relative (scaled by the unperturbed tune) nonlinear tuneshift.

Equation (5.33) represents the most general expression for the nonlinear tuneshift due to even multipole errors. It is worthwhile to note that the invariant $\mathcal{J}$ is more general as compared to the linear action invariant J [see expression (2.70)].

5.3 The Method of Multiple Scales

In the *method of multiple time-scales* (see e.g. [Bender (1978)], [Kevorkian (1996)], [Murdock (1991)], [Nayfeh (1981)]), the single time variable is replaced by an infinite sequence of independent time-scales. The latter describe slow variations of dependent quantities, characteristic for the process under investigation.

To reveal the basic features of the method of multiple time-scales, we consider again the model equation (5.2) in the smooth focusing approximation

$$\frac{\mathrm{d}^2 X}{\mathrm{d}\chi^2} + X = \sum_{n=1}^{\infty} \epsilon^n a_{2n+1} X^{2n+1}. \tag{5.34}$$

We define the different time-scales

$$\tau_0 = \chi, \qquad \tau_1 = \epsilon\chi, \qquad \tau_2 = \epsilon^2\chi, \qquad \ldots, \qquad \tau_n = \epsilon^n\chi, \qquad \ldots, \tag{5.35}$$

and seek a uniform approximate solution to equation (5.34) in the form

$$X(\chi, \tau_1, \tau_2, \ldots) = \sum_{n=0}^{\infty} \epsilon^n X_n(\chi, \tau_1, \tau_2, \ldots). \tag{5.36}$$

In addition, we replace the derivative with respect to χ by

$$\frac{\mathrm{d}}{\mathrm{d}\chi} = \sum_{n=0}^{\infty} \epsilon^n \widehat{\mathcal{D}}_n, \tag{5.37}$$

where

$$\widehat{\mathcal{D}}_n = \frac{\partial}{\partial \tau_n}. \tag{5.38}$$

Equating coefficients of equal powers in ϵ, the order-by-order analysis through $O(\epsilon^2)$ yields

$$\widehat{\mathcal{D}}_0^2 X_0 + X_0 = 0, \tag{5.39}$$

$$\widehat{\mathcal{D}}_0^2 X_1 + 2\widehat{\mathcal{D}}_0\widehat{\mathcal{D}}_1 X_0 + X_1 = a_3 X_0^3, \tag{5.40}$$

$$\widehat{\mathcal{D}}_0^2 X_2 + 2\widehat{\mathcal{D}}_0\widehat{\mathcal{D}}_1 X_1 + \left(\widehat{\mathcal{D}}_1^2 + 2\widehat{\mathcal{D}}_0\widehat{\mathcal{D}}_2\right) X_0 + X_2 = 3a_3 X_0^2 X_1 + a_5 X_0^5. \tag{5.41}$$

The general solution of the zero-order equation (5.39) can be written as

$$X_0 = \mathcal{A}e^{-i\chi} + c.c., \tag{5.42}$$

where the amplitude $\mathcal{A}$ depends on the slower scales

$$\mathcal{A} = \mathcal{A}(\tau_1, \tau_2, \ldots) = |\mathcal{A}(\tau_1, \tau_2, \ldots)| e^{-i\varphi_a(\tau_1, \tau_2, \ldots)}. \tag{5.43}$$

Note that the functional dependence of $\mathcal{A}$ on τ_1, τ_2, ... cannot be determined in zero order because the differential operator $\widehat{\mathcal{D}}_0$ acts with respect to χ alone.

To avoid a secular term (linearly dependent on χ) in the first-order correction X_1, it is necessary to eliminate all contributions in equation (5.40) that are proportional to $e^{\pm i\chi}$. Exact cancellation of these contributions can be performed by using the degree of freedom provided by the generic dependence (in zero order) of the amplitude $\mathcal{A}$ on the independent time-scales τ_n $(n \geq 1)$. This leads to the so called *solvability condition*

$$\widehat{\mathcal{D}}_1 \mathcal{A} = \mathcal{G}_1(\mathcal{A}, \mathcal{A}^*) = \frac{3ia_3}{2} |\mathcal{A}|^2 \mathcal{A}. \tag{5.44}$$

Obviously

$$\widehat{\mathcal{D}}_1 |\mathcal{A}| = 0, \qquad \varphi_a = -\frac{3a_3}{2} |\mathcal{A}|^2 \tau_1 + \psi_a(\tau_2, \tau_3, \dots). \tag{5.45}$$

With equation (5.45) obeyed, the regular solution for X_1 obtained from equation (5.40) reads as

$$X_1 = -\frac{a_3}{8} \mathcal{A}^3 e^{-3i\chi} + \mathcal{B} e^{-i\chi} + c.c.. \tag{5.46}$$

Here $\mathcal{B}$ is the amplitude of the solution to the homogeneous equation which together with $\mathcal{A}$ depends on the slower time scales.

To eliminate secular terms in second order, the following solvability condition must be imposed

$$\widehat{\mathcal{D}}_2 \mathcal{A} + \widehat{\mathcal{D}}_1 \mathcal{B} = \mathcal{G}_2(\mathcal{A}, \mathcal{A}^*) + \frac{3ia_3}{2} \left(\mathcal{A}^2 \mathcal{B}^* + 2|\mathcal{A}|^2 \mathcal{B} \right), \tag{5.47}$$

where

$$\mathcal{G}_2(\mathcal{A}, \mathcal{A}^*) = 5i \left(\frac{3a_3^2}{16} + a_5 \right) |\mathcal{A}|^4 \mathcal{A}. \tag{5.48}$$

Equation (5.47) together with its complex conjugate equation may be regarded as linear, first-order differential equations for the τ_1 dependence of the first-order amplitude $\mathcal{B}$. Some simple manipulations yield

$$\widehat{\mathcal{D}}_1 (\mathcal{A}\mathcal{B}^* + \mathcal{A}^*\mathcal{B}) + \widehat{\mathcal{D}}_2 |\mathcal{A}|^2 = 0, \tag{5.49}$$

$$\widehat{\mathcal{D}}_1 (\mathcal{A}^*\mathcal{B} - \mathcal{A}\mathcal{B}^*) - 2i|\mathcal{A}|^2 \widehat{\mathcal{D}}_2 \varphi_a$$

$$= 10i \left(\frac{3a_3^2}{16} + a_5 \right) |\mathcal{A}|^6 + 3ia_3 |\mathcal{A}|^2 (\mathcal{A}\mathcal{B}^* + \mathcal{A}^*\mathcal{B}). \tag{5.50}$$

The above equations can be solved in principle for the combinations $\mathcal{A}\mathcal{B}^* + \mathcal{A}^*\mathcal{B}$ and $\mathcal{A}^*\mathcal{B} - \mathcal{A}\mathcal{B}^*$. These however, do not give the explicit dependence of $\mathcal{A}$ on τ_2. A common way to fix the τ_2 dependence of $\mathcal{A}$ is to require that the solution be constructed solely out of periodic functions. For that purpose, we brake equations (5.49) and (5.50) into

$$\widehat{\mathcal{D}}_2 |\mathcal{A}|^2 = 0, \qquad \widehat{\mathcal{D}}_1 (\mathcal{A}\mathcal{B}^* + \mathcal{A}^*\mathcal{B}) = 0, \tag{5.51}$$

$$\widehat{\mathcal{D}}_1 (\mathcal{A}^*\mathcal{B} - \mathcal{A}\mathcal{B}^*) = 0. \tag{5.52}$$

Thus the argument φ_a of the zero-order amplitude $\mathcal{A}$ can be determined which shows that through $O(\epsilon^2)$, the dependence of $\mathcal{A}$ on τ_1 and τ_2 is factored into a product of terms, each depending on one time-scale only. This pattern is preserved in higher orders as well, suffices the requirement that the solvability conditions are solved solely in terms of periodic functions is imposed order-by-order. With the choice that the solution retains its periodic nature, equations (5.44), (5.51) and (5.52) lead to

$$\widehat{\mathcal{D}}_1 (\mathcal{A}^*\mathcal{B}) = 0, \qquad \widehat{\mathcal{D}}_1 \mathcal{B} = \frac{3ia_3}{2} |\mathcal{A}|^2 \mathcal{B}.$$

Therefore, the dependence of the phase φ_b of $\mathcal{B}$ on τ_1 is the same as that of φ_a, which implies that in second order the solvability conditions leave $\mathcal{B}$ free. In particular, $\mathcal{B} = 0$ is allowed. To see that, we construct the quantity

$$\mathcal{C} = \left[\widehat{\mathcal{D}}_1, \widehat{\mathcal{D}}_2\right] \mathcal{A}, \tag{5.53}$$

where

$$\left[\widehat{\mathcal{D}}_1, \widehat{\mathcal{D}}_2\right] = \widehat{\mathcal{D}}_1 \widehat{\mathcal{D}}_2 - \widehat{\mathcal{D}}_2 \widehat{\mathcal{D}}_1 \tag{5.54}$$

denotes the commutator of the two operators $\widehat{\mathcal{D}}_1$ and $\widehat{\mathcal{D}}_2$. Since the sought-for dependence of $\mathcal{A}$ on τ_1 and τ_2 should be appropriately smooth, commutativity of derivatives ($\mathcal{C} \equiv 0$) is automatically guaranteed. This yields a second-order linear differential equation for the amplitude $\mathcal{B}$ (and $\mathcal{B}^*$) with respect to τ_1. It is trivial to verify that a possible solution is $\mathcal{B} = 0$.

Setting the free amplitude $\mathcal{B}$ to zero, the mutual consistency condition $\mathcal{C} \equiv 0$ of the first-order and the second-order solvability conditions can be rewritten as

$$\mathcal{C} = \widehat{\mathcal{D}}_1 \mathcal{G}_2(\mathcal{A}, \mathcal{A}^*) - \widehat{\mathcal{D}}_2 \mathcal{G}_1(\mathcal{A}, \mathcal{A}^*) = \{\mathcal{G}_2(\mathcal{A}, \mathcal{A}^*), \mathcal{G}_1(\mathcal{A}, \mathcal{A}^*)\} \equiv 0. \tag{5.55}$$

Here $\{\mathcal{G}_2, \mathcal{G}_1\}$ denotes the *Lie brackets* of $\mathcal{G}_2$ and $\mathcal{G}_1$. By definition, the Lie bracket of two n-dimensional vector-valued functions $\mathbf{F}(\mathbf{z})$ and $\mathbf{G}(\mathbf{z})$

depending on a vector variable $\mathbf{z}$ is given by the expression

$$\{\mathbf{F}(\mathbf{z}), \mathbf{G}(\mathbf{z})\}_i = \sum_{k=1}^{n} \left(G_k \frac{\partial F_i}{\partial z_k} - F_k \frac{\partial G_i}{\partial z_k} \right) \qquad 1 \le i \le n. \tag{5.56}$$

In our case

$$\mathbf{F} = (\mathcal{G}_2, \mathcal{G}_2^*), \qquad \mathbf{G} = (\mathcal{G}_1, \mathcal{G}_1^*), \qquad \mathbf{z} = (\mathcal{A}, \mathcal{A}^*).$$

It is straightforward to check by direct substitution of (5.44) and (5.48) that indeed

$$\{\mathcal{G}_2, \mathcal{G}_1\} = -\{\mathcal{G}_1, \mathcal{G}_2\} \equiv 0.$$

The consistency requirement of the order-by-order solvability conditions opens a possibility to reconstruct the full dependence of the zero-order amplitude $\mathcal{A}$ on χ. By virtue of (5.37), we can write

$$\frac{\mathrm{d}\mathcal{A}}{\mathrm{d}\chi} = \epsilon \widehat{\mathcal{D}}_1 \mathcal{A} + \epsilon^2 \widehat{\mathcal{D}}_2 \mathcal{A} + \cdots .$$

Since the n-th order solvability condition provides the dependence of $\mathcal{A}$ on the time-scale τ_n making use of equations (5.44) and (5.47), we obtain

$$\frac{\mathrm{d}\mathcal{A}}{\mathrm{d}\chi} = \frac{3i\epsilon a_3}{2} |\mathcal{A}|^2 \mathcal{A} + 5i\epsilon^2 \left(\frac{3a_3^2}{16} + a_5 \right) |\mathcal{A}|^4 \mathcal{A} + \cdots . \tag{5.57}$$

Equation (5.57) is known as the *reconstituted equation* [Nayfeh (1985)].

It is worthwhile to note that the method of multiple scales yields the same result as the RG method. Indeed, identifying the zero-order amplitude $\mathcal{A}$ with the renormalized amplitude $\mathcal{E}_R$ [see equation (5.14)], we see that the reconstituted equation (5.57) and the RG equation (5.18) coincide.

5.4 Renormalization Group Analysis of Hill's Equation

As an example of the application of the RG method to non-autonomous systems, we will study the *Hill*'s equation (2.2) discussed in chapter 2. Since the focusing coefficient $G(\theta)$ is a periodic function of θ with period $2\pi/N$ (see the discussion in chapter 2), we can write

$$G(\theta) = n_0^2 + \Omega + \sum_{n \neq 0} G_n e^{inN\theta}, \tag{5.58}$$

where

$$G_n = \frac{N}{2\pi} \int_0^{2\pi/N} \mathrm{d}\theta G(\theta) e^{-inN\theta}.$$

Here n_0 is some integer and $0 \leq \Omega < 2n_0 + 1$. We wish to determine the "fractional" part Ω in a way that the solution of *Hill*'s equation is stable in the case when a resonance between the fundamental harmonic $e^{-in_0\theta}$ and a particular harmonic of the *Fourier* expansion of $G(\theta)$ is possible. The *Hill*'s equation can be cast to the form

$$\frac{\mathrm{d}^2 u}{\mathrm{d}\tau^2} + u = -\epsilon \left(\omega + \sum_{n \neq 0} g_n e^{in\sigma\tau} \right) u, \tag{5.59}$$

where

$$\tau = n_0 \theta, \qquad g_n = \frac{G_n}{n_0^2}, \qquad \sigma = \frac{N}{n_0}, \tag{5.60}$$

$$\omega = \frac{\Omega}{n_0^2} = \omega_1 + \epsilon\omega_2 + \epsilon^2\omega_3 + \cdots, \tag{5.61}$$

and ϵ is a formal small parameter accounting for the order of the *Fourier* components of the focusing strength. Expanding u as in the example of the previous paragraph

$$u = \mathcal{A} e^{-i\tau} + c.c. + \epsilon u_1 + \epsilon^2 u_2 + \cdots, \tag{5.62}$$

to the leading first order we obtain the perturbation equation

$$\left(\frac{\mathrm{d}^2}{\mathrm{d}\tau^2} + 1 \right) u_1 = -\omega_1 \mathcal{A} e^{-i\tau} - \mathcal{A}^* \sum_{n \neq 0} g_n e^{i(n\sigma + 1)\tau} + c.c.. \tag{5.63}$$

Let us now assume that there exists an integer m_0 such that $m_0\sigma = -2$. Then the general solution to equation (5.63) reads as

$$u_1 = -(\omega_1 \mathcal{A} + \mathcal{A}^* g_{m_0}) \mathcal{P}_1(\tau; \tau_0) e^{-i\tau} + \sum_{n \neq 0}{}' \frac{\mathcal{A}^* g_n}{n\sigma(n\sigma + 2)} e^{i(n\sigma+1)\tau} + c.c., \tag{5.64}$$

where $\mathcal{P}_1(\tau; \tau_0)$ is given by equations (5.6) and (5.10), and $\sum'$ implies exclusion of the harmonic m_0 from the summation. Similar perturbation calculations yield higher order solutions

$$u_2 = -\left[\left(|g_{m_0}|^2 - \omega_1^2\right)\mathcal{P}_2(\tau;\tau_0) + \left(\omega_2 + \sum_{n\neq 0}{}' \frac{|g_n|^2}{n\sigma(n\sigma+2)}\right)\mathcal{P}_1(\tau;\tau_0)\right]$$

$$\times \mathcal{A}e^{-i\tau} - \left(\omega_1\mathcal{A}^* + \mathcal{A}g_{m_0}^*\right)\sum_{n\neq 0}{}' g_n \mathcal{Q}_n(\tau;\tau_0)e^{i(n\sigma+1)\tau}$$

$$+\omega_1\mathcal{A}^*\sum_{n\neq 0}{}' \frac{g_n}{n^2\sigma^2(n\sigma+2)^2}e^{i(n\sigma+1)\tau}$$

$$+\mathcal{A}\sum_{m\neq 0}{}' \sum_{n\neq m} \frac{g_m^* g_n}{\sigma^2 m(n-m)(m\sigma+2)[(n-m)\sigma-2]}e^{i[(n-m)\sigma-1]\tau} + c.c., \tag{5.65}$$

$$u_3 = \left[-\left(\omega_3 + \omega_1\sum_{n\neq 0}{}' \frac{|g_n|^2}{n^2\sigma^2(n\sigma+2)^2}\right)\mathcal{A}\mathcal{P}_1(\tau;\tau_0)\right.$$

$$+(\omega_1\mathcal{A} + \mathcal{A}^* g_{m_0})\sum_{n\neq 0}{}' |g_n|^2\mathcal{R}_n(\tau;\tau_0) + \omega_1\left(|g_{m_0}|^2 - \omega_1^2\right)\mathcal{A}\mathcal{P}_3'(\tau;\tau_0)$$

$$+g_{m_0}\left(|g_{m_0}|^2 - \omega_1^2\right)\mathcal{A}^*\mathcal{P}_3(\tau;\tau_0) + 2\omega_1\omega_2\mathcal{A}\mathcal{P}_2(\tau;\tau_0)$$

$$+(\omega_1\mathcal{A} - \mathcal{A}^* g_{m_0})\mathcal{P}_2(\tau;\tau_0)\sum_{n\neq 0}{}' \frac{|g_n|^2}{n\sigma(n\sigma+2)}$$

$$\left. - \mathcal{A}\mathcal{P}_1(\tau;\tau_0)\sum_{n\neq 0}{}' \sum_{m\neq n} \frac{g_m g_n^* g_{m-n}^*}{\sigma^2 n(m-n)(n\sigma+2)[(m-n)\sigma-2]}\right] e^{-i\tau} + c.c., \tag{5.66}$$

where higher-harmonic terms in equation (5.66) are not listed explicitly. Moreover, the secular function $\mathcal{P}_2(\tau;\tau_0)$ is given by equations (5.9) and (5.11), while the other secular functions $\mathcal{Q}_n(\tau;\tau_0)$, $\mathcal{P}_3(\tau;\tau_0)$, $\mathcal{P}_3'(\tau;\tau_0)$ and $\mathcal{R}_n(\tau;\tau_0)$ satisfy the equations

$$\left[\frac{\mathrm{d}^2}{\mathrm{d}\tau^2} + 2i(n\sigma+1)\frac{\mathrm{d}}{\mathrm{d}\tau} - n\sigma(n\sigma+2)\right]\mathcal{Q}_n = \widehat{\mathcal{L}}_n(\tau)\mathcal{Q}_n = \mathcal{P}_1, \tag{5.67}$$

$$\widehat{\mathcal{L}}_1(\tau)\mathcal{P}_3' = \mathcal{P}_2, \qquad \widehat{\mathcal{L}}_1(\tau)\mathcal{P}_3 = \mathcal{P}_2^*, \qquad \widehat{\mathcal{L}}_1(\tau)\mathcal{R}_n = \mathcal{Q}_n^*. \tag{5.68}$$

For completeness we present the secular terms $\mathcal{Q}_n(\tau;\tau_0)$, $\mathcal{P}_3(\tau;\tau_0)$, $\mathcal{P}_3'(\tau;\tau_0)$ and $\mathcal{R}_n(\tau;\tau_0)$ in their explicit form

$$\mathcal{P}_3(\tau;\tau_0) = -\frac{i(\tau-\tau_0)^3}{48}, \tag{5.69}$$

$$\mathcal{P}_3'(\tau;\tau_0) = -\frac{i(\tau-\tau_0)^3}{48} - \frac{(\tau-\tau_0)^2}{16} + \frac{i(\tau-\tau_0)}{16}, \tag{5.70}$$

$$\mathcal{Q}_n(\tau;\tau_0) = -\frac{i(\tau-\tau_0)}{2n\sigma(n\sigma+2)} + \frac{n\sigma+1}{n^2\sigma^2(n\sigma+2)^2}, \tag{5.71}$$

$$\mathcal{R}_n(\tau;\tau_0) = -\frac{(\tau-\tau_0)^2}{8n\sigma(n\sigma+2)} + \frac{i(n^2\sigma^2+6n\sigma+4)}{8n^2\sigma^2(n\sigma+2)^2}(\tau-\tau_0). \tag{5.72}$$

Performing a renormalization transformation similar to what has been done in the previous paragraph, for the renormalized amplitude $\mathcal{A}_R$ we obtain the RG equation

$$\begin{aligned}\frac{\mathrm{d}\mathcal{A}_R}{\mathrm{d}\tau} &= -i\left[\frac{\epsilon\omega_1}{2} + \frac{\epsilon^2}{2}\left(\omega_2 - \frac{\omega_1^2}{4}\right) + \frac{\epsilon^3}{2}\left(\omega_3 + \frac{\omega_1^3}{8} - \frac{\omega_1\omega_2}{2}\right)\right]\mathcal{A}_R \\ &- \frac{i\epsilon^2}{2}\left\{\frac{|g_{m_0}|^2}{4} + \sum_{n\neq 0}{}' \frac{|g_n|^2}{n\sigma(n\sigma+2)} - \epsilon\omega_1\left[\frac{|g_{m_0}|^2}{8} + \frac{1}{2}\sum_{n\neq 0}{}' \frac{(n\sigma+4)|g_n|^2}{n\sigma(n\sigma+2)^2}\right.\right. \\ &\left.\left. - \frac{1}{\omega_1}\sum_{n\neq 0}{}' \sum_{m\neq n} \frac{g_m g_n^* g_{m-n}^*}{\sigma^2 n(m-n)(n\sigma+2)[(m-n)\sigma-2]}\right]\right\}\mathcal{A}_R \\ &- \frac{i\epsilon}{2}g_{m_0}\left[1 - \epsilon^2\sum_{n\neq 0}{}' \frac{(n\sigma+1)|g_n|^2}{n^2\sigma^2(n\sigma+2)^2}\right]\mathcal{A}_R^*.\end{aligned} \tag{5.73}$$

Note that the expression in the square brackets of the first term on the right-hand-side of the RG equation (5.73) represents the third-order expansion of $\sqrt{1+\epsilon\omega}-1$.

Having obtained the RG equation two important conclusions are now in order. Let us first consider the case when there is no m_0, such that $m_0 = -2n_0/N$, which means that there is no resonance occurring between the fundamental harmonic and any particular harmonic of $G(\theta)$. Then the solution of the *Hill's* equation is stable and can be written approximately as

$$u = Ae^{-i\nu\theta} + c.c., \tag{5.74}$$

where A is a constant amplitude and the betatron tune ν is given by the expression

$$\nu = n_0\sqrt{1+\epsilon\omega} + \frac{\epsilon^2 n_0}{2}\sum_{n\neq 0}\frac{|g_n|^2}{n\sigma(n\sigma+2)} - \frac{\epsilon^3 n_0}{2}\left[\frac{\omega_1}{2}\sum_{n\neq 0}\frac{(n\sigma+4)|g_n|^2}{n\sigma(n\sigma+2)^2}\right.$$

$$\left. - \sum_{n\neq 0}\sum_{m\neq n}\frac{g_m g_n^* g_{m-n}^*}{\sigma^2 n(m-n)(n\sigma+2)[(m-n)\sigma-2]}\right]. \tag{5.75}$$

Suppose that the largest harmonic in the *Fourier* expansion of $G(\theta)$ is the resonant one with a harmonic number $m_0 = \pm 2n_0/N$ and all other harmonics are small, so that can be neglected. Then the RG equation (5.73) reads as

$$\frac{\mathrm{d}\mathcal{A}_R}{\mathrm{d}\tau} = -i\left[\sqrt{1+\epsilon\omega} - 1 + \frac{\epsilon^2|g_{m_0}|^2}{16}\left(3 - \frac{7\epsilon\omega_1}{4}\right)\right]\mathcal{A}_R$$

$$-\frac{i\epsilon}{2}g_{m_0}\left(1 - \frac{3\epsilon^2|g_{m_0}|^2}{64}\right)\mathcal{A}_R^*, \tag{5.76}$$

from which we obtain

$$\frac{\mathrm{d}^2\mathcal{A}_R}{\mathrm{d}\tau^2} + \left[\sqrt{1+\epsilon\omega} - 1 + \frac{\epsilon^2|g_{m_0}|^2}{16}\left(3 - \frac{7\epsilon\omega_1}{4}\right)\right]^2\mathcal{A}_R$$

$$= \frac{\epsilon^2|g_{m_0}|^2}{4}\left(1 - \frac{3\epsilon^2|g_{m_0}|^2}{64}\right)^2\mathcal{A}_R. \tag{5.77}$$

The solution to equation (5.77) is stable provided

$$\sqrt{1+\epsilon\omega} - 1 + \frac{\epsilon^2|g_{m_0}|^2}{16}\left(3 - \frac{7\epsilon\omega_1}{4}\right) > \frac{\epsilon|g_{m_0}|}{2}\left(1 - \frac{3\epsilon^2|g_{m_0}|^2}{64}\right), \tag{5.78}$$

which represents the well-known transition curves for the *Mathieu* equation that separate stable from unstable solutions.

5.5 Renormalization Group Reduction of Nonlinear Resonances

In this paragraph we will show how the RG approach works when applied to study nonlinear resonances. In order to demonstrate the method, we consider one-dimensional betatron oscillations in the presence of a sextupole perturbation. The Hamiltonian can be written in terms of the normalized canonical variables (2.38) as

$$H(X, P; \theta) = \frac{\dot{\chi}(\theta)}{2}\left(P^2 + X^2\right) + \frac{\epsilon \mathcal{S}(\theta)}{3} X^3, \tag{5.79}$$

where

$$\mathcal{S}(\theta) = \frac{\lambda_0(\theta)\beta^{3/2}(\theta)}{2R^2}, \tag{5.80}$$

and ϵ is a formal small parameter.

In what follows, it will prove convenient to perform a canonical transformation defined by the generating function

$$\mathcal{F}_2\left(X, \widetilde{P}; \theta\right) = -\frac{\tan\psi}{2}\widetilde{P}^2 + \frac{X\widetilde{P}}{\cos\psi} - \frac{\tan\psi}{2}X^2, \tag{5.81}$$

where the old and the new canonical variables are related via a rotation transformation

$$X = \widetilde{X}\cos\psi + \widetilde{P}\sin\psi, \qquad P = -\widetilde{X}\sin\psi + \widetilde{P}\cos\psi, \tag{5.82}$$

and

$$\psi(\theta) = \chi(\theta) - \nu\theta. \tag{5.83}$$

The new Hamiltonian acquires the form

$$\widetilde{H}\left(\widetilde{X}, \widetilde{P}; \theta\right) = \frac{\nu}{2}\left(\widetilde{P}^2 + \widetilde{X}^2\right) + \frac{\epsilon \mathcal{S}(\theta)}{3}\left[\widetilde{X}\cos\psi(\theta) + \widetilde{P}\sin\psi(\theta)\right]^3. \tag{5.84}$$

We now assume that the unperturbed betatron tune ν is close to a third order resonance

$$\nu = \nu_0 + \delta, \qquad \nu_0 = \frac{m_0}{3}, \tag{5.85}$$

where m_0 is an integer and δ is the resonance detuning. The *Hamilton*'s equations of motion are

$$\frac{\mathrm{d}\widetilde{X}}{\mathrm{d}\theta} = \nu\widetilde{P} + \epsilon\mathcal{S}\left(\widetilde{X}\cos\psi + \widetilde{P}\sin\psi\right)^2 \sin\psi, \tag{5.86}$$

$$\frac{\mathrm{d}\widetilde{P}}{\mathrm{d}\theta} = -\nu\widetilde{X} - \epsilon\mathcal{S}\left(\widetilde{X}\cos\psi + \widetilde{P}\sin\psi\right)^2 \cos\psi, \tag{5.87}$$

or in an alternative form

$$\frac{\mathrm{d}\widetilde{X}}{\mathrm{d}\theta} = (\nu_0 + \delta)\widetilde{P} + \epsilon\sum_n \left(a_n\widetilde{X}^2 + b_n\widetilde{P}^2 + 2c_n\widetilde{X}\widetilde{P}\right)e^{in\theta}, \tag{5.88}$$

$$\frac{\mathrm{d}\widetilde{P}}{\mathrm{d}\theta} = -(\nu_0 + \delta)\widetilde{X} - \epsilon\sum_n \left(d_n\widetilde{X}^2 + c_n\widetilde{P}^2 + 2a_n\widetilde{X}\widetilde{P}\right)e^{in\theta}, \tag{5.89}$$

where we have introduced the *Fourier* expansions

$$\mathcal{S}(\theta)\cos^2\psi(\theta)\sin\psi(\theta) = \sum_n a_n e^{in\theta}, \qquad \mathcal{S}(\theta)\sin^3\psi(\theta) = \sum_n b_n e^{in\theta}, \tag{5.90}$$

$$\mathcal{S}(\theta)\sin^2\psi(\theta)\cos\psi(\theta) = \sum_n c_n e^{in\theta}, \qquad \mathcal{S}(\theta)\cos^3\psi(\theta) = \sum_n d_n e^{in\theta}. \tag{5.91}$$

Following the prescription of the RG method, we again adopt the naive perturbation expansion

$$\widetilde{X} = Ae^{-i\nu_0\theta} + c.c. + \epsilon\widetilde{X}_1 + \epsilon^2\widetilde{X}_2 + \dots, \tag{5.92}$$

$$\widetilde{P} = -iAe^{-i\nu_0\theta} + c.c. + \epsilon\widetilde{P}_1 + \epsilon^2\widetilde{P}_2 + \dots. \tag{5.93}$$

In addition, we assume that

$$\delta = \epsilon\delta_1 + \epsilon^2\delta_2 + \dots. \tag{5.94}$$

The first-order perturbation equations are

$$\frac{\mathrm{d}\widetilde{X}_1}{\mathrm{d}\theta} = \nu_0\widetilde{P}_1 - i\delta_1 Ae^{-i\nu_0\theta} + |A|^2\sum_n (a_n + b_n)e^{in\theta}$$

$$+A^2\sum_n (a_n - b_n - 2ic_n)e^{i(n-2\nu_0)\theta} + c.c., \tag{5.95}$$

$$\frac{\mathrm{d}\widetilde{P}_1}{\mathrm{d}\theta} = -\nu_0 \widetilde{X}_1 - \delta_1 A e^{-i\nu_0\theta} - |A|^2 \sum_n (c_n + d_n) e^{in\theta}$$

$$-A^2 \sum_n (d_n - c_n - 2ia_n) e^{i(n-2\nu_0)\theta} + c.c.. \tag{5.96}$$

Before proceeding further, we note that for the n-th order the perturbation equations have the general form

$$\frac{\mathrm{d}\widetilde{X}_n}{\mathrm{d}\theta} = \nu_0 \widetilde{P}_n + \alpha_n(\theta), \qquad \frac{\mathrm{d}\widetilde{P}_n}{\mathrm{d}\theta} = -\nu_0 \widetilde{X}_n + \beta_n(\theta), \tag{5.97}$$

where the functions $\alpha_n(\theta)$ and $\beta_n(\theta)$ are known from previous lower-order calculations. The system of equations (5.97) is equivalent to the two equations

$$\frac{\mathrm{d}^2\widetilde{X}_n}{\mathrm{d}\theta^2} + \nu_0^2 \widetilde{X}_n = \frac{\mathrm{d}\alpha_n}{\mathrm{d}\theta} + \nu_0\beta_n, \qquad \frac{\mathrm{d}^2\widetilde{P}_n}{\mathrm{d}\theta^2} + \nu_0^2 \widetilde{P}_n = \frac{\mathrm{d}\beta_n}{\mathrm{d}\theta} - \nu_0\alpha_n \tag{5.98}$$

for $\widetilde{X}_n$ and $\widetilde{P}_n$ alone. Thus for the first order, we obtain

$$\frac{\mathrm{d}^2\widetilde{X}_1}{\mathrm{d}\theta^2} + \nu_0^2 \widetilde{X}_1 = -2\nu_0\delta_1 A e^{-i\nu_0\theta} + |A|^2 \sum_n \mathcal{A}_n e^{in\theta} + A^2 \sum_n \mathcal{B}_n e^{i(n-2\nu_0)\theta} + c.c., \tag{5.99}$$

$$\frac{\mathrm{d}^2\widetilde{P}_1}{\mathrm{d}\theta^2} + \nu_0^2 \widetilde{P}_1 = 2i\nu_0\delta_1 A e^{-i\nu_0\theta} + |A|^2 \sum_n \mathcal{C}_n e^{in\theta} + A^2 \sum_n \mathcal{D}_n e^{i(n-2\nu_0)\theta} + c.c., \tag{5.100}$$

where

$$\mathcal{A}_n = in(a_n + b_n) - \nu_0(c_n + d_n), \tag{5.101}$$

$$\mathcal{B}_n = i(n - 2\nu_0)(a_n - b_n - 2ic_n) - \nu_0(d_n - c_n - 2ia_n), \tag{5.102}$$

$$\mathcal{C}_n = -\nu_0(a_n + b_n) - in(c_n + d_n), \tag{5.103}$$

$$\mathcal{D}_n = -\nu_0(a_n - b_n - 2ic_n) - i(n - 2\nu_0)(d_n - c_n - 2ia_n), \tag{5.104}$$

The solutions to the first-order perturbation equations (5.99) and (5.100) can be found in a straightforward manner to be

$$\widetilde{X}_1 = \left(-\frac{2\delta_1}{\nu_0} A + \frac{\mathcal{B}^*_{m_0}}{\nu_0^2} A^{*2} \right) \mathcal{P}_1(\nu_0\theta; \nu_0\theta_0) e^{-i\nu_0\theta}$$

$$-|A|^2 \sum_n \frac{\mathcal{A}_n e^{in\theta}}{n^2 - \nu_0^2} - A^2 \sum_{n \neq m_0} \frac{\mathcal{B}_n e^{i(n-2\nu_0)\theta}}{n^2 - 4\nu_0 n + 3\nu_0^2} + c.c., \tag{5.105}$$

$$\widetilde{P}_1 = \left(\frac{2i\delta_1}{\nu_0} A + \frac{\mathcal{D}_{m_0}^*}{\nu_0^2} A^{*2} \right) \mathcal{P}_1(\nu_0\theta; \nu_0\theta_0) e^{-i\nu_0\theta}$$

$$-|A|^2 \sum_n \frac{\mathcal{C}_n e^{in\theta}}{n^2 - \nu_0^2} - A^2 \sum_{n \neq m_0} \frac{\mathcal{D}_n e^{i(n-2\nu_0)\theta}}{n^2 - 4\nu_0 n + 3\nu_0^2} + c.c., \tag{5.106}$$

The RG equation for the renormalized amplitude A_R can be written as

$$\frac{\mathrm{d}A_R}{\mathrm{d}\theta} = -i\epsilon\delta_1 A_R - i\epsilon\mathcal{D}_R A_R^{*2}, \tag{5.107}$$

where

$$\mathcal{D}_R = \frac{1}{4\pi} \int_0^{2\pi} \mathrm{d}\theta \mathcal{S}(\theta) e^{i[m_0\theta + 3\psi(\theta)]}. \tag{5.108}$$

It can be directly checked that the invariant ($\mathrm{d}\mathcal{H}_R/\mathrm{d}\theta \equiv 0$)

$$\mathcal{H}_R = a_R\delta_1 |A_R|^2 + \frac{a_R}{3} \left(\mathcal{D}_R^* A_R^3 + \mathcal{D}_R A_R^{*3} \right) = const, \tag{5.109}$$

which can be interpreted as the *resonant Hamiltonian*, is a formal solution to the RG equation(s)[1] (5.107). Here a_R is an arbitrary real constant. As a matter of fact, if we define the canonical equations of motion

$$\frac{\mathrm{d}A_R}{\mathrm{d}\theta} = -i\frac{\partial\mathcal{H}_R}{\partial A_R^*}, \qquad \frac{\mathrm{d}A_R^*}{\mathrm{d}\theta} = i\frac{\partial\mathcal{H}_R}{\partial A_R} \tag{5.110}$$

with $a_R = \epsilon$, the RG equation(s) are recovered within the framework of a Hamiltonian formalism. Note that, if we define

$$A_R = \sqrt{J} e^{i\alpha}, \tag{5.111}$$

the resonant Hamiltonian (5.109) can be transformed to

$$\mathcal{H}_R = \epsilon\delta_1 J + \epsilon\frac{2|\mathcal{D}_R|}{3} J^{3/2} \cos(3\alpha - \varphi_\mathcal{D}), \tag{5.112}$$

where $\varphi_\mathcal{D}$ is the argument of the driving term coefficient $\mathcal{D}_R$. The latter after appropriate rescaling can be cast to a form coinciding with the

[1]The RG equation (5.107) automatically implies a second equation for A_R^*.

Hamiltonian (3.24). By using the invariant (5.109) it is possible to derive a second-order differential equation for $|A_R|^2$ alone, that is

$$\frac{\mathrm{d}^2|A_R|^2}{\mathrm{d}\theta^2} + 9\epsilon^2\delta_1^2|A_R|^2 = 9\epsilon\delta_1\mathcal{H}_R + 6\epsilon^2|\mathcal{D}_R|^2|A_R|^4. \tag{5.113}$$

A similar equation was derived and studied in paragraphs 3.3 and 3.4, hence results already obtained spread on equation (5.113) automatically.

The remarkable feature of the RG approach, when applied to investigate continuous Hamiltonian systems, is that the RG reduction leads to amplitude equations that preserve the Hamiltonian structure of the original dynamical equations. This property remains valid in every order. Although calculations become more and more tedious, in principle it is possible to proceed to arbitrary high order. Nevertheless higher order contributions do not add much new, it is worthwhile to note that in second (and higher) order(s) the RG equation (and respectively the resonant Hamiltonian) includes terms responsible for the nonlinear tuneshift, as well as second-order (and higher-order) corrections to the resonant driving term.

5.6 Reduction of Nonlinear Resonances Using the Method of Multiple Scales

In paragraph 5.3 it was shown that the method of multiple scales when applied to autonomous dynamical systems yields the same result as the RG method. It is instructive to demonstrate the efficiency of the method of multiple scales to study nonlinear resonances of betatron oscillations.

Starting with the *Hamilton*'s equations of motion in the form (5.88) and (5.89) and adopting the perturbation expansions (5.92) – (5.94), we obtain the first-order perturbation equations

$$\widehat{\mathcal{D}}_0\widetilde{X}_1 + \widehat{\mathcal{D}}_1\widetilde{X}_0 = \nu_0\widetilde{P}_1 + \delta_1\widetilde{P}_0 + |A|^2\sum_n(a_n + b_n)e^{in\theta}$$

$$+A^2\sum_n(a_n - b_n - 2ic_n)e^{i(n-2\nu_0)\theta} + c.c., \tag{5.114}$$

$$\widehat{\mathcal{D}}_0\widetilde{P}_1 + \widehat{\mathcal{D}}_1\widetilde{P}_0 = -\nu_0\widetilde{X}_1 - \delta_1\widetilde{X}_0 - |A|^2\sum_n(c_n + d_n)e^{in\theta}$$

$$-A^2 \sum_n (d_n - c_n - 2ia_n) e^{i(n-2\nu_0)\theta} + c.c.. \tag{5.115}$$

Here $\left(\widetilde{X}_0, \widetilde{P}_0\right)$ is the zero-order solution

$$\widetilde{X}_0 = A e^{-i\nu_0\theta} + c.c., \qquad \widetilde{P}_0 = -iAe^{-i\nu_0\theta} + c.c.,$$

$\widehat{\mathcal{D}}_n$ are the operators defined in (5.38), and the amplitude A is allowed to depend on the slower scales. Similar to the corresponding expressions obtained in the previous paragraph, the general form of the n-th order perturbation equations can be written as

$$\widehat{\mathcal{D}}_0^2 \widetilde{X}_n + \nu_0^2 \widetilde{X}_n = \widehat{\mathcal{D}}_0 \alpha_n + \nu_0 \beta_n, \qquad \widehat{\mathcal{D}}_0^2 \widetilde{P}_n + \nu_0^2 \widetilde{P}_n = \widehat{\mathcal{D}}_0 \beta_n - \nu_0 \alpha_n. \tag{5.116}$$

In first order these become

$$\widehat{\mathcal{D}}_0^2 \widetilde{X}_1 + \nu_0^2 \widetilde{X}_1 = 2i\nu_0 \left(\widehat{\mathcal{D}}_1 + i\delta_1\right) A e^{-i\nu_0\theta} + |A|^2 \sum_n \mathcal{A}_n e^{in\theta}$$

$$+ A^2 \sum_n \mathcal{B}_n e^{i(n-2\nu_0)\theta} + c.c., \tag{5.117}$$

$$\widehat{\mathcal{D}}_0^2 \widetilde{P}_1 + \nu_0^2 \widetilde{P}_1 = 2\nu_0 \left(\widehat{\mathcal{D}}_1 - i\delta_1\right) A e^{-i\nu_0\theta} + |A|^2 \sum_n \mathcal{C}_n e^{in\theta}$$

$$+ A^2 \sum_n \mathcal{D}_n e^{i(n-2\nu_0)\theta} + c.c., \tag{5.118}$$

To eliminate secular terms in the first-order solution $\left(\widetilde{X}_1, \widetilde{P}_1\right)$, the solvability condition

$$\widehat{\mathcal{D}}_1 A = -i\delta_1 A - i\mathcal{D}_R A^{*2} \tag{5.119}$$

must be imposed.

Up to first order the reconstituted equation for the amplitude A coincides with the RG equation (5.107). One may proceed further in obtaining the first-order contribution, which includes the solution of the homogeneous equation. The latter is specified again by a free first-order amplitude B. In second order, it can be shown that B remains undetermined and can be set to zero, thus assuring the mutual consistency of the first and second-order solvability conditions. This is left to the reader as an exercise.

5.7 Renormalization Group Reduction of Hamilton's Equations of Motion

In this paragraph, we apply the RG method to investigate the long-time behavior, that survives appropriate averaging and/or factorizing of rapidly oscillating contributions to the nonlinear dynamics of particles in accelerators and storage rings. The results thus obtained will be further utilized to study the problem of periodic crossing of a nonlinear resonance. For the sake of simplicity in what follows, we consider one-dimensional betatron motion governed by a Hamiltonian

$$H(J, \alpha; \theta) = H_0(J) + \Delta V(J, \alpha; \theta), \tag{5.120}$$

written in action-angle variables [see equation (4.1)]. Here $H_0(J)$ is the integrable part of the Hamiltonian, Δ is a formal small parameter, while $V(J, \alpha; \theta)$ is the perturbation periodic in the angle variable

$$V(J, \alpha; \theta) = \sum_{m \neq 0} \mathcal{V}_m(J; \theta) e^{im\alpha}. \tag{5.121}$$

Let us consider the solution to *Hamilton*'s equations of motion

$$\frac{\mathrm{d}\alpha}{\mathrm{d}\theta} = \nu(J) + \Delta \frac{\partial V}{\partial J}, \qquad \frac{\mathrm{d}J}{\mathrm{d}\theta} = -\Delta \frac{\partial V}{\partial \alpha} \tag{5.122}$$

for a small perturbation parameter Δ [Tzenov (2002a)]. Similar to equation (4.8), $\nu(J)$ denotes the unperturbed betatron tune

$$\nu(J) = \frac{\mathrm{d}H_0}{\mathrm{d}J}.$$

Following the basic idea of the RG method already discussed in preceding paragraphs, we introduce the naive perturbation expansion

$$\alpha = \alpha_0 + \Delta\alpha_1 + \Delta^2\alpha_2 + \dots, \qquad J = J_0 + \Delta J_1 + \Delta^2 J_2 + \dots. \tag{5.123}$$

The lowest-order perturbation equations have the trivial solution

$$\alpha_0 = \nu(\mathcal{A})\theta + \varphi, \qquad J_0 = \mathcal{A}, \tag{5.124}$$

where $\mathcal{A}$ and φ are constant amplitude and phase, respectively. We write the first-order perturbation equations as

$$\frac{\mathrm{d}\alpha_1}{\mathrm{d}\theta} = \gamma(\mathcal{A}) J_1 + \frac{\partial}{\partial \mathcal{A}} V(\mathcal{A}, \alpha_0; \theta), \qquad \frac{\mathrm{d}J_1}{\mathrm{d}\theta} = -\frac{\partial}{\partial \alpha_0} V(\mathcal{A}, \alpha_0; \theta), \tag{5.125}$$

where the nonlinearity coefficient $\gamma(J)$ is given by

$$\gamma(J) = \frac{\mathrm{d}^2 H_0}{\mathrm{d}J^2}. \tag{5.126}$$

Assuming that the angular harmonics $\mathcal{V}_m(J;\theta)$ are periodic in θ, we can expand them in a *Fourier* series

$$\mathcal{V}_m(J;\theta) = \sum_{\mu=-\infty}^{\infty} \mathcal{V}_m(J;\mu) e^{i\mu\nu_m\theta}. \tag{5.127}$$

If the original system (5.120) is far from primary resonances of the form

$$n_R\nu(J) + \mu\nu_R = 0, \tag{5.128}$$

we can solve the first-order perturbation equations (5.125), yielding the result

$$\alpha_1(\theta) = i\gamma(\mathcal{A}) \sum_{m\neq 0} \sum_{\mu=-\infty}^{\infty} \frac{m\mathcal{V}_m(\mathcal{A};\mu)}{(m\nu+\mu\nu_m)^2} e^{i(m\nu+\mu\nu_m)\theta} e^{im\varphi}$$

$$-i \sum_{m\neq 0} \sum_{\mu=-\infty}^{\infty} \frac{\mathcal{V}'_m(\mathcal{A};\mu)}{m\nu+\mu\nu_m} e^{i(m\nu+\mu\nu_m)\theta} e^{im\varphi}, \tag{5.129}$$

$$J_1(\theta) = -\sum_{m\neq 0} \sum_{\mu=-\infty}^{\infty} \frac{m\mathcal{V}_m(\mathcal{A};\mu)}{m\nu+\mu\nu_m} e^{i(m\nu+\mu\nu_m)\theta} e^{im\varphi}, \tag{5.130}$$

where the prime implies differentiation with respect to $\mathcal{A}$. The second-order perturbation equations have the form

$$\frac{\mathrm{d}\alpha_2}{\mathrm{d}\theta} = \frac{\gamma'(\mathcal{A})}{2} J_1^2 + \gamma(\mathcal{A}) J_2 + J_1 \frac{\partial^2}{\partial \mathcal{A}^2} V(\mathcal{A},\alpha_0;\theta) + \alpha_1 \frac{\partial^2}{\partial\alpha_0 \partial\mathcal{A}} V(\mathcal{A},\alpha_0;\theta), \tag{5.131}$$

$$\frac{\mathrm{d}J_2}{\mathrm{d}\theta} = -J_1 \frac{\partial^2}{\partial\alpha_0\partial\mathcal{A}} V(\mathcal{A},\alpha_0;\theta) - \alpha_1 \frac{\partial^2}{\partial\alpha_0^2} V(\mathcal{A},\alpha_0;\theta), \tag{5.132}$$

The solution to equation (5.132) reads as

$$J_2(\theta) = i\gamma(\mathcal{A})(\theta-\theta_0) \sum_{m\neq 0} \sum_{\mu=-\infty}^{\infty} \frac{m^3 |\mathcal{V}_m(\mathcal{A};\mu)|^2}{(m\nu+\mu\nu_m)^2}$$

$$-i(\theta - \theta_0) \sum_{m \neq 0} \sum_{\mu=-\infty}^{\infty} \frac{m^2}{m\nu + \mu\nu_m} \frac{\mathrm{d}}{\mathrm{d}\mathcal{A}} |\mathcal{V}_m(\mathcal{A}; \mu)|^2 + \text{oscillating terms.} \tag{5.133}$$

For the purpose of regularization, we add *ad hoc* a small imaginary quantity $-i\gamma$ to $m\nu + \mu\nu_m$ in the denominators of equation (5.133). Taking into account the limit $\gamma \to 0$ and utilizing

$$\pi\Re(x; y) = \frac{x}{x^2 + y^2}, \qquad \lim_{x \to 0} \Re(x; y) = \delta(y), \tag{5.134}$$

we rewrite equation (5.133) in the form

$$J_2(\theta) = -2\pi\gamma(\mathcal{A})(\theta - \theta_0) \sum_{m>0} \frac{m^3}{\nu_m} \frac{\partial}{\partial a} |\mathcal{V}_m(\mathcal{A}; a)|^2 \bigg|_{a = -\left[\frac{m\nu}{\nu_m}\right]}$$

$$+2\pi(\theta - \theta_0) \sum_{m>0} \frac{m^2}{\nu_m} \frac{\mathrm{d}}{\mathrm{d}\mathcal{A}} \left| \mathcal{V}_m\left(\mathcal{A}; -\left[\frac{m\nu}{\nu_m}\right]\right) \right|^2 + \text{oscillating terms,} \tag{5.135}$$

where $[z]$ implies the integer part of z. Note that the formal procedure presented above is equivalent to making use of the well-known *Plemelj* formula

$$\frac{1}{\omega} = \mathcal{P}\left(\frac{1}{\omega}\right) + i\pi\delta(\omega), \tag{5.136}$$

where $\mathcal{P}$ denotes the principal value. Renormalizing the amplitude $\mathcal{A}$ in a way similar to what has been done in preceding paragraphs, for the renormalized amplitude $\mathcal{A}_R$ we obtain the RG equation

$$\frac{\mathrm{d}\mathcal{A}_R}{\mathrm{d}\theta} = 2\pi\Delta^2 \sum_{m>0} \frac{m^2}{\nu_m} \frac{\mathrm{d}}{\mathrm{d}\mathcal{A}_R} \left| \mathcal{V}_m\left(\mathcal{A}_R; -\left[\frac{m\nu}{\nu_m}\right]\right) \right|^2$$

$$-2\pi\Delta^2\gamma(\mathcal{A}_R) \sum_{m>0} \frac{m^3}{\nu_m} \frac{\partial}{\partial a} |\mathcal{V}_m(\mathcal{A}_R; a)|^2 \bigg|_{a = -\left[\frac{m\nu}{\nu_m}\right]}. \tag{5.137}$$

One may proceed further with solving equation (5.131), and as a result a RG equation for the phase φ_R can be derived [Tzenov (2002a)]. It is worthwhile to mention that the RG equations for the amplitude $\mathcal{A}_R$ and the phase φ_R are decoupled in the case when the system is far from primary resonances of the form (5.128).

As an example to demonstrate the theory just developed, we consider the simplest example of a one-and-a-half-degree-of-freedom dynamical system exhibiting chaotic motion, which was already studied in paragraph 3.7.

The Hamiltonian (3.109) can be cast to the form considered in the present paragraph by means of a simple canonical transformation, which shifts the angle variable as $\alpha \to \alpha + \xi \sin \nu_m \theta$ and leaves the action unchanged. As a result, the integrable part of the Hamiltonian and the perturbation become

$$H_0(J) = \epsilon J + \mathcal{H}_s(J), \qquad V(J, \alpha; \theta) = V(J) \cos(\alpha + \xi \sin \nu_m \theta), \quad (5.138)$$

where $\xi = \nu_{12}/\nu_m$ and without loss of generality $\alpha_{m0} = -\pi/2$ (which is always possible to achieve by appropriate choice of the initial time θ_0). The modulation of the resonant phase (or the unperturbed betatron tune) is known to cause a weak instability induced by modulational layers (see e.g. [Lichtenberg (1982)] and references therein). This phenomenon is usually referred to as *modulational diffusion.* Since

$$\nu(J) = \epsilon + \frac{\mathrm{d}\mathcal{H}_s}{\mathrm{d}J}, \qquad V_1(J; \mu) = \frac{1}{2} V(J) \mathcal{J}_\mu(\mu), \qquad (5.139)$$

where $\mathcal{J}_n(z)$ is the *Bessel* function of order n, the RG equation (5.137) can be rewritten as

$$\frac{\mathrm{d}\mathcal{A}_R}{\mathrm{d}\theta} = \frac{\pi \Delta^2}{2\nu_m} \left\{ \frac{\partial}{\partial \mathcal{A}_R} \left[V^2(\mathcal{A}_R) \mathcal{J}^2_{\left[\frac{\nu}{\nu_m}\right]}(\xi) \right] \right.$$

$$\left. - \gamma(\mathcal{A}_R) V^2(\mathcal{A}_R) \frac{\partial}{\partial a} \left[\mathcal{J}_a^2(\xi) \right] \Big|_{a = -\left[\frac{\nu}{\nu_m}\right]} \right\} \qquad (5.140)$$

Instead of solving the RG equation (5.140), we address the important problem of computing the so-called *Liapunov exponent.* It represents the exponential rate of divergence of nearby solutions (trajectories) of equation (5.140) surrounding a given solution. Let $\mathcal{A}_R(J_0; \theta)$ and $\mathcal{A}_R(J_0 + \Delta J_0; \theta)$ be two solutions of the RG equation (5.140), corresponding to the initial conditions J_0 and $J_0 + \Delta J_0$, respectively. The time evolution of the deviation

$$\mathcal{W}(J_0; \theta) = \mathcal{A}_R(J_0 + \Delta J_0; \theta) - \mathcal{A}_R(J_0; \theta) \qquad (5.141)$$

is governed by the equation

$$\frac{\mathrm{d}\mathcal{W}}{\mathrm{d}\theta} = \mathcal{M}(\mathcal{A}_R(\theta)) \mathcal{W}, \qquad (5.142)$$

where

$$\mathcal{M}(\mathcal{A}_R(\theta)) = \frac{\pi \Delta^2}{2\nu_m} \frac{\partial}{\partial \mathcal{A}_R} \left\{ \frac{\partial}{\partial \mathcal{A}_R} \left[V^2(\mathcal{A}_R) \mathcal{J}^2_{\left[\frac{\nu}{\nu_m}\right]}(\xi) \right] \right.$$

$$-\gamma(\mathcal{A}_R)V^2(\mathcal{A}_R)\,\frac{\partial}{\partial a}\left[\mathcal{J}_a^2(\xi)\right]\bigg|_{a=-\left[\frac{\nu}{\nu_m}\right]}\Bigg\}\Bigg|_{\mathcal{A}_R=\mathcal{A}_R(J_0;\theta)} \tag{5.143}$$

If for $\mathcal{A}_R(J_0;\theta)$, we use the approximate solution

$$\mathcal{A}_R(J_0;\theta) = J_0 + O(\Delta^2),$$

equation (5.142) can be integrated at once yielding the approximate *Liapunov* exponent

$$\sigma(J_0;\mathcal{W}) = \mathcal{M}(J_0). \tag{5.144}$$

Chapter 6

Transfer Maps

6.1 Introduction

In the preceding chapters, we argued that once the Hamiltonian governing the motion of a single particle is properly defined, we can formally write the corresponding *Hamilton*'s equations of motion. The solution of the *Hamilton*'s equations with specified initial conditions gives us the complete information about the beam. In most cases of practical interest an analytical solution to the equations of motion is hopeless to find, so as numerical methods have to be employed. Since all numerical techniques for solving differential equations involve discretization schemes anyway, it is natural to pose the question about the possibility of substitution of the *Hamilton*'s equations of motion with a mapping.

The study of Hamiltonian systems (not only such systems, but also a large variety of dynamical systems) can often be accomplished by using one-turn discrete maps, which represents a great conceptual and computational simplification. For a complicated system, such as a large storage ring with a large number of nonlinear elements, it is impractical to extract an exact one-turn map even though analytical forms of Hamiltonians for the individual elements are known.

A useful tool to study the stability of multidimensional Hamiltonian systems is the so-called *Poincare return map* [Poincare (1992)], [Lichtenberg (1982)], [Wiggins (1990)]. Let us consider first the case of an autonomous system with D degrees of freedom, and let Ω be the $(2D-1)$-dimensional surface of constant energy on which the motion is confined. Next, we choose a point $\mathbf{z}_0$ on a periodic orbit Γ with a period T. There exists a $(2D-2)$-dimensional surface Σ (called *Poincare section*) transversely cutting through the orbit Γ at the point $\mathbf{z}_0$. Any orbit beginning in a sufficiently small neighbourhood Λ of $\mathbf{z}_0$ in the *Poincare* section Σ returns to Σ after a time t_Γ close to the period T. The map restricted to Λ which defines the time evolution is called the *Poincare* return map. The choice of the *Poincare* section Σ is to a large extent optional. Like the entire Hamiltonian flow in $2D$-dimensional space, the *Poincare* return map is symplectic, which as we will see later, implies that its Jacobian matrix is a symplectic matrix.

In the case of a non autonomous Hamiltonian system with explicit time dependence, the standard procedure is to extend the phase space to $(2D+1)$-dimensional one, in which the time is a coordinate along with the position $\mathbf{x}$ and the momentum $\mathbf{p}$ of the particle. Suppose that the Hamiltonian is periodic in time with a period T. Then the *Poincare* section is defined as the set of all points with $t = t_0(\text{mod}T)$. The *Poincare* return map is defined now on the full *Poincare* section as the time evolution of $(\mathbf{x}, \mathbf{p})$ over a time T, and is obviously symplectic.

We already encountered in chapter 2 an example of a *Poincare* return map in the non autonomous case, which arose naturally in the theory of linear stability under the name of transfer matrix. In the present chapter, we will discuss the possible extension to the case where nonlinear multipole elements are significant and necessary to be taken into account.

6.2 Nonlinear Transfer Maps of Betatron Motion

We begin with the two-dimensional model describing betatron oscillations in a plane transversal to the particle orbit. Combining equations (2.38), (3.1) and (3.2), we write the Hamiltonian as

$$H(X, P_x, Z, P_z; \theta) = \frac{\dot{\chi}_x(\theta)}{2}(P_x^2 + X^2) + \frac{\dot{\chi}_z(\theta)}{2}(P_z^2 + Z^2) + V(X, Z; \theta), \tag{6.1}$$

where

$$V(X, Z; \theta) = \sum_{I=2}^{\infty} \sum_{\substack{k,m=0 \\ k+m=I}}^{I} b_{km}^{(I)}(\theta) X^k Z^m, \tag{6.2}$$

$$b_{km}^{(I)}(\theta) = \beta_x^{k/2}(\theta) \beta_z^{m/2}(\theta) a_{k0m0}^{(I)}(\theta). \tag{6.3}$$

In order to build the iterative transfer map, we formally solve the *Hamilton*'s equations of motion

$$\frac{\mathrm{d}U}{\mathrm{d}\theta} = \dot{\chi}_u P_u, \qquad \frac{\mathrm{d}P_u}{\mathrm{d}\theta} = -\dot{\chi}_u U - \frac{\partial V}{\partial U}, \tag{6.4}$$

where as usual $u = (x, z)$. By defining the state-vector

$$\mathbf{z} = \begin{pmatrix} X \\ P_x \\ Z \\ P_z \end{pmatrix}, \tag{6.5}$$

we can rewrite equation (6.4) in a vector form

$$\frac{\mathrm{d}\mathbf{z}}{\mathrm{d}\theta} = \widehat{\mathbf{K}}(\theta)\mathbf{z} + \mathbf{F}(\mathbf{z}; \theta). \tag{6.6}$$

Here

$$\widehat{\mathbf{K}} = \begin{pmatrix} 0 & \dot{\chi}_x & 0 & 0 \\ -\dot{\chi}_x & 0 & 0 & 0 \\ 0 & 0 & 0 & \dot{\chi}_z \\ 0 & 0 & -\dot{\chi}_z & 0 \end{pmatrix}, \qquad \mathbf{F} = \begin{pmatrix} 0 \\ -\frac{\partial V}{\partial X} \\ 0 \\ -\frac{\partial V}{\partial Z} \end{pmatrix}. \tag{6.7}$$

Next we perform a linear transformation defined as

$$\mathbf{z} = \widehat{\mathcal{R}}(\theta; \theta_0)\boldsymbol{\xi},$$

where the matrix $\widehat{\mathcal{R}}(\theta; \theta_0)$ is a solution to the linear part of equation (6.6)

$$\frac{\mathrm{d}\widehat{\mathcal{R}}(\theta; \theta_0)}{\mathrm{d}\theta} = \widehat{\mathbf{K}}(\theta)\widehat{\mathcal{R}}(\theta; \theta_0)$$

with a supplementary initial condition $\widehat{\mathcal{R}}(\theta_0; \theta_0) = \widehat{\mathbf{I}}$ for some initial θ_0. The equation for the transformed state-vector $\boldsymbol{\xi}$ can be written as follows

$$\frac{\mathrm{d}\boldsymbol{\xi}}{\mathrm{d}\theta} = \widehat{\mathcal{R}}^{-1}(\theta; \theta_0)\mathbf{F}(\mathbf{z}; \theta), \qquad \boldsymbol{\xi}(\theta_0) = \mathbf{z}_0. \tag{6.8}$$

Equation (6.8) can be solved directly yielding the result

$$\mathbf{z}(\theta) = \widehat{\mathcal{R}}(\theta; \theta_0)\mathbf{z}_0 + \int\limits_{\theta_0}^{\theta} \mathrm{d}\tau \widehat{\mathcal{R}}(\theta; \theta_0)\widehat{\mathcal{R}}^{-1}(\tau; \theta_0)\mathbf{F}(\mathbf{z}(\tau); \tau). \tag{6.9}$$

It can be easily checked that the matrix $\widehat{\mathcal{R}}(\theta;\theta_0)$ of fundamental solutions to the unperturbed problem is of the form [compare with equation (2.45)]

$$\widehat{\mathcal{R}} = \begin{pmatrix} \cos\Delta\chi_x(\theta) & \sin\Delta\chi_x(\theta) & 0 & 0 \\ -\sin\Delta\chi_x(\theta) & \cos\Delta\chi_x(\theta) & 0 & 0 \\ 0 & 0 & \cos\Delta\chi_z(\theta) & \sin\Delta\chi_z(\theta) \\ 0 & 0 & -\sin\Delta\chi_z(\theta) & \cos\Delta\chi_z(\theta) \end{pmatrix}, \tag{6.10}$$

where

$$\Delta\chi_u(\theta) = \chi_u(\theta) - \chi_u(\theta_0), \qquad (u = x, z). \tag{6.11}$$

As a matter of fact, equation (6.9) is only a formal solution to *Hamilton*'s equations of motion, being an equivalent integral representation of the latter. It can be solved iteratively starting with the initial approximation

$$\mathbf{z}^{(0)}(\theta) = \widehat{\mathcal{R}}(\theta;\theta_0)\mathbf{z}_0.$$

Then the n-th approximation will be

$$\mathbf{z}^{(n)}(\theta) = \widehat{\mathcal{R}}(\theta;\theta_0)\left[\mathbf{z}^{(n-1)}(\theta_0) + \int\limits_{\theta_0}^{\theta} \mathrm{d}\tau \widehat{\mathcal{R}}^{-1}(\tau;\theta_0)\mathbf{F}\Big(\mathbf{z}^{(n-1)}(\tau);\tau\Big)\right]. \tag{6.12}$$

The idea of solving equation (6.9) recursively by successive approximations is not far reaching, because the problems of small resonant denominators and secular terms encountered at consecutive orders still remain. Moreover, the recursive solution (6.12) is not exactly symplectic. Instead, we should seek a reasonable approximation able to provide high accuracy and simplicity in utilizing the result, and in addition leading to conservation of symplecticity. Such an approximation is the so-called *thin lens approximation.*

Let us assume that a certain nonlinearity is concentrated in a single point azimuthally located at θ_1

$$\mathbf{F}(\mathbf{z};\theta) = \mathbf{G}(\mathbf{z};\theta)\delta(\theta - \theta_1). \tag{6.13}$$

It can be proved[1] that such point exists and the equivalent strength of the

[1] According to the mean value theorem, the integral on the right-hand-side of equation (6.9) is equal to

$$\frac{l}{R}\widehat{\mathcal{R}}^{-1}\Big(\overline{\theta};\theta_0\Big)\mathbf{F}\Big(\mathbf{z}\Big(\overline{\theta}\Big);\overline{\theta}\Big),$$

where $\overline{\theta}$ is the azimuthal position for which the mean value of the function under the nonlinearity is

$$\mathbf{G}(\mathbf{z};\theta) = \frac{l}{R}\mathbf{F}(\mathbf{z};\theta), \tag{6.14}$$

where l is the length of the multipole element. Shifting the initial reference point at θ_1 ($\theta_0 = \theta_1$), we apply expression (6.9) in a small ϵ-interval $\theta \in (\theta_0 - \epsilon, \theta_0 + \epsilon)$ around θ_0 and then take the limit $\epsilon \to 0$. Thus, we obtain the *kick map*

$$\mathbf{z}_{kick} = \mathbf{z}_0 + \frac{l}{R}\mathbf{F}(\mathbf{z}_0;\theta_0). \tag{6.15}$$

In order to obtain the *rotation map* in between successive kicks, we apply once again expression (6.9) in the interval $\theta \in (\theta_0 + \epsilon, \theta_0 + 2\pi - \epsilon)$

$$\mathbf{z}_{rot} = \widehat{\mathcal{R}}_t \mathbf{z}_{kick}, \tag{6.16}$$

where $\widehat{\mathcal{R}}_t = \widehat{\mathcal{R}}(\theta_0 + 2\pi;\theta_0)$. In explicit form

$$\widehat{\mathcal{R}}_t = \begin{pmatrix} \cos 2\pi\nu_x & \sin 2\pi\nu_x & 0 & 0 \\ -\sin 2\pi\nu_x & \cos 2\pi\nu_x & 0 & 0 \\ 0 & 0 & \cos 2\pi\nu_z & \sin 2\pi\nu_z \\ 0 & 0 & -\sin 2\pi\nu_z & \cos 2\pi\nu_z \end{pmatrix}. \tag{6.17}$$

Combining equations (6.15) and (6.16), we finally arrive at the one-turn transfer map

$$\mathbf{z}' = \widehat{\mathcal{R}}_t \left[\mathbf{z} + \frac{l}{R}\mathbf{F}(\mathbf{z};\theta_0)\right], \tag{6.18}$$

where $\mathbf{z}$ is the state-vector before the kick has been applied, while $\mathbf{z}'$ is its value prior to the next application of the nonlinear kick. An alternate notation for the one-turn transfer map, which we will use extensively in what follows is

$$\mathbf{z}_{n+1} = \widehat{\mathcal{R}}_t \left[\mathbf{z}_n + \frac{l}{R}\mathbf{F}(\mathbf{z}_n;\theta_0)\right], \tag{6.19}$$

---

integral is attained. On the other hand, substituting equation (6.13) into (6.9) for the value of the integral, we obtain

$$\widehat{\mathcal{R}}^{-1}(\theta_1;\theta_0)\mathbf{G}(\mathbf{z}(\theta_1);\theta_1).$$

Comparing these two expressions, we deduce

$$\mathbf{G}(\mathbf{z}(\theta_1);\theta_1) = \frac{l}{R}\widehat{\mathcal{R}}(\theta_1;\theta_0)\widehat{\mathcal{R}}^{-1}\left(\overline{\theta};\theta_0\right)\mathbf{F}\left(\mathbf{z}\left(\overline{\theta}\right);\overline{\theta}\right).$$

It suffices to choose $\theta_1 = \overline{\theta}$, which yields equation (6.14) automatically.

where $\mathbf{z}_n$ denotes the value of the state-vector after the n-th successive revolution.

So far, we built the so-called *one-turn map* in the ideal case of a linear machine with a single nonlinearity. Real machines include many nonlinear elements (and nonlinear errors in bending magnets and quadrupoles) located at different places all along the machine circumference. Therefore, the question of how to construct the one-turn map for a real machine is of utmost importance for practical applications. Consider the portion between the k-th and $(k+1)$-st nonlinear element. The transfer map for that portion is given by

$$\mathbf{z}' = \widehat{\mathcal{M}}^{(k)} \circ \mathbf{z} = \widehat{\mathcal{R}}_k \left[\mathbf{z} + \frac{l_k}{R} \mathbf{F}_k(\mathbf{z}; \theta_{k-1}) \right], \tag{6.20}$$

where $\widehat{\mathcal{R}}_k = \widehat{\mathcal{R}}(\theta_k; \theta_{k-1})$. The one-turn map, called the *Poincare* map at section $\theta = \theta_0$, is given by the composition of the single-element maps

$$\mathbf{z}' = \cdots \circ \widehat{\mathcal{M}}^{(k)} \circ \widehat{\mathcal{M}}^{(k-1)} \circ \cdots \circ \widehat{\mathcal{M}}^{(2)} \circ \widehat{\mathcal{M}}^{(1)} \circ \mathbf{z}. \tag{6.21}$$

A map $\widehat{\mathcal{M}} : \mathbf{R}^4 \to \mathbf{R}^4$ is called *symplectic* if its Jacobian $\widehat{\mathcal{J}}_{\mathcal{M}}(\mathbf{z})$ with elements

$$(\mathcal{J}_{\mathcal{M}})_{kl}(\mathbf{z}) = \frac{\partial \mathcal{M}_k(\mathbf{z})}{\partial z_l} \tag{6.22}$$

is a symplectic matrix, that is the following matrix identity

$$\widehat{\mathcal{J}}_{\mathcal{M}}(\mathbf{z}) \widehat{\mathbf{J}} \widehat{\mathcal{J}}_{\mathcal{M}}^T(\mathbf{z}) = \widehat{\mathbf{J}} \tag{6.23}$$

holds for every $\mathbf{z}$. Here the upper index T denotes matrix transposition, and $\widehat{\mathbf{J}}$ is the fundamental symplectic matrix, given in explicit form by

$$\widehat{\mathbf{J}} = \begin{pmatrix} 0 & 1 & 0 & 0 \\ -1 & 0 & 0 & 0 \\ 0 & 0 & 0 & 1 \\ 0 & 0 & -1 & 0 \end{pmatrix}. \tag{6.24}$$

The map $\widehat{\mathcal{M}}$ is *area preserving* if its Jacobian (6.22) has a determinant equal to one, i.e.

$$\det\left(\widehat{\mathcal{J}}_{\mathcal{M}}\right) = 1. \tag{6.25}$$

It is easy to verify that the condition for area preservation (6.25) is equivalent to the symplectic condition (6.23). One can straightforwardly check by direct calculation that the single-element map (6.20) is area preserving and therefore symplectic. This property is a direct consequence of the Hamiltonian character of the flow represented by the map $\widehat{\mathcal{M}}^{(k)}$. Because the one-turn map (6.21) is composed of symplectic single-element maps, it is symplectic as well[2].

An important remark is now in order. The transfer map is well defined only if the nonlinear term $\mathbf{F}$ has the form specified by equation (6.7), namely only the second and the fourth components are nonzero and depend on X and Z only. In this case only the transfer map derived by following the prescription of the present paragraph is symplectic. It is however possible to "symplectify" the transfer map even in the case when the coordinate and momentum variables are not separated [see the Hamiltonian (6.1)], but the momentum variables enter the nonlinear term $\mathbf{F}$ as well. This will be discussed in subsequent paragraphs of this chapter.

6.3 Linear Transfer Maps

Suppose that the accelerator lattice is composed of linear magnetic elements (dipoles and quadrupoles) only. Alternatively, we may assume that since the nonlinear perturbations are small enough, they can be neglected, so that the linear part of the *Hamilton* equations of motion (6.4) describes the dynamics of single particles with good accuracy. This means that the potential function V given by equation (6.2) is a homogeneous polynomial of second order in the variables X and Z. In this case the single-element map (6.20) represents a linear transformation from $\mathbf{z}$ to $\mathbf{z}'$, and can be

[2]Let us first show that if $\widehat{\mathbf{M}}_1$ and $\widehat{\mathbf{M}}_2$ are two symplectic matrices, their product is also a symplectic matrix. We have

$$\widehat{\mathbf{M}}_1\widehat{\mathbf{M}}_2\hat{\mathbf{J}}\widehat{\mathbf{M}}_2^T\widehat{\mathbf{M}}_1^T = \widehat{\mathbf{M}}_1\hat{\mathbf{J}}\widehat{\mathbf{M}}_1^T = \hat{\mathbf{J}}.$$

This can be easily extended by induction for any finite product of symplectic matrices. Denoting

$$\mathbf{z}^{(1)} = \widehat{\mathcal{M}}^{(1)} \circ \mathbf{z}, \qquad \mathbf{z}^{(2)} = \widehat{\mathcal{M}}^{(2)} \circ \mathbf{z}^{(1)}, \qquad \dots,$$

we can write the Jacobian for the one-turn map as

$$(\mathcal{J}_{\mathcal{M}})_{mn} = \cdots \frac{\partial \mathcal{M}_m^{(k)}}{\partial z_{l_{k-1}}^{(k-1)}} \frac{\partial \mathcal{M}_{l_{k-1}}^{(k-1)}}{\partial z_{l_{k-2}}^{(k-2)}} \cdots \frac{\partial \mathcal{M}_{l_2}^{(2)}}{\partial z_{l_1}^{(1)}} \frac{\partial \mathcal{M}_{l_1}^{(1)}}{\partial z_n}.$$

Since each matrix in the product on the right-hand-side of the above expression is symplectic, the Jacobian $\widehat{\mathcal{J}}_{\mathcal{M}}$ of the full one-turn map is a symplectic matrix as well.

written explicitly as

$$\mathbf{z}' = \widehat{\mathbf{M}}^{(k)}\mathbf{z},$$

where $\widehat{\mathbf{M}}^{(k)}$ is a symplectic matrix. Thus, the one-turn map (6.21) becomes a simple multiplication of matrices

$$\mathbf{z}' = \widehat{\mathbf{M}}\mathbf{z} = \ldots \widehat{\mathbf{M}}^{(k)}\widehat{\mathbf{M}}^{(k-1)} \ldots \widehat{\mathbf{M}}^{(2)}\widehat{\mathbf{M}}^{(1)}\mathbf{z}. \tag{6.26}$$

Here the one-turn transfer matrix $\widehat{\mathbf{M}}$ being a product of symplectic matrices, is symplectic as well. Therefore the study of linear transfer maps reduces to the analysis of properties of symplectic matrices.

Following *Dragt* (see [Dragt (1982)]), we review some basic properties of symplectic matrices. Let us first define the fundamental $2n \times 2n$ symplectic matrix [compare with equation (6.24)] as

$$\widehat{\mathbf{J}} = \begin{pmatrix} \widehat{\mathbf{J}}_2 & \widehat{\mathbf{O}}_2 & \ldots & \widehat{\mathbf{O}}_2 \\ \widehat{\mathbf{O}}_2 & \widehat{\mathbf{J}}_2 & \ldots & \widehat{\mathbf{O}}_2 \\ \ldots & \ldots & \ldots & \ldots \\ \widehat{\mathbf{O}}_2 & \widehat{\mathbf{O}}_2 & \ldots & \widehat{\mathbf{J}}_2 \end{pmatrix}, \tag{6.27}$$

where $\widehat{\mathbf{O}}_2$ denotes the 2×2 null matrix and

$$\widehat{\mathbf{J}}_2 = \begin{pmatrix} 0 & 1 \\ -1 & 0 \end{pmatrix}, \tag{6.28}$$

is the fundamental 2×2 symplectic matrix. Generalizing equation (6.23), we have that a $2n \times 2n$ matrix $\widehat{\mathbf{M}}$ is symplectic if

$$\widehat{\mathbf{M}}\widehat{\mathbf{J}}\widehat{\mathbf{M}}^T = \widehat{\mathbf{J}}. \tag{6.29}$$

It is easy to check the following properties of the fundamental symplectic matrix

$$\widehat{\mathbf{J}}^2 = -\widehat{\mathbf{I}}, \qquad \Longrightarrow \qquad \widehat{\mathbf{J}}^{-1} = -\widehat{\mathbf{J}}, \tag{6.30}$$

$$\det\left(\widehat{\mathbf{J}}\right) = 1, \tag{6.31}$$

$$\widehat{\mathbf{J}}^T = -\widehat{\mathbf{J}}. \tag{6.32}$$

Here $\widehat{\mathbf{I}}$ denotes the $2n \times 2n$ identity matrix. From equations (6.30) and (6.32), it follows that

$$\widehat{\mathbf{J}}^T = \widehat{\mathbf{J}}^{-1}. \tag{6.33}$$

Note that on the grounds of equation (6.30), $\widehat{\mathbf{J}}$ can be regarded as the matrix analogue of the imaginary unit. Combining the definition (6.29) with the identity (6.33), it is straightforward to show that the matrix $\widehat{\mathbf{J}}$ is symplectic by itself. It should be stressed emphatically that the dimensionality of symplectic matrices is *even* by definition. Taking the determinant of both sides of equation (6.29) and making use of (6.31), we obtain

$$\det\left(\widehat{\mathbf{M}}\right) = \pm 1. \tag{6.34}$$

As a matter of fact, it can be shown [Barut (1987)], [Hammermesh (1962)] that the determinant of a symplectic matrix always equals $+1$. Furthermore, it is trivial to check that the identity matrix $\widehat{\mathbf{I}}$ is symplectic. From the relation

$$\widehat{\mathbf{M}}^T = -\widehat{\mathbf{J}}\widehat{\mathbf{M}}^{-1}\widehat{\mathbf{J}}$$

[which can be easily derived using equations (6.29) and (6.30)], an alternate definition of a symplectic matrix follows, namely

$$\widehat{\mathbf{M}}^T\widehat{\mathbf{J}}\widehat{\mathbf{M}} = -\widehat{\mathbf{J}}\widehat{\mathbf{M}}^{-1}\widehat{\mathbf{J}}^2\widehat{\mathbf{M}} = \widehat{\mathbf{J}}\widehat{\mathbf{M}}^{-1}\widehat{\mathbf{M}} = \widehat{\mathbf{J}}. \tag{6.35}$$

Moreover

$$\widehat{\mathbf{M}}^{-1}\widehat{\mathbf{J}}\left(\widehat{\mathbf{M}}^{-1}\right)^T = -\widehat{\mathbf{M}}^{-1}\widehat{\mathbf{J}}\widehat{\mathbf{J}}^{-1}\widehat{\mathbf{M}}\widehat{\mathbf{J}}^{-1} = \widehat{\mathbf{J}}, \tag{6.36}$$

which asserts that if $\widehat{\mathbf{M}}$ is a symplectic matrix, so is its inverse $\widehat{\mathbf{M}}^{-1}$.

To this end, we have shown that:

(1) The identity matrix $\widehat{\mathbf{I}}$ is symplectic.
(2) If $\widehat{\mathbf{M}}$ is symplectic, its inverse $\widehat{\mathbf{M}}^{-1}$ is symplectic as well.
(3) If $\widehat{\mathbf{M}}$ and $\widehat{\mathbf{N}}$ are two symplectic matrices, their product $\widehat{\mathbf{M}}\widehat{\mathbf{N}}$ is also a symplectic matrix.

Therefore, the set of all $2n \times 2n$ symplectic matrices form a *group* [Barut (1987)], [Hammermesh (1962)], called the *symplectic group* and usually denoted by $Sp(2n)$ for any particular value of n.

An *eigenvector* $\mathbf{v}$ of any matrix $\widehat{\mathbf{M}}$ satisfies the equation

$$\widehat{\mathbf{M}}\mathbf{v} = \lambda\mathbf{v}, \tag{6.37}$$

where λ is called an *eigenvalue* correspondent to the eigenvector $\mathbf{v}$. Equation (6.37) represents a linear system for the components of $\mathbf{v}$ and has a nontrivial solution ($\mathbf{v} \neq 0$) if and only if

$$\mathcal{P}(\lambda) = \det\left(\widehat{\mathbf{M}} - \lambda\widehat{\mathbf{I}}\right) = 0. \tag{6.38}$$

The function $\mathcal{P}(\lambda)$ is called *characteristic polynomial*, while equation (6.38) is called the *secular equation* for the matrix $\widehat{\mathbf{M}}$. If the matrix $\widehat{\mathbf{M}}$ is real, the coefficients of the characteristic polynomial are real as well, therefore the eigenvalues must be real or must occur in complex conjugate pairs λ, λ^*. Since the lowest order monomial $a_0\lambda^0$ entering the characteristic polynomial equals $\det\widehat{\mathbf{M}}$, it follows that a symplectic matrix (with $\det\widehat{\mathbf{M}} = 1$) cannot have $\lambda = 0$ as an eigenvalue. For a symplectic matrix $\widehat{\mathbf{M}}$, we also have

$$\widehat{\mathbf{J}}^{-1}\left(\widehat{\mathbf{M}} - \lambda\widehat{\mathbf{I}}\right)\widehat{\mathbf{J}} = \left(\widehat{\mathbf{M}}^{-1} - \lambda\widehat{\mathbf{I}}\right)^T = -\lambda\left(\widehat{\mathbf{M}} - \lambda^{-1}\widehat{\mathbf{I}}\right)^T\left(\widehat{\mathbf{M}}^{-1}\right)^T. \tag{6.39}$$

Taking the determinant of both sides of equation (6.39), we obtain the important relation

$$\mathcal{P}(\lambda) = \lambda^{2n}\mathcal{P}\left(\frac{1}{\lambda}\right). \tag{6.40}$$

It confirms that if λ is an eigenvalue of a symplectic matrix, so is the reciprocal $1/\lambda$. Consequently, the eigenvalues of a symplectic matrix form reciprocal pairs.

We can rewrite equation (6.40) in the form

$$\lambda^{-n}\mathcal{P}(\lambda) = \lambda^{n}\mathcal{P}\left(\lambda^{-1}\right), \tag{6.41}$$

which further displays the symmetry between λ and $1/\lambda$. Another function $\mathcal{Q}(\lambda)$ can be defined as

$$\mathcal{Q}(\lambda) = \lambda^{-n}\mathcal{P}(\lambda), \tag{6.42}$$

which evidently possesses the same zeroes as $\mathcal{P}(\lambda)$. From equation (6.41) it follows that $\mathcal{Q}(\lambda)$ has the symmetry property

$$\mathcal{Q}(\lambda) = \mathcal{Q}\left(\frac{1}{\lambda}\right). \tag{6.43}$$

Apart from the fact that the eigenvalues of a symplectic matrix form reciprocal pairs, it shows in addition that these reciprocal pairs must occur with the same multiplicity. In other words, if the root λ_0 has multiplicity m, so must the root $1/\lambda_0$.

If either $+1$ or -1 is an eigenvalue, then this eigenvalue must have *even* multiplicity. By noting that $(+1)^{-1} = 1$ and $(-1)^{-1} = -1$, this property follows trivially from the above statement concerning the reciprocal pairs of eigenvalues with the same multiplicity possessed by each member of the pair.

Summarizing the properties of the eigenvalues of a real symplectic matrix, we have shown that:

(1) They must be real or occur in complex conjugate pairs.
(2) They form reciprocal pairs, and each member of the pair must have the same multiplicity.
(3) If either ± 1 is an eigenvalue, it must have even multiplicity.

These conditions impose strong restriction on the possible eigenvalues of a real symplectic matrix. For a 2×2 symplectic matrix the (only two) eigenvalues λ_1 and λ_2 must satisfy

$$\lambda_1 \lambda_2 = 1. \tag{6.44}$$

Equation (6.44) combined with the conditions just enumerated above yields five possible cases:

(1) Hyperbolic case: $\lambda_1 > 1$ and $0 < \lambda_2 < 1$. To show that the dynamics is *unstable*, we perform the linear transformation

$$\mathbf{z} = \widehat{\mathbf{A}}\mathbf{u} \tag{6.45}$$

and cast the map (6.26) into the simple decoupled form

$$u'_k = \lambda_k u_k, \qquad k = 1, 2. \tag{6.46}$$

Here the matrix $\widehat{\mathbf{A}}$ is such that the similarity transformation

$$\left(\widehat{\mathbf{A}}^{-1}\widehat{\mathbf{M}}\widehat{\mathbf{A}}\right)_{ij} = \lambda_i \delta_{ij} \tag{6.47}$$

diagonalizes the symplectic matrix $\widehat{\mathbf{M}}$. The matrix $\widehat{\mathbf{A}}$ exists, and as can be easily verified, it is composed of the eigenvectors of $\widehat{\mathbf{M}}$. The first component u_1 of the new state vector $\mathbf{u}$ stretches (while at the same time u_2 shrinks) turn by turn

$$u_k(n) = \lambda_k^n u_k(0), \qquad k = 1, 2, \tag{6.48}$$

which leads to unbound phase trajectories.

(2) Inversion hyperbolic case (unstable): $\lambda_1 < -1$ and $-1 < \lambda_2 < 0$.

(3) Elliptic case (stable): $\lambda_1 = e^{i\phi}$ and $\lambda_2 = e^{-i\phi}$. Note that from equation (6.44) it follows that if the eigenvalues of a 2×2 symplectic matrix occur in a complex conjugate pair, they must lie on the unit circle in the complex plane.
(4) Parabolic case (generally linearly unstable): $\lambda_1 = \lambda_2 = +1$.
(5) Inversion parabolic case (generally linearly unstable): $\lambda_1 = \lambda_2 = -1$.

Another simple example is that of a 4×4 symplectic matrix ($n = 2$). Here, we have to deal with four possible eigenvalues and take into account the conditions and reasoning analogous to the 2×2 case. Due to the larger degree of freedom, more combinations among the eigenvalues are possible which can be grouped in two basic configurations.

Generic configurations:

(1) All eigenvalues are complex and off the unit circle in the complex plane. Note that all eigenvalues can be obtained from a single one

$$\lambda_1 = ae^{i\phi}, \tag{6.49}$$

where to be more precise, we have assumed that $a < 1$ and $0 < \phi < \pi$. Using the operations of complex conjugation and taking reciprocals, we obtain

$$\lambda_2 = ae^{-i\phi}, \qquad \lambda_3 = \frac{1}{a}e^{-i\phi}, \qquad \lambda_4 = \frac{1}{a}e^{i\phi}. \tag{6.50}$$

This case is unstable.
(2) All eigenvalues are real, off the unit circle and of the same sign

$$\lambda_1 = a, \qquad \lambda_2 = b, \qquad \lambda_3 = \frac{1}{a}, \qquad \lambda_4 = \frac{1}{b}, \tag{6.51}$$

where $ab > 0$. This case is unstable.
(3) All eigenvalues are real, off the unit circle and of different signs

$$\lambda_1 = a, \qquad \lambda_2 = b, \qquad \lambda_3 = \frac{1}{a}, \qquad \lambda_4 = \frac{1}{b}, \tag{6.52}$$

where $ab < 0$. This case is unstable.
(4) Two eigenvalues are complex and confined to the unit circle in the complex plane[3]

$$\lambda_1 = e^{i\phi}, \qquad \lambda_2 = e^{-i\phi}, \tag{6.53}$$

[3]The fact that the only two complex eigenvalues of a real 4×4 symplectic matrix are not only complex conjugate, but in addition must also occur in a reciprocal pair, automatically implies that they lie on the unit circle in the complex plane.

and two eigenvalues are real

$$\lambda_3 = a, \qquad \lambda_4 = \frac{1}{a}, \tag{6.54}$$

This case is unstable.

(5) All eigenvalues are complex and confined to the unit circle

$$\lambda_1 = e^{i\phi_1}, \qquad \lambda_2 = e^{i\phi_2}, \qquad \lambda_3 = e^{-i\phi_1}, \qquad \lambda_4 = e^{-i\phi_2}, \tag{6.55}$$

Eigenvalues form reciprocal pairs which are also complex conjugate. This case is stable.

Note that the generic configurations form a two-parameter family, that is to say, they are fully described by two parameters, amplitude and phase, two amplitudes, or two phases of the corresponding eigenvalues for each of the cases enumerated above [see also equation (6.61)]. When these parameters vary transitions between generic configurations can occur by passage through a degenerate configuration.

Degenerate configurations:

(1) Two eigenvalues are equal $\lambda_1 = \lambda_2 = 1$, and two eigenvalues are real and positive $(a > 0)$

$$\lambda_3 = a, \qquad \lambda_4 = \frac{1}{a}. \tag{6.56}$$

This configuration occurs in transition between generic cases (2) and (4). It is unstable.

(2) Two eigenvalues are equal $\lambda_1 = \lambda_2 = -1$, and two eigenvalues are real and positive $(a > 0)$

$$\lambda_3 = a, \qquad \lambda_4 = \frac{1}{a}. \tag{6.57}$$

This configuration occurs in transition between generic cases (3) and (4). It is unstable.

(3) Two eigenvalues are equal $\lambda_1 = \lambda_2 = 1$, and two eigenvalues are confined to the unit circle

$$\lambda_3 = e^{i\phi}, \qquad \lambda_4 = e^{-i\phi}. \tag{6.58}$$

This configuration occurs in transition between generic cases (4) and (5). It is generally linearly unstable.

(4) Two eigenvalues are equal $\lambda_1 = \lambda_2 = 1$, and two eigenvalues are equal $\lambda_3 = \lambda_4 = -1$. This configuration occurs in transition between generic cases (3) and (5), or (3) and (4), or (4) and (5). It also occurs between degenerate cases (2) and (3). The underlying dynamics described by this configuration is generally linearly unstable.

(5) Two pairs of equal eigenvalues confined to the unit circle

$$\lambda_1 = \lambda_2 = e^{i\phi}, \qquad \lambda_3 = \lambda_4 = e^{-i\phi}. \tag{6.59}$$

This configuration occurs in transition between generic cases (1) and (5). The condition that guarantees the existence of this case is $b = -\cos\phi$ and $c = \cos^2\phi$ in equation (6.61). There is however no sufficient condition to guarantee that such a transition is possible. Stability is also undetermined in absence of further conditions.

(6) Two pairs of equal and real eigenvalues

$$\lambda_1 = \lambda_2 = a, \qquad \lambda_3 = \lambda_4 = \frac{1}{a}. \tag{6.60}$$

This configuration occurs in transition between generic cases (1) and (2). It is unstable.

(7) All eigenvalues are equal and have value ± 1. This configuration occurs in transition between generic cases (1), (2), (4) and (5) as well as between degenerate cases (1), (3), (5) and (6). It is generally linearly unstable.

It is natural to expect that the determination of the four eigenvalues of a 4×4 symplectic matrix would in general require the solution of a quartic equation. However, thanks to the symplectic condition matters simplify considerably. Using the identity (6.40), we obtain

$$\mathcal{P}(\lambda) = \lambda^4 + 4b\lambda^3 + 2(2c+1)\lambda^2 + 4b\lambda + 1, \tag{6.61}$$

where

$$b = \frac{1}{16}[\mathcal{P}(1) - \mathcal{P}(-1)], \qquad c = \frac{1}{8}[\mathcal{P}(1) + \mathcal{P}(-1)] - 1. \tag{6.62}$$

Thus, $\mathcal{Q}(\lambda)$ can be written in the form

$$\mathcal{Q}(\lambda) = \left(\lambda + \frac{1}{\lambda}\right)^2 + 4b\left(\lambda + \frac{1}{\lambda}\right) + 4c. \tag{6.63}$$

The eigenvalues can be found from the simple algebraic expressions

$$\lambda = \mathcal{W} \pm \sqrt{\mathcal{W}^2 - 1}, \tag{6.64}$$

where

$$\mathcal{W} = -b \pm \sqrt{b^2 - c}. \tag{6.65}$$

Let us suppose that $\widehat{\mathbf{M}}$ is a real symplectic matrix close to the identity matrix $\widehat{\mathbf{I}}$. Then $\widehat{\mathbf{M}}$ can be expressed as

$$\widehat{\mathbf{M}} = \exp\left(\epsilon \widehat{\mathbf{B}}\right) = \sum_{n=0}^{\infty} \frac{\epsilon^n}{n!} \widehat{\mathbf{B}}^n, \tag{6.66}$$

where ϵ is a small parameter and $\widehat{\mathbf{B}}$ is a real matrix. To understand the implication of equation (6.66) and further reveal the properties of $\widehat{\mathbf{B}}$, we insert (6.66) into the symplectic condition (6.29). Then the relation

$$\sum_{\substack{k,l=0 \\ k+l=n}}^{n} \frac{1}{k!l!} \widehat{\mathbf{B}}^k \widehat{\mathbf{J}} \left(\widehat{\mathbf{B}}^T\right)^l = 0 \tag{6.67}$$

must be satisfied for every order $n \geq 1$. For $n = 1$, we find

$$\widehat{\mathbf{B}}\widehat{\mathbf{J}} + \widehat{\mathbf{J}}\widehat{\mathbf{B}}^T = 0. \tag{6.68}$$

This relation is consistent with equation (6.67) for all orders for $n \geq 2$. To verify this assertion, it suffices to rewrite the left-hand-side of equation (6.67) in the form

$$\frac{\widehat{\mathbf{B}}^n \widehat{\mathbf{J}}}{n!} \sum_{k=0}^{n} (-1)^k \binom{n}{k},$$

where $\binom{n}{k}$ is the binomial coefficient, and use of (6.68) has been made. The fact that the sum in the above expression is identically zero completes the proof of (6.67). Next, we write $\widehat{\mathbf{B}}$ in the form

$$\widehat{\mathbf{B}} = \widehat{\mathbf{J}}\widehat{\mathbf{S}}, \tag{6.69}$$

where $\widehat{\mathbf{S}}$ is a real matrix. Note that the representation (6.69) is always possible, since $\widehat{\mathbf{J}}^{-1}$ exists and equals $-\widehat{\mathbf{J}}$ [see equation (6.30)]. Upon substituting (6.69) into (6.68), we obtain

$$\widehat{\mathbf{S}}^T = \widehat{\mathbf{S}}, \tag{6.70}$$

which implies that $\widehat{\mathbf{S}}$ is a symmetric matrix. Conversely, suppose that $\widehat{\mathbf{M}}$ is any matrix of the form

$$\widehat{\mathbf{M}} = \exp\left(\epsilon \widehat{\mathbf{J}} \widehat{\mathbf{S}}\right) \tag{6.71}$$

with $\widehat{\mathbf{S}}$ real and symmetric. Then the matrix $\widehat{\mathbf{M}}$ is symplectic. Taking into account

$$\widehat{\mathbf{M}}^T = \exp\left(-\epsilon \widehat{\mathbf{S}} \widehat{\mathbf{J}}\right),$$

we find

$$\widehat{\mathbf{M}} \widehat{\mathbf{J}} \widehat{\mathbf{M}}^T = \exp\left(\epsilon \widehat{\mathbf{J}} \widehat{\mathbf{S}}\right) \widehat{\mathbf{J}} \exp\left(-\epsilon \widehat{\mathbf{S}} \widehat{\mathbf{J}}\right)$$

$$= \exp\left(\epsilon \widehat{\mathbf{J}} \widehat{\mathbf{S}}\right) \widehat{\mathbf{J}} \exp\left(-\epsilon \widehat{\mathbf{S}} \widehat{\mathbf{J}}\right) \widehat{\mathbf{J}}^{-1} \widehat{\mathbf{J}} = \exp\left(\epsilon \widehat{\mathbf{J}} \widehat{\mathbf{S}}\right) \exp\left(-\epsilon \widehat{\mathbf{J}} \widehat{\mathbf{S}}\right) \widehat{\mathbf{J}} = \widehat{\mathbf{J}}. \tag{6.72}$$

A set of matrices is said to form a *Lie algebra* if is satisfies the following properties:

(1) If the matrix $\widehat{\mathbf{A}}$ is an element of the Lie algebra, then so is the matrix $a\widehat{\mathbf{A}}$, where a is any scalar.
(2) If the matrices $\widehat{\mathbf{A}}$ and $\widehat{\mathbf{B}}$ belong to the Lie algebra, then so does their sum $\widehat{\mathbf{A}} + \widehat{\mathbf{B}}$.
(3) If two matrices $\widehat{\mathbf{A}}$ and $\widehat{\mathbf{B}}$ are in the Lie algebra, then so is their commutator

$$\left[\widehat{\mathbf{A}}, \widehat{\mathbf{B}}\right] = \widehat{\mathbf{A}} \widehat{\mathbf{B}} - \widehat{\mathbf{B}} \widehat{\mathbf{A}}. \tag{6.73}$$

It can be easily verified that matrices represented as $\widehat{\mathbf{J}} \widehat{\mathbf{S}}$, where $\widehat{\mathbf{S}}$ is symmetric form a Lie algebra.

Let $\widehat{\mathbf{A}}$ be a given matrix and $\widehat{\mathbf{B}}$ an arbitrary matrix. We associate with $\widehat{\mathbf{A}}$ an operator $\widehat{\mathcal{A}}$ (which acts on matrices) according to the rule

$$\widehat{\mathcal{A}} \widehat{\mathbf{B}} = \left[\widehat{\mathbf{A}}, \widehat{\mathbf{B}}\right]. \tag{6.74}$$

Next, we define the matrix

$$\widehat{\mathbf{D}}(\epsilon) = \exp\left(\epsilon \widehat{\mathbf{A}}\right) \widehat{\mathbf{B}} \exp\left(-\epsilon \widehat{\mathbf{A}}\right), \qquad \widehat{\mathbf{D}}(0) = \widehat{\mathbf{B}}, \tag{6.75}$$

where $\widehat{\mathbf{A}}$ and $\widehat{\mathbf{B}}$ are given matrices and ϵ is a parameter. By differentiation, we find

$$\frac{\mathrm{d}\widehat{\mathbf{D}}}{\mathrm{d}\epsilon} = \widehat{\mathbf{A}}\widehat{\mathbf{D}} - \widehat{\mathbf{D}}\widehat{\mathbf{A}} = \widehat{\mathcal{A}}\widehat{\mathbf{D}}.$$

This differential equation can be easily solved. Taking into account the boundary condition (6.75), we obtain

$$\widehat{\mathbf{D}}(\epsilon) = \exp\left(\epsilon\widehat{\mathcal{A}}\right)\widehat{\mathbf{B}}. \tag{6.76}$$

There is a remarkably close connection between the concept of a Lie algebra and that of a group, which is based on a fundamental theorem by *Campbell, Baker* and *Hausdorff*. The *Campbell-Baker-Hausdorff* theorem states the following:

Let $\widehat{\mathbf{A}}$ *and* $\widehat{\mathbf{B}}$ *be any two matrices (or operators in general), and let* α *and* β *be parameters. Then, we can formally write*

$$\exp\left(\alpha\widehat{\mathbf{A}}\right)\exp\left(\beta\widehat{\mathbf{B}}\right) = \exp\left(\widehat{\mathbf{C}}\right), \tag{6.77}$$

where

$$\widehat{\mathbf{C}} = \alpha\widehat{\mathbf{A}} + \beta\widehat{\mathbf{B}} + \frac{\alpha\beta}{2}\left[\widehat{\mathbf{A}},\widehat{\mathbf{B}}\right] + \frac{\alpha^2\beta}{12}\left[\widehat{\mathbf{A}},\left[\widehat{\mathbf{A}},\widehat{\mathbf{B}}\right]\right] + \frac{\alpha\beta^2}{12}\left[\widehat{\mathbf{B}},\left[\widehat{\mathbf{B}},\widehat{\mathbf{A}}\right]\right] + \ldots. \tag{6.78}$$

Depending on the properties of the matrices $\widehat{\mathbf{A}}$ and $\widehat{\mathbf{B}}$, the series for $\widehat{\mathbf{C}}$ may or may not converge. The general form of all the coefficients on the right-hand-side of equation (6.78) is not known, however summation of the series to all orders in α and the first few orders in β is possible. Through first order in β the result is

$$\widehat{\mathbf{C}} = \alpha\widehat{\mathbf{A}} + \beta\alpha\left[1 - \exp\left(-\alpha\widehat{\mathcal{A}}\right)\right]^{-1}\widehat{\mathcal{A}}\widehat{\mathbf{B}} + O(\beta^2). \tag{6.79}$$

The first step in the proof of the *Campbell-Baker-Hausdorff* theorem is to verify that the sought-for matrix $\widehat{\mathbf{C}}(\alpha,\beta)$ satisfies the differential equation

$$\frac{\partial\widehat{\mathbf{C}}}{\partial\beta} = \left[1 - \exp\left(-\widehat{\mathcal{C}}\right)\right]^{-1}\widehat{\mathcal{C}}\widehat{\mathbf{B}}. \tag{6.80}$$

Differentiation of the left-hand-side of equation (6.77) yields

$$\frac{\partial}{\partial\beta}\exp\left(\alpha\widehat{\mathbf{A}}\right)\exp\left(\beta\widehat{\mathbf{B}}\right) = \exp\left(\widehat{\mathbf{C}}\right)\widehat{\mathbf{B}}. \tag{6.81}$$

Because the argument of the exponential on the right-hand-side of equation (6.77) is an operator, the computation of its derivative should be handled with care. Through first order in $\Delta\beta$, we obtain

$$\exp\left[\widehat{\mathbf{C}}(\alpha,\beta+\Delta\beta)\right] = \exp\left[\widehat{\mathbf{C}}(\alpha,\beta) + \Delta\beta\frac{\partial}{\partial\beta}\widehat{\mathbf{C}}(\alpha,\beta)\right]$$

$$= \sum_{n=0}^{\infty}\frac{1}{n!}\left[\widehat{\mathbf{C}}(\alpha,\beta) + \Delta\beta\frac{\partial}{\partial\beta}\widehat{\mathbf{C}}(\alpha,\beta)\right]^{n}.$$

Expanding the power series and retaining zero- and first-order terms, we find that

$$\frac{\partial}{\partial\beta}\exp\left(\widehat{\mathbf{C}}\right) = \sum_{n=1}^{\infty}\sum_{m=0}^{n-1}\frac{1}{n!}\widehat{\mathbf{C}}^{m}\left(\frac{\partial\widehat{\mathbf{C}}}{\partial\beta}\right)\widehat{\mathbf{C}}^{n-m-1}, \tag{6.82}$$

where the definition of the derivative of an operator $\widehat{\mathbf{D}}(\beta)$

$$\frac{\mathrm{d}\widehat{\mathbf{D}}}{\mathrm{d}\beta} = \lim_{\Delta\beta\to 0}\frac{\widehat{\mathbf{D}}(\beta+\Delta\beta) - \widehat{\mathbf{D}}(\beta)}{\Delta\beta}$$

has been used and particular attention has been paid to the possibility that $\widehat{\mathbf{C}}$ and $\partial\widehat{\mathbf{C}}/\partial\beta$ may not commute. Next, we change the order of summation and rewrite equation (6.82) in the form

$$\frac{\partial}{\partial\beta}\exp\left(\widehat{\mathbf{C}}\right) = \sum_{m=0}^{\infty}\sum_{l=0}^{\infty}\frac{1}{(l+m+1)!}\widehat{\mathbf{C}}^{m}\left(\frac{\partial\widehat{\mathbf{C}}}{\partial\beta}\right)\widehat{\mathbf{C}}^{l}. \tag{6.83}$$

The series (6.83) has an integral representation

$$\sum_{m=0}^{\infty}\sum_{l=0}^{\infty}\frac{1}{(l+m+1)!}\widehat{\mathbf{C}}^{m}\left(\frac{\partial\widehat{\mathbf{C}}}{\partial\beta}\right)\widehat{\mathbf{C}}^{l}$$

$$= \int_{0}^{1}\mathrm{d}\lambda\exp\left[(1-\lambda)\widehat{\mathbf{C}}\right]\left(\frac{\partial\widehat{\mathbf{C}}}{\partial\beta}\right)\exp\left(\lambda\widehat{\mathbf{C}}\right),$$

which can be verified in a straightforward manner by expanding out the two exponentials and integrating term by term with the due account of the identity

$$\int_0^1 d\lambda \lambda^l (1-\lambda)^m = \frac{m!}{(l+1)(l+2)\dots(l+m+1)}.$$

Using equation (6.76), we obtain

$$\int_0^1 d\lambda \exp\left[(1-\lambda)\widehat{\mathbf{C}}\right]\left(\frac{\partial\widehat{\mathbf{C}}}{\partial\beta}\right)\exp\left(\lambda\widehat{\mathbf{C}}\right) = \exp\left(\widehat{\mathbf{C}}\right)\int_0^1 d\lambda \exp\left(-\lambda\widehat{\mathcal{C}}\right)\left(\frac{\partial\widehat{\mathbf{C}}}{\partial\beta}\right)$$

$$= \exp\left(\widehat{\mathbf{C}}\right)\widehat{\mathcal{C}}^{-1}\left[1-\exp\left(-\widehat{\mathcal{C}}\right)\right]\left(\frac{\partial\widehat{\mathbf{C}}}{\partial\beta}\right).$$

The last expression has been obtained by expanding $\exp\left(-\lambda\widehat{\mathcal{C}}\right)$ in a power series, integrating term by term, and resumming the final result. Therefore

$$\frac{\partial}{\partial\beta}\exp\left(\widehat{\mathbf{C}}\right) = \exp\left(\widehat{\mathbf{C}}\right)\widehat{\mathcal{C}}^{-1}\left[1-\exp\left(-\widehat{\mathcal{C}}\right)\right]\left(\frac{\partial\widehat{\mathbf{C}}}{\partial\beta}\right). \tag{6.84}$$

Comparing (6.81) and (6.84), we finally arrive at equation (6.80).

To prove equation (6.78), we expand $\widehat{\mathbf{C}}(\alpha,\beta)$ as

$$\widehat{\mathbf{C}}(\alpha,\beta) = \sum_{m,n=0}^{\infty} \alpha^m \beta^n \widehat{\mathbf{C}}_{mn},$$

and substitute the expansion into equation (6.80). Observing that

$$\widehat{\mathbf{C}}_{10} = \widehat{\mathbf{A}}, \qquad \widehat{\mathbf{C}}_{01} = \widehat{\mathbf{B}}, \tag{6.85}$$

and equating coefficients of like powers $\alpha^m\beta^n$, we obtain the series (6.78).

The proof of equation (6.79) now follows directly from the differential equation (6.80). Suppressing the dependence on α, the Taylor expansion of $\widehat{\mathbf{C}}$ with respect to β can be written as

$$\widehat{\mathbf{C}}(\beta) = \widehat{\mathbf{C}}(0) + \beta\widehat{\mathbf{C}}'(0) + \frac{\beta^2}{2}\widehat{\mathbf{C}}''(0) + \dots. \tag{6.86}$$

The quantity $\widehat{\mathbf{C}}(0)$ has been determined already in equation (6.85), where it was noted that it is equal to $\alpha\widehat{\mathbf{A}}$. To find $\widehat{\mathbf{C}}'(0)$, we evaluate the right-hand-side of equation (6.80) for β set equal to zero. The result is

$$\widehat{\mathbf{C}}'(0) = \alpha\left[1-\exp\left(-\alpha\widehat{\mathcal{A}}\right)\right]^{-1}\widehat{\mathcal{A}}\widehat{\mathbf{B}}. \tag{6.87}$$

Insertion of (6.85) and (6.87) into the Taylor expansion (6.86) yields the desired result (6.79).

We have just proved the remarkable fact that the matrix $\widehat{\mathbf{C}}$ defined by equation (6.77) belongs to the *Lie* algebra generated by $\widehat{\mathbf{A}}$ and $\widehat{\mathbf{B}}$. That is, we have shown that $\widehat{\mathbf{C}}$ is a sum of elements formed only from $\widehat{\mathbf{A}}$ and $\widehat{\mathbf{B}}$ and their multiple commutators. From the *Campbell-Baker-Hausdorff* theorem it follows that given any *Lie* algebra of matrices, there exists a corresponding *Lie group.* The rule for multiplication of any two group elements is contained within the *Lie* algebra and can be extended to the corresponding *Lie* group. Consider all matrices of the form

$$\widehat{\mathbf{G}}(\alpha) = \exp\left(\alpha\widehat{\mathbf{L}}\right) \tag{6.88}$$

with $\widehat{\mathbf{L}}$ being a member of the *Lie* algebra. Because

$$\widehat{\mathbf{G}}_2 = \exp\left(\widehat{\mathbf{L}}_2\right) = \exp\left(\alpha\widehat{\mathbf{L}}\right)\exp\left(\beta\widehat{\mathbf{L}}_1\right)$$

for α and β sufficiently small, it follows that $\widehat{\mathbf{G}}_2$ bears the same properties as $\widehat{\mathbf{G}}$. In addition

$$\widehat{\mathbf{G}}(0) = \widehat{\mathbf{I}}, \qquad \widehat{\mathbf{G}}^{-1}(\alpha) = \widehat{\mathbf{G}}(-\alpha).$$

Therefore, matrices of the form (6.88), at least those sufficiently close to the identity, form a group. Once the group has been constructed near the identity, it can be further extended to a global group by successively multiplying the $\widehat{\mathbf{G}}$'s already obtained. Based on equation (6.71) and on the fact that matrices of the form $\widehat{\mathbf{J}}\widehat{\mathbf{S}}$ (with $\widehat{\mathbf{S}}$ symmetric) are elements of a *Lie* algebra, it follows that symplectic matrices near the identity form a *Lie* group.

Let $\widehat{\mathbf{M}}$ be any real nonsingular matrix. Then $\widehat{\mathbf{M}}$ can be represented uniquely as

$$\widehat{\mathbf{M}} = \widehat{\mathbf{P}}\widehat{\mathbf{O}}, \tag{6.89}$$

where $\widehat{\mathbf{P}}$ is a real positive definite symmetric matrix, and $\widehat{\mathbf{O}}$ is a real orthogonal matrix [Gantmacher (1959)][4]. Suppose that $\widehat{\mathbf{M}}$ is symplectic. Then

[4]Since

$$\widehat{\mathbf{M}}\widehat{\mathbf{M}}^T = \widehat{\mathbf{P}}\widehat{\mathbf{O}}\widehat{\mathbf{O}}^T\widehat{\mathbf{P}}^T = \widehat{\mathbf{P}}^2,$$

we can set

$$\widehat{\mathbf{P}} = \sqrt{\widehat{\mathbf{M}}\widehat{\mathbf{M}}^T}.$$

the symplectic condition (6.29) can be written in alternate form

$$\widehat{\mathbf{M}} = \widehat{\mathbf{J}}\left(\widehat{\mathbf{M}}^T\right)^{-1}\widehat{\mathbf{J}}^{-1}$$

Using the polar decomposition (6.89), we obtain

$$\widehat{\mathbf{P}}\widehat{\mathbf{O}} = \widehat{\mathbf{J}}\widehat{\mathbf{P}}^{-1}\widehat{\mathbf{J}}^{-1}\widehat{\mathbf{J}}\widehat{\mathbf{O}}\widehat{\mathbf{J}}^{-1}. \qquad (6.90)$$

It can be easily checked by direct computation that the matrix $\widehat{\mathbf{J}}\widehat{\mathbf{P}}^{-1}\widehat{\mathbf{J}}^{-1}$ is real, symmetric and positive definite, while the matrix $\widehat{\mathbf{J}}\widehat{\mathbf{O}}\widehat{\mathbf{J}}^{-1}$ is real and orthogonal. Due to the uniqueness of the polar decomposition, equation (6.90) implies the relations

$$\widehat{\mathbf{P}} = \widehat{\mathbf{J}}\widehat{\mathbf{P}}^{-1}\widehat{\mathbf{J}}^{-1}, \qquad \widehat{\mathbf{O}} = \widehat{\mathbf{J}}\widehat{\mathbf{O}}\widehat{\mathbf{J}}^{-1}. \qquad (6.91)$$

From the above equations (6.91) it follows that the matrices $\widehat{\mathbf{P}}$ and $\widehat{\mathbf{O}}$ apart from being symmetric and orthogonal, respectively, they are also symplectic.

We will now be seeking a suitable representation for $\widehat{\mathbf{P}}$ and $\widehat{\mathbf{O}}$, similar to the one displayed by equation (6.71). Since $\widehat{\mathbf{O}}$ is real, orthogonal and its determinant is equal to $+1$ (recall that $\widehat{\mathbf{O}}$ is also symplectic), it can be represented in exponential form

$$\widehat{\mathbf{O}} = \exp\left(\widehat{\mathbf{A}}\right), \qquad (6.92)$$

where $\widehat{\mathbf{A}}$ is a real antisymmetric matrix

$$\widehat{\mathbf{A}}^T = -\widehat{\mathbf{A}}. \qquad (6.93)$$

Next, we insert the representation (6.92) into the symplectic condition (6.91). The latter acquires the form

$$\widehat{\mathbf{O}} = \exp\left(\widehat{\mathbf{A}}\right) = \exp\left(\widehat{\mathbf{J}}\widehat{\mathbf{A}}\widehat{\mathbf{J}}^{-1}\right). \qquad (6.94)$$

Since the matrix under the second exponential in equation (6.94) is real and antisymmetric provided $\widehat{\mathbf{A}}$ is real and antisymmetric itself, and because $\widehat{\mathbf{A}}$

Here

$$\left(\det\widehat{\mathbf{P}}\right)^2 = \left(\det\widehat{\mathbf{M}}\right)^2 \neq 0, \qquad \widehat{\mathbf{O}} = \widehat{\mathbf{P}}^{-1}\widehat{\mathbf{M}}.$$

What is left is to verify that the matrix $\widehat{\mathbf{O}}$ is orthogonal. We have

$$\widehat{\mathbf{O}}\widehat{\mathbf{O}}^T = \widehat{\mathbf{P}}^{-1}\widehat{\mathbf{M}}\widehat{\mathbf{M}}^T\widehat{\mathbf{P}}^{-1} = \widehat{\mathbf{P}}^{-1}\widehat{\mathbf{P}}^2\widehat{\mathbf{P}}^{-1} = \widehat{\mathbf{I}}.$$

Clearly, both matrices $\widehat{\mathbf{P}}$ and $\widehat{\mathbf{O}}$ are uniquely determined by construction.

is uniquely determined, from equation (6.94) we must have

$$\widehat{\mathbf{A}}\widehat{\mathbf{J}} = \widehat{\mathbf{J}}\widehat{\mathbf{A}}. \tag{6.95}$$

This implies that $\widehat{\mathbf{A}}$ commutes with $\widehat{\mathbf{J}}$. Taking into account that $\widehat{\mathbf{A}}$ is antisymmetric, the condition (6.95) can also be written in the form

$$\widehat{\mathbf{A}}\widehat{\mathbf{J}} + \widehat{\mathbf{J}}\widehat{\mathbf{A}}^T = 0. \tag{6.96}$$

According to an earlier argument [compare (6.96) with equation (6.68)], the matrix $\widehat{\mathbf{A}}$ can be written as a product

$$\widehat{\mathbf{A}} = \widehat{\mathbf{J}}\widehat{\mathbf{S}}_c,$$

where $\widehat{\mathbf{S}}_c$ is a real symmetric matrix. Moreover, since $\widehat{\mathbf{A}}$ commutes with $\widehat{\mathbf{J}}$, then $\widehat{\mathbf{S}}_c$ must commute with $\widehat{\mathbf{J}}$ as well

$$\widehat{\mathbf{S}}_c\widehat{\mathbf{J}} = \widehat{\mathbf{J}}\widehat{\mathbf{S}}_c. \tag{6.97}$$

Thus, we have showed that the matrix $\widehat{\mathbf{O}}$ can be represented according to the relation

$$\widehat{\mathbf{O}} = \exp\left(\widehat{\mathbf{J}}\widehat{\mathbf{S}}_c\right), \tag{6.98}$$

where $\widehat{\mathbf{S}}_c$ is a real symmetric matrix which commutes with $\widehat{\mathbf{J}}$.

Recall that according to the polar decomposition theorem the matrix $\widehat{\mathbf{P}}$ is real, symmetric and positive definite. Therefore, it can be written in the form

$$\widehat{\mathbf{P}} = \exp\left(\widehat{\mathbf{B}}\right), \tag{6.99}$$

where $\widehat{\mathbf{B}}$ is unique, real and symmetric

$$\widehat{\mathbf{B}}^T = \widehat{\mathbf{B}}. \tag{6.100}$$

Inserting the representation (6.99) into the symplectic condition (6.91), and using a reasoning similar to the one already applied for the matrix $\widehat{\mathbf{A}}$, we conclude that $\widehat{\mathbf{B}}$ anticommutes with $\widehat{\mathbf{J}}$, namely

$$\widehat{\mathbf{B}}\widehat{\mathbf{J}} = -\widehat{\mathbf{J}}\widehat{\mathbf{B}}. \tag{6.101}$$

Since $\widehat{\mathbf{B}}$ is symmetric, equation (6.101) can be rewritten in the form

$$\widehat{\mathbf{B}}\widehat{\mathbf{J}} + \widehat{\mathbf{J}}\widehat{\mathbf{B}}^T = 0. \tag{6.102}$$

Consequently,

$$\widehat{\mathbf{B}} = \widehat{\mathbf{J}}\widehat{\mathbf{S}}_a,$$

where $\widehat{\mathbf{S}}_a$ is a real symmetric matrix. However, in this case $\widehat{\mathbf{S}}_a$ anticommutes with $\widehat{\mathbf{J}}$

$$\widehat{\mathbf{S}}_a\widehat{\mathbf{J}} = -\widehat{\mathbf{J}}\widehat{\mathbf{S}}_a. \tag{6.103}$$

In summary, we have shown that $\widehat{\mathbf{P}}$ has the representation

$$\widehat{\mathbf{P}} = \exp\left(\widehat{\mathbf{J}}\widehat{\mathbf{S}}_a\right), \tag{6.104}$$

where $\widehat{\mathbf{S}}_a$ is a real symmetric matrix which anticommutes with $\widehat{\mathbf{J}}$.

Combining now equations (6.89), (6.98) and (6.104), we obtain

$$\widehat{\mathbf{M}} = \exp\left(\widehat{\mathbf{J}}\widehat{\mathbf{S}}_a\right)\exp\left(\widehat{\mathbf{J}}\widehat{\mathbf{S}}_c\right). \tag{6.105}$$

Thus, we have shown that a symplectic matrix of the most general form can be represented as a product of two exponentials whose arguments are elements of a special type, belonging to the symplectic group *Lie* algebra.

It is interesting to study the properties of commuting and anticommuting of a symmetric matrix $\widehat{\mathbf{S}}$ with an arbitrary nonsingular matrix $\widehat{\mathbf{F}}$. We write $\widehat{\mathbf{S}}$ as a sum of two matrices

$$\widehat{\mathbf{S}} = \widehat{\mathbf{S}}_a + \widehat{\mathbf{S}}_c, \tag{6.106}$$

where $\widehat{\mathbf{S}}_a$ is the part anticommuting with $\widehat{\mathbf{F}}$, while $\widehat{\mathbf{S}}_c$ is the one that commutes with $\widehat{\mathbf{F}}$. Multiplying both sides of equation (6.106) by $\widehat{\mathbf{F}}$ and taking into account that $\widehat{\mathbf{S}}_a$ and $\widehat{\mathbf{S}}_c$ anticommutes and commutes with $\widehat{\mathbf{F}}$ respectively, we obtain

$$\widehat{\mathbf{F}}^{-1}\widehat{\mathbf{S}}\widehat{\mathbf{F}} = -\widehat{\mathbf{S}}_a + \widehat{\mathbf{S}}_c, \tag{6.107}$$

From the two equations (6.106) and (6.107), we find

$$\widehat{\mathbf{S}}_a = \frac{1}{2}\left(\widehat{\mathbf{S}} - \widehat{\mathbf{F}}^{-1}\widehat{\mathbf{S}}\widehat{\mathbf{F}}\right), \qquad \widehat{\mathbf{S}}_c = \frac{1}{2}\left(\widehat{\mathbf{S}} + \widehat{\mathbf{F}}^{-1}\widehat{\mathbf{S}}\widehat{\mathbf{F}}\right), \tag{6.108}$$

If in addition the condition that both $\widehat{\mathbf{S}}_a$ and $\widehat{\mathbf{S}}_c$ are symmetric is imposed, it is straightforward to verify that the matrix $\widehat{\mathbf{F}}$ must be of a special type obeying the relation

$$\widehat{\mathbf{F}}\widehat{\mathbf{F}}^T = \sigma\widehat{\mathbf{I}}, \tag{6.109}$$

where σ is any scalar. Clearly, the fundamental symplectic matrix $\widehat{\mathbf{J}}$ satisfies equation (6.109) with $\sigma = 1$.

Let us now address the topic of a suitable basis for the symplectic group *Lie* algebra. First, we observe that matrices of the form $\widehat{\mathbf{J}}\widehat{\mathbf{S}}_c$ constitute a *Lie* algebra by themselves because the commutator of any two matrices $\widehat{\mathbf{J}}\widehat{\mathbf{S}}_c^{(1)}$ and $\widehat{\mathbf{J}}\widehat{\mathbf{S}}_c^{(2)}$ is again a matrix of the form $\widehat{\mathbf{J}}\widehat{\mathbf{S}}_c$. Further, the commutator of a matrix of the form $\widehat{\mathbf{J}}\widehat{\mathbf{S}}_c$ with a matrix of the form $\widehat{\mathbf{J}}\widehat{\mathbf{S}}_a$ is a matrix of the form $\widehat{\mathbf{J}}\widehat{\mathbf{S}}_a$. Finally, the commutator of any two matrices of the form $\widehat{\mathbf{J}}\widehat{\mathbf{S}}_a$ is a matrix of the form $\widehat{\mathbf{J}}\widehat{\mathbf{S}}_c$.

In the 2×2 case of $Sp(2)$, the most general symmetric matrix $\widehat{\mathbf{S}}$ can be written as

$$\widehat{\mathbf{S}} = \begin{pmatrix} a & b \\ b & c \end{pmatrix}. \tag{6.110}$$

The matrix $\widehat{\mathbf{S}}$ commutes with $\widehat{\mathbf{J}}_2$ [see equation (6.28)] if

$$a = c, \qquad b = 0.$$

Therefore, the most general form of $\widehat{\mathbf{S}}_c$ in the 2×2 case is simply proportional to the identity matrix

$$\widehat{\mathbf{S}}_c = a\widehat{\mathbf{I}}.$$

Consequently,

$$\widehat{\mathbf{J}}_2\widehat{\mathbf{S}}_c = a\widehat{\mathbf{J}}_2. \tag{6.111}$$

This implies that a convenient choice for the first element of the basis for the *Lie* algebra is $\widehat{\mathbf{J}}_2$ itself.

We now study the matrix $\widehat{\mathbf{S}}_a$. The requirement that $\widehat{\mathbf{J}}_2$ anticommute with the matrix $\widehat{\mathbf{S}}$ of the form (6.110) gives the restriction

$$c = -a.$$

Therefore $\widehat{\mathbf{S}}_a$ is of the general form

$$\widehat{\mathbf{S}}_a = a\begin{pmatrix} 1 & 0 \\ 0 & -1 \end{pmatrix} + b\begin{pmatrix} 0 & 1 \\ 1 & 0 \end{pmatrix},$$

and consequently $\widehat{\mathbf{J}}_2\widehat{\mathbf{S}}_a$ can be written in the form

$$\widehat{\mathbf{J}}_2\widehat{\mathbf{S}}_a = a\begin{pmatrix} 0 & -1 \\ -1 & 0 \end{pmatrix} + b\begin{pmatrix} 1 & 0 \\ 0 & -1 \end{pmatrix}. \tag{6.112}$$

Upon combining the results of equations (6.111) and (6.112), we are finally in a position to construct the convenient basis for the *Lie* algebra of $Sp(2)$. This basis consists of the matrices

$$\widehat{\mathbf{D}}_1 = \begin{pmatrix} 0 & 1 \\ -1 & 0 \end{pmatrix}, \qquad \widehat{\mathbf{D}}_2 = \begin{pmatrix} 0 & 1 \\ 1 & 0 \end{pmatrix}, \qquad \widehat{\mathbf{D}}_3 = \begin{pmatrix} 1 & 0 \\ 0 & -1 \end{pmatrix}. \tag{6.113}$$

It is straightforward to check that they satisfy the commutation rules

$$\left[\widehat{\mathbf{D}}_1, \widehat{\mathbf{D}}_2\right] = 2\widehat{\mathbf{D}}_3, \qquad \left[\widehat{\mathbf{D}}_2, \widehat{\mathbf{D}}_3\right] = -2\widehat{\mathbf{D}}_1, \qquad \left[\widehat{\mathbf{D}}_3, \widehat{\mathbf{D}}_1\right] = 2\widehat{\mathbf{D}}_2. \tag{6.114}$$

We use the basis elements $\widehat{\mathbf{D}}_i$ to evaluate the right-hand-side of equation (6.105). A symplectic 2×2 matrix of the most general form can be written as

$$\widehat{\mathbf{M}} = \exp\left(\alpha_2 \widehat{\mathbf{D}}_2 + \alpha_3 \widehat{\mathbf{D}}_3\right) \exp\left(\alpha_1 \widehat{\mathbf{D}}_1\right), \tag{6.115}$$

where α_i are arbitrary coefficients. Thus, we have found a complete parameterization of $Sp(2)$.

Since the matrix $\widehat{\mathbf{D}}_1$ plays the role of the imaginary unit for matrices $\left(\widehat{\mathbf{D}}_1^2 = -\widehat{\mathbf{I}}\right)$, the second exponential on the right-hand-side of equation (6.115) is easily evaluated, i.e.,

$$\exp\left(\alpha_1 \widehat{\mathbf{D}}_1\right) = \widehat{\mathbf{I}} \cos\alpha_1 + \widehat{\mathbf{D}}_1 \sin\alpha_1. \tag{6.116}$$

In order to evaluate the first exponential $\exp\left(\alpha_2 \widehat{\mathbf{D}}_2 + \alpha_3 \widehat{\mathbf{D}}_3\right)$, we note that

$$\left(\alpha_2 \widehat{\mathbf{D}}_2 + \alpha_3 \widehat{\mathbf{D}}_3\right)^2 = (\alpha_2^2 + \alpha_3^2)\widehat{\mathbf{I}}. \tag{6.117}$$

Consequently, we obtain

$$\exp\left(\alpha_2 \widehat{\mathbf{D}}_2 + \alpha_3 \widehat{\mathbf{D}}_3\right) = \widehat{\mathbf{I}} \cosh\Delta + \frac{1}{\Delta}\left(\alpha_2 \widehat{\mathbf{D}}_2 + \alpha_3 \widehat{\mathbf{D}}_3\right) \sinh\Delta, \tag{6.118}$$

where

$$\Delta = \sqrt{\alpha_2^2 + \alpha_3^2}. \tag{6.119}$$

An alternate parameterization of the symplectic group $Sp(2)$ follows from equation (6.71). Suppose now that the transfer matrix has the form suggested by equation (6.71), where $\widehat{\mathbf{S}}$ is a symmetric matrix given by equation (6.110). The exponential representation (6.71) of $\widehat{\mathbf{M}}$ can be written in equivalent form

$$\widehat{\mathbf{M}} = \exp\left(\mu\widehat{\mathbf{C}}_2\right),$$

where

$$\mu^2 = \det\widehat{\mathbf{S}} = ac - b^2, \qquad \widehat{\mathbf{C}}_2 = \begin{pmatrix} \alpha & \beta \\ -\gamma & -\alpha \end{pmatrix},$$

$$\alpha = \frac{b}{\mu}, \qquad \beta = \frac{c}{\mu}, \qquad \gamma = \frac{a}{\mu}.$$

Note that the matrix $\widehat{\mathbf{C}}_2$ thus constructed has the property

$$\widehat{\mathbf{C}}_2^2 = -\widehat{\mathbf{I}}.$$

Consequently, the transfer matrix $\widehat{\mathbf{M}}$ can be evaluated explicitly in a manner similar to that applied to obtain equation (6.116). The result is

$$\widehat{\mathbf{M}} = \widehat{\mathbf{I}}\cos\mu + \widehat{\mathbf{C}}_2\sin\mu. \tag{6.120}$$

In the representation (6.120) one can recognize immediately the expression (2.54) for the transfer matrix obtained earlier.

6.4 The Henon Map

The simplest nontrivial example of a polynomial transfer map is the so-called *Henon map.* It can describe the horizontal betatron oscillations in an accelerator possessing a FODO-cell structure with a single thin sextupole. The two-dimensional *Henon* map can be obtained from equation (6.18) in the case when the potential V [see equation (6.2)] contains a single localized cubic nonlinearity. In explicit form it can be written as

$$X_{n+1} = X_n\cos\omega + \left(P_n - \mathcal{S}X_n^2\right)\sin\omega, \tag{6.121}$$

$$P_{n+1} = -X_n\sin\omega + \left(P_n - \mathcal{S}X_n^2\right)\cos\omega, \tag{6.122}$$

where

$$\omega = 2\pi\nu, \qquad \mathcal{S} = \frac{l\lambda_0(\theta_0)\beta^{3/2}(\theta_0)}{2R^3}. \tag{6.123}$$

In the previous chapter, we discussed the essence of the RG method and worked out a few examples, which demonstrate its power to handle a number of problems arising in the theory of continuous dynamical systems. In the present paragraph, we will show that the RG method can be successfully applied to study discrete dynamical systems, and as a particular example, we consider the *Henon* map. The latter can be further simplified by eliminating the canonical momentum variable P. Multiplying equation (6.121) by $\cos\omega$, multiplying equation (6.122) by $-\sin\omega$, and summing up the two equations, we obtain

$$X_{n+1}\cos\omega - P_{n+1}\sin\omega = X_n. \tag{6.124}$$

Substitution of the recursion relation (6.124) into equation (6.121) yields a second-order difference equation for X alone

$$\widehat{\mathcal{L}}X_n = X_{n+1} - 2X_n\cos\omega + X_{n-1} = -\epsilon\mathcal{S}X_n^2\sin\omega. \tag{6.125}$$

Here ϵ is a formal small parameter introduced for convenience to take into account the perturbative character of the sextupole nonlinearity.

Next we consider an asymptotic solution of the map (6.125) for small ϵ by means of the RG method. The naive perturbation expansion

$$X_n = X_n^{(0)} + \epsilon X_n^{(1)} + \epsilon^2 X_n^{(2)} + \dots \tag{6.126}$$

when substituted into equation (6.125), yields the perturbation equations order by order

$$\widehat{\mathcal{L}}X_n^{(0)} = 0, \tag{6.127}$$

$$\widehat{\mathcal{L}}X_n^{(1)} = -\mathcal{S}X_n^{(0)\mathbf{2}}\sin\omega, \tag{6.128}$$

$$\widehat{\mathcal{L}}X_n^{(2)} = -2\mathcal{S}X_n^{(0)}X_n^{(1)}\sin\omega, \tag{6.129}$$

$$\widehat{\mathcal{L}}X_n^{(3)} = -\mathcal{S}\left(X_n^{(1)\mathbf{2}} + 2X_n^{(0)}X_n^{(2)}\right)\sin\omega, \tag{6.130}$$

$$\widehat{\mathcal{L}}X_n^{(4)} = -2\mathcal{S}\left(X_n^{(0)}X_n^{(3)} + X_n^{(1)}X_n^{(2)}\right)\sin\omega. \tag{6.131}$$

Solving equation (6.127) for the zeroth-order contribution, we obtain the obvious result

$$X_n^{(0)} = Ae^{i\omega n} + c.c., \qquad P_n^{(0)} = iAe^{i\omega n} + c.c., \tag{6.132}$$

where A is a complex integration constant. By virtue of (6.132) the first-order perturbation equation (6.128) becomes

$$\widehat{\mathcal{L}}X_n^{(1)} = -\mathcal{S}\Big(A^2 e^{2i\omega n} + 2|A|^2 + c.c.\Big)\sin\omega. \tag{6.133}$$

Its solution can be written as

$$X_n^{(1)} = -\mathcal{S}|A|^2 \cot\frac{\omega}{2} + \frac{\sin\omega}{2}\mathcal{S}_1 A^2 e^{2i\omega n} + c.c., \tag{6.134}$$

where

$$\mathcal{S}_1 = \frac{\mathcal{S}}{\cos\omega - \cos 2\omega}. \tag{6.135}$$

To avoid resonant secular terms, we assume in addition that

$$\omega \neq 2k\pi, \qquad \omega \neq \frac{2\pi}{3} + 2k\pi, \qquad \omega \neq \frac{4\pi}{3} + 2k\pi, \tag{6.136}$$

where k is an integer.

The second-order perturbation equation (6.129) becomes

$$\widehat{\mathcal{L}}X_n^{(2)} = -2\mathcal{F}\sin\omega|A|^2 A e^{i\omega n} - \sin^2\omega\mathcal{S}\mathcal{S}_1 A^3 e^{3i\omega n} + c.c.,$$

where

$$\mathcal{F} = \mathcal{S}\bigg(\frac{\mathcal{S}_1}{2}\sin\omega - \mathcal{S}\cot\frac{\omega}{2}\bigg). \tag{6.137}$$

Its solution can be readily obtained in the form

$$X_n^{(2)} = in\mathcal{F}|A|^2 A e^{i\omega n} + \frac{\sin^2\omega}{2}\mathcal{S}_1\mathcal{S}_2 A^3 e^{3i\omega n} + c.c., \tag{6.138}$$

where

$$\mathcal{S}_2 = \frac{\mathcal{S}}{\cos\omega - \cos 3\omega}, \qquad \omega \neq (2k+1)\pi, \qquad \omega \neq (2k+1)\frac{\pi}{2}. \tag{6.139}$$

Continuing further with third-order calculations, we note that

$$\widehat{\mathcal{L}}X_n^{(3)} = -\mathcal{S}\Big(\Sigma_0|A|^4 + \Sigma_2|A|^2 A^2 e^{2i\omega n} + \Sigma_4 A^4 e^{4i\omega n} + c.c.\Big)\sin\omega.$$

Here the following notations

$$\Sigma_0 = \mathcal{S}^2\cot^2\frac{\omega}{2} + \frac{\mathcal{S}_1^2}{2}\sin^2\omega, \tag{6.140}$$

$$\Sigma_2 = \mathcal{S}_1\Big(\mathcal{S}_2\sin\omega - \mathcal{S}\cot\frac{\omega}{2}\Big)\sin\omega + 2in\mathcal{F}, \tag{6.141}$$

$$\Sigma_4 = \frac{\mathcal{S}_1}{4}(\mathcal{S}_1 + 4\mathcal{S}_2)\sin^2\omega \tag{6.142}$$

have been introduced. The solution to the third-order perturbation equation is found in a straightforward manner to be

$$X_n^{(3)} = -\frac{\mathcal{S}}{2}\Sigma_0|A|^4\cot\frac{\omega}{2} + (\mathcal{B} + in\mathcal{S}_1\mathcal{F}\sin\omega)|A|^2A^2e^{2i\omega n}$$

$$+\frac{\sin\omega}{2}\mathcal{S}_3\Sigma_4A^4e^{4i\omega n} + c.c., \tag{6.143}$$

where

$$\mathcal{B} = \frac{\mathcal{S}_1^2}{2}\left(\mathcal{S}_2\sin\omega - \mathcal{S}\cot\frac{\omega}{2}\right)\sin^2\omega - \frac{\mathcal{S}_1^2\mathcal{F}}{\mathcal{S}}\sin\omega\sin 2\omega, \tag{6.144}$$

$$\mathcal{S}_3 = \frac{\mathcal{S}}{\cos\omega - \cos 4\omega}, \qquad \omega \neq \frac{2\pi}{5} + 2k\pi, \qquad \omega \neq \frac{4\pi}{5} + 2k\pi. \tag{6.145}$$

Finally, retaining terms proportional to the fundamental harmonic $e^{i\omega n}$ only in the fourth-order perturbation equation

$$\widehat{\mathcal{L}}X_n^{(4)} = (\mathcal{C} - 2in\mathcal{F}^2\sin\omega)|A|^4Ae^{i\omega n} + c.c.,$$

where

$$\mathcal{C} = -2\mathcal{S}\left(\mathcal{B} + \frac{\mathcal{S}_1^2\mathcal{S}_2}{4}\sin^3\omega - \frac{\mathcal{S}\Sigma_0}{2}\cot\frac{\omega}{2}\right)\sin\omega, \tag{6.146}$$

we obtain the secular fourth-order contribution to the fundamental harmonic in the form

$$X_n^{(4)} = \left(in\mathcal{D} - \frac{n^2\mathcal{F}^2}{2}\right)|A|^4Ae^{i\omega n} + c.c.. \tag{6.147}$$

Here the coefficient $\mathcal{D}$ is given by the expression

$$\mathcal{D} = -\frac{\mathcal{C} + \mathcal{F}^2\cos\omega}{2\sin\omega}. \tag{6.148}$$

For the sake of completeness, let us write the remaining part of the exact fourth-order solution containing third and fifth harmonics

$$X_n^{(4)} = \left(\mathcal{B}_3 + \frac{3in}{2}\mathcal{S}_1\mathcal{S}_2\mathcal{F}\sin^2\omega\right)|A|^2A^3e^{3i\omega n} + \mathcal{C}_5A^5e^{5i\omega n} + c.c.. \tag{6.149}$$

The coefficients $\mathcal{B}_3$ and $\mathcal{C}_5$ are given by the expressions

$$\mathcal{B}_3 = \mathcal{S}_2\left(\mathcal{B} + \frac{\mathcal{S}_3\Sigma_4}{2}\sin\omega - \frac{\mathcal{S}\mathcal{S}_1\mathcal{S}_2}{2}\sin^2\omega\cot\frac{\omega}{2}\right)\sin\omega$$

$$-\frac{3\mathcal{S}_1\mathcal{S}_2^2\mathcal{F}}{2\mathcal{S}}\sin^2\omega\cot\frac{\omega}{2}, \tag{6.150}$$

and

$$\mathcal{C}_5 = \frac{\mathcal{S}_4}{2}\left(\mathcal{S}_3\Sigma_4 + \frac{\mathcal{S}_1^2\mathcal{S}_2}{2}\sin^2\omega\right)\sin^2\omega. \tag{6.151}$$

Analogously to equations (6.135), (6.139) and (6.145), the coefficient $\mathcal{S}_4$ can be written as

$$\mathcal{S}_4 = \frac{\mathcal{S}}{\cos\omega - \cos 5\omega}, \qquad \omega \neq \frac{\pi}{3} + 2k\pi. \tag{6.152}$$

Close inspection of the naive perturbation solution starting from the second-order result (6.138) shows that secular terms (proportional to n, n^2, etc.) are present. To remove these terms, we define the renormalization transformation $A \to \widetilde{A}(n)$ by collecting all terms proportional to the fundamental harmonic $e^{i\omega n}$

$$\widetilde{A}(n) = \left[1 + i\epsilon^2 n\mathcal{F}|A|^2 + \epsilon^4\left(in\mathcal{D} - \frac{n^2\mathcal{F}^2}{2}\right)|A|^4\right]A. \tag{6.153}$$

Solving perturbatively equation (6.153) for A in terms of $\widetilde{A}(n)$, we obtain

$$A = \left[1 - i\epsilon^2 n\mathcal{F}\left|\widetilde{A}(n)\right|^2 + O(\epsilon^3)\right]\widetilde{A}(n). \tag{6.154}$$

A discrete version of the RG equation can be defined by considering the difference

$$\widetilde{A}(n+1) - \widetilde{A}(n) = i\epsilon^2\mathcal{F}|A|^2 A + \epsilon^4\left(i\mathcal{D} - \frac{\mathcal{F}^2}{2} - n\mathcal{F}^2\right)|A|^4 A. \tag{6.155}$$

Substituting the expression for A in terms of $\widetilde{A}(n)$ [see equation (6.154)] into the right-hand-side of the above equation (6.155), we can eliminate the secular terms up to $O(\epsilon^4)$. The result is

$$\widetilde{A}(n+1) = \left[1 + i\epsilon^2\mathcal{F}\left|\widetilde{A}(n)\right|^2 + \epsilon^4\left(i\mathcal{D} - \frac{\mathcal{F}^2}{2}\right)\left|\widetilde{A}(n)\right|^4\right]\widetilde{A}(n). \tag{6.156}$$

This naive RG map does not preserve the symplectic symmetry of the original system and does not have a *constant of motion.* To recover the symplectic symmetry, we regularize [Goto (2001)] the naive RG map (6.156) by noting that the coefficient in the square brackets multiplying $\widetilde{A}(n)$ can be exponentiated as

$$\widetilde{A}(n+1) = \widetilde{A}(n) \exp\left[i\widetilde{\omega}\left(\left|\widetilde{A}(n)\right|\right)\right], \tag{6.157}$$

where

$$\widetilde{\omega}\left(\left|\widetilde{A}(n)\right|\right) = \epsilon^2 \mathcal{F}\left|\widetilde{A}(n)\right|^2 + \epsilon^4 \mathcal{D}\left|\widetilde{A}(n)\right|^4. \tag{6.158}$$

Although the renormalization procedure described above may seem somewhat artificial, it holds in all orders. By extracting a symplectic implicit map in terms of the real part and the argument (phase) of the renormalized amplitude $\widetilde{A}(n)$, a partial proof (up to fourth order) of this assertion will be presented in the next paragraph. It is clear now that the regularized RG map (6.157) possesses the obvious integral of motion

$$\left|\widetilde{A}(n+1)\right| = \left|\widetilde{A}(n)\right| = \sqrt{\frac{\mathcal{J}}{2}}. \tag{6.159}$$

It is important to note that the secular terms encountered in higher harmonics $\left(e^{2i\omega n},\ e^{3i\omega n},\ \text{etc.}\right)$ can be summed up to give renormalized amplitudes respectively, which expressed in terms of $\widetilde{A}(n)$ do not contain secular terms. This means that once the amplitude of the fundamental harmonic is renormalized, problems with divergences in higher harmonics are being fixed automatically. To see that, let us write the amplitude of the second harmonic

$$A_2 = \epsilon\left[\frac{\mathcal{S}_1}{2}\sin\omega + \epsilon^2(\mathcal{B} + in\mathcal{S}_1\mathcal{F}\sin\omega)|A|^2\right]A^2, \tag{6.160}$$

which by virtue of equation (6.154) acquires the form

$$A_2 = \epsilon\left[\frac{\mathcal{S}_1}{2}\sin\omega + \epsilon^2\mathcal{B}\left|\widetilde{A}(n)\right|^2\right]\widetilde{A}^2(n). \tag{6.161}$$

In analogy, the amplitude of the third harmonic

$$A_3 = \epsilon^2\left[\frac{\mathcal{S}_1\mathcal{S}_2}{2}\sin^2\omega + \epsilon^2\left(\mathcal{B}_3 + \frac{3in}{2}\mathcal{S}_1\mathcal{S}_2\mathcal{F}\sin^2\omega\right)|A|^2\right]A^3 \tag{6.162}$$

can be renormalized. The result is

$$A_3 = \epsilon^2 \left[\frac{\mathcal{S}_1 \mathcal{S}_2}{2} \sin^2 \omega + \epsilon^2 \mathcal{B}_3 \left| \widetilde{A}(n) \right|^2 \right] \widetilde{A}^3(n). \tag{6.163}$$

Proceeding in a similar way as above, we can represent the canonical conjugate momentum P_n according to

$$P_n = i\widetilde{B}(n) e^{i\omega n} + c.c. + \text{higher harmonics}, \tag{6.164}$$

where

$$\widetilde{B}(n+1) = \widetilde{B}(n) \exp \left[i\widetilde{\omega}\left(\left| \widetilde{A}(n) \right| \right) \right]. \tag{6.165}$$

Using now the relation (6.124) between the canonical conjugate variables (X_n, P_n), we can express the renormalized amplitude $\widetilde{B}(n)$ in terms of $\widetilde{A}(n)$ as

$$\widetilde{B}(n) = \frac{i\widetilde{A}(n)}{\sin \omega} \left[e^{-i(\omega + \widetilde{\omega})} - \cos \omega \right]. \tag{6.166}$$

In addition, the sextupole nonlinearity shifts the closed orbit by a constant (in normalized coordinates) value

$$X_{co} = -\frac{\epsilon \mathcal{S} \mathcal{J}}{2} \left[1 + \frac{\epsilon^2}{4} \Sigma_0 \mathcal{J} + O(\epsilon^3) \right] \cot \frac{\omega}{2}, \qquad P_{co} = -X_{co} \tan \frac{\omega}{2}, \tag{6.167}$$

which is a property common for all odd multipole nonlinearities.

Neglecting higher harmonics and iterating the regularized RG maps (6.157) and (6.165), we can write the renormalized solution of the *Henon* map

$$X_n^{(1)} = X_{co} + \sqrt{2\mathcal{J}} \cos \psi(\mathcal{J}; n), \tag{6.168}$$

$$P_n^{(1)} = P_{co} + \sqrt{2\mathcal{J}} \left[\alpha_H^{(1)}(\mathcal{J}) \cos \psi(\mathcal{J}; n) - \beta_H^{(1)}(\mathcal{J}) \sin \psi(\mathcal{J}; n) \right], \tag{6.169}$$

where

$$\psi(\mathcal{J}; n) = [\omega + \widetilde{\omega}(\mathcal{J})] n + \widetilde{\phi}, \tag{6.170}$$

$$\alpha_H^{(1)}(\mathcal{J}) = \frac{\cos \omega - \cos [\omega + \widetilde{\omega}(\mathcal{J})]}{\sin \omega}, \qquad \beta_H^{(1)}(\mathcal{J}) = \frac{\sin [\omega + \widetilde{\omega}(\mathcal{J})]}{\sin \omega}. \tag{6.171}$$

It is easy to see that the integral of motion $\mathcal{J}$ has the form of a *generalized Courant-Snyder invariant*, which can be written as

$$2\mathcal{J} = \left(X^{(1)} - X_{co}\right)^2 + \frac{\left[P^{(1)} - P_{co} - \alpha_H^{(1)}(\mathcal{J})\left(X^{(1)} - X_{co}\right)\right]^2}{\beta_H^{(1)\mathbf{2}}(\mathcal{J})}. \quad (6.172)$$

It is important to emphasize that equation (6.172) represents a transcendental equation for the invariant $\mathcal{J}$ as a function of the canonical variables (X, P), because the coefficients α_H and β_H depend on $\mathcal{J}$ themselves. Note also that the sextupole nonlinearity gives rise to a nonlinear tune shift $\widetilde{\omega}$, leading to the distortion of the invariant curves [being circles in normalized phase space (X, P)][5] even in the approximation when the contribution of the first harmonic only is taken into account. Further distortion of the phase-space trajectories is introduced by higher harmonics.

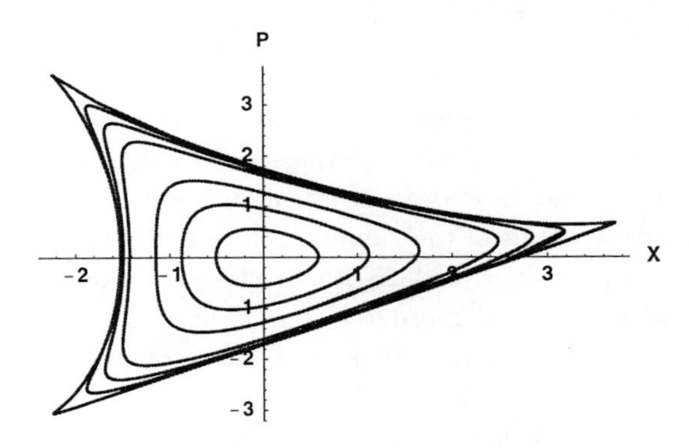

Fig. 6.1 Phase portrait of the Henon map obtained from equations (6.173) - (6.177) near the third-order resonance with $\nu = 0.323$. Here, $\mathcal{J}$ takes values ranging from $\mathcal{J} = 0.15$ (inner contour) to $\mathcal{J} = \mathcal{J}_{max} = 3.85$ (outer contour).

Taking into account all harmonics up to the fifth one, we can write the

[5]Note that $\alpha_H = 0$ and $\beta_H = 1$ for $\widetilde{\omega} = 0$.

renormalized fourth-order solution of the *Henon* map in the form

$$X_n = X_n^{(1)} + \sum_{M=2}^{5} \mathcal{X}_M \cos M\psi(\mathcal{J}; n), \tag{6.173}$$

$$P_n = P_n^{(1)} + \sum_{M=2}^{5} \mathcal{X}_M \left[\alpha_H^{(M)} \cos M\psi(\mathcal{J}; n) - \beta_H^{(M)} \sin M\psi(\mathcal{J}; n) \right]. \tag{6.174}$$

The amplitudes $\mathcal{X}_M$ of various harmonics are given by the expressions

$$\mathcal{X}_2 = \frac{\epsilon \mathcal{J}}{2} \left(\mathcal{S}_1 \sin\omega + \epsilon^2 \mathcal{B} \mathcal{J} \right), \qquad \mathcal{X}_4 = \frac{1}{4} \epsilon^3 \mathcal{S}_3 \Sigma_4 \mathcal{J}^2 \sin\omega, \tag{6.175}$$

$$\mathcal{X}_3 = \frac{\epsilon^2 \mathcal{J} \sqrt{\mathcal{J}}}{2\sqrt{2}} \left(\mathcal{S}_1 \mathcal{S}_2 \sin^2\omega + \epsilon^2 \mathcal{B}_3 \mathcal{J} \right), \qquad \mathcal{X}_5 = \frac{1}{2\sqrt{2}} \epsilon^4 \mathcal{C}_5 \mathcal{J}^2 \sqrt{\mathcal{J}}. \tag{6.176}$$

Furthermore, similar to equation (6.171), the generalized $\alpha_H^{(M)}$ and $\beta_H^{(M)}$ functions can be written as

$$\alpha_H^{(M)}(\mathcal{J}) = \frac{\cos\omega - \cos M[\omega + \widetilde{\omega}(\mathcal{J})]}{\sin\omega}, \qquad \beta_H^{(M)}(\mathcal{J}) = \frac{\sin M[\omega + \widetilde{\omega}(\mathcal{J})]}{\sin\omega}. \tag{6.177}$$

To conclude this paragraph, we present illustrative numerical results revealing the stability properties of the Henon map. We use the analytical renormalized solution expressed by equations (6.173) - (6.177) to construct the phase portrait of the Henon map near the third-order resonance with $\nu = 0.323$ (also $\nu = 0.34525$), near the fourth-order resonance with $\nu = 0.24$, and near the fifth-order resonance with $\nu = 0.19$. All calculations are performed for a relatively large value of the sextupole strength corresponding to $\mathcal{S} = 0.1$.

In Figure 6.1 the phase portrait of the Henon map near the third-order resonance with $\nu = 0.323$ is depicted. It shows the stability region and the invariant curves for different values of the modulus squared of the renormalized amplitude $\widetilde{A}$ [see equation (6.159)]. As the value of the invariant $\mathcal{J}$ increases, it reaches a value $\mathcal{J}_{max}$ above which the phase trajectories begin to intersect. This is due to the fact that the perturbation renormalization technique, valid for unperturbed betatron tunes sufficiently far from resonances, does not work well in this region, and the reduction procedure near resonances developed in the next paragraph should be employed. However, the quantity $\mathcal{J}_{max}$ is closely related to the *dynamic aperture*. For the case where $\nu = 0.323$ it is approximately $\mathcal{J}_{max} = 3.85$. Figure 6.2 represents

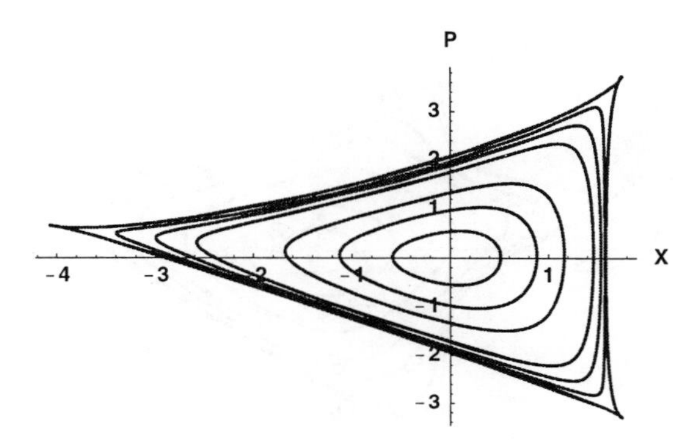

Fig. 6.2 Phase portrait of the Henon map obtained from equations (6.173) - (6.177) near the third-order resonance with $\nu = 0.34525$. Here, $\mathcal{J}$ takes values ranging from $\mathcal{J} = 0.15$ (inner contour) to $\mathcal{J} = \mathcal{J}_{max} = 3.85$ (outer contour).

the phase portrait of the Henon map for $\nu = 0.34525$. The value of $\mathcal{J}_{max}$ is the same as in the case where $\nu = 0.323$.

Figures 6.3 and 6.4 show the phase portrait of the Henon map near the fourth-order and the fifth-order resonances where $\nu = 0.24$ and $\nu = 0.19$, respectively. The dynamic aperture in the case of the fourth-order resonance is approximately $\mathcal{J}_{max} = 19.01$, while in the case of the fifth-order resonance it is $\mathcal{J}_{max} = 28.5$.

6.5 Resonance Structure of the Henon Map

The solution to the first-order perturbation equation (6.133) was obtained in the form (6.134) assuming that the unperturbed betatron tune ν is far from the third order resonance $3\nu = 1$. It is important to study the properties of the *Henon* map near a nonlinear resonance by means of the RG method. In what follows, we will demonstrate how the RG reduction of the *Henon* map works near the one-third resonance, however a similar procedure can be performed for all other resonances.

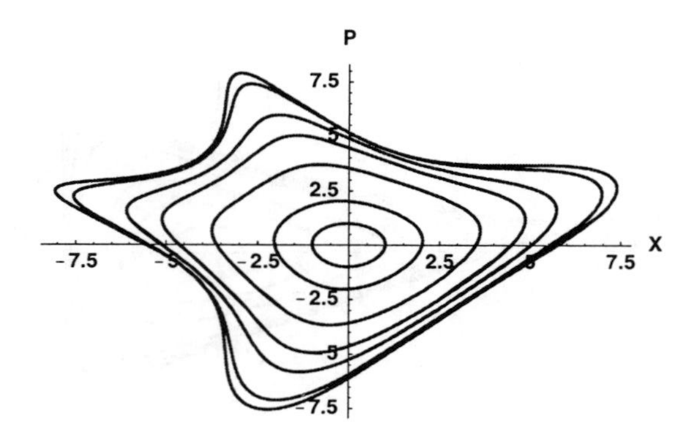

Fig. 6.3 Phase portrait of the Henon map obtained from equations (6.173) - (6.177) near the fourth-order resonance with $\nu = 0.24$. Here, $\mathcal{J}$ takes values ranging from $\mathcal{J} = 0.5$ (inner contour) to $\mathcal{J} = \mathcal{J}_{max} = 19.01$ (outer contour).

Let us assume [similar to equations (5.85) and (5.94)] that

$$\omega = \omega_0 + \epsilon\delta_1 + \epsilon^2\delta_2 + \dots, \qquad \omega_0 = \frac{2\pi}{3}. \tag{6.178}$$

Equation (6.125) can be written in alternate form as

$$\widehat{\mathcal{L}}_0 X_n = X_{n+1} - 2X_n \cos\omega_0 + X_{n-1} = 2X_n(\cos\omega - \cos\omega_0) - \epsilon\mathcal{S}X_n^2 \sin\omega. \tag{6.179}$$

The perturbation expansion (6.126) when substituted into equation (6.179), yields the perturbation equations

$$\widehat{\mathcal{L}}_0 X_n^{(0)} = 0, \tag{6.180}$$

$$\widehat{\mathcal{L}}_0 X_n^{(1)} = -\frac{\sqrt{3}}{2}\Big(2\delta_1 X_n^{(0)} + \mathcal{S}X_n^{(0)\mathbf{2}}\Big), \tag{6.181}$$

$$\widehat{\mathcal{L}}_0 X_n^{(2)} = -\sqrt{3}\delta_1 X_n^{(1)} + \left(\frac{\delta_1^2}{2} - \sqrt{3}\delta_2\right)X_n^{(0)} + \mathcal{S}\left(\frac{\delta_1}{2}X_n^{(0)\mathbf{2}} - \sqrt{3}X_n^{(0)}X_n^{(1)}\right). \tag{6.182}$$

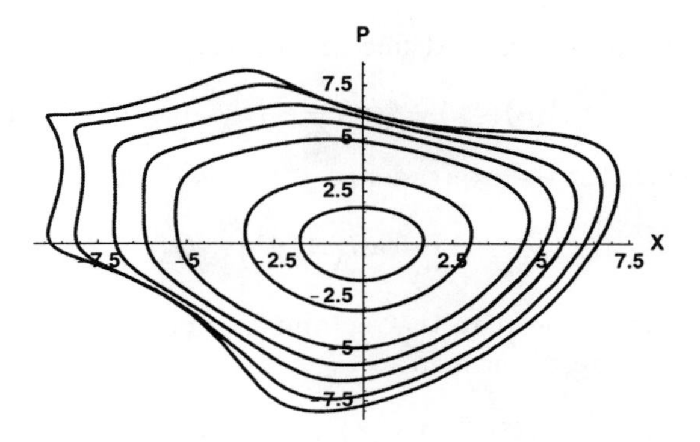

Fig. 6.4 Phase portrait of the Henon map obtained from equations (6.173) - (6.177) near the fifth-order resonance with $\nu = 0.19$. Here, $\mathcal{J}$ takes values ranging from $\mathcal{J} = 1.5$ (inner contour) to $\mathcal{J} = \mathcal{J}_{max} = 28.5$ (outer contour).

By noting that $2\omega_0 = 2\pi - \omega_0$, these can be solved yielding the result

$$X_n^{(0)} = Ae^{i\omega_0 n} + c.c., \qquad P_n^{(0)} = iAe^{i\omega_0 n} + c.c., \tag{6.183}$$

$$X_n^{(1)} = -\frac{\mathcal{S}\sqrt{3}}{3}|A|^2 + in\mathcal{G}(A, A^*)e^{i\omega_0 n} + c.c., \tag{6.184}$$

$$X_n^{(2)} = \mathcal{A} + in\mathcal{B} + \left(in\mathcal{C} + n^2\mathcal{D}\right)e^{i\omega_0 n} + c.c., \tag{6.185}$$

where

$$\mathcal{G}(A, A^*) = \delta_1 A + \frac{\mathcal{S}}{2}A^{*\mathbf{2}}, \tag{6.186}$$

$$\mathcal{A} = \frac{2\delta_1}{3}\mathcal{S}|A|^2, \qquad \mathcal{B} = \frac{\sqrt{3}\mathcal{S}}{3}(\mathcal{G}^* A - \mathcal{G}A^*), \tag{6.187}$$

$$\mathcal{D} = \frac{1}{2}(-\delta_1\mathcal{G} + \mathcal{S}\mathcal{G}^* A^*), \qquad \mathcal{C} = -\frac{\Sigma + \mathcal{D}}{\sqrt{3}}, \tag{6.188}$$

$$\Sigma = \left(\frac{\delta_1^2}{2} - \sqrt{3}\delta_2 \right) A + \frac{\delta_1 \mathcal{S}}{2} A^{*\mathbf{2}} + \mathcal{S}^2 |A|^2 A. \tag{6.189}$$

Proceeding as before, we define the renormalized amplitude

$$\widetilde{A}(n) = A + i\epsilon n \mathcal{G} + \epsilon^2 \left(in\mathcal{C} + n^2 \mathcal{D} \right). \tag{6.190}$$

Taking into account the expression

$$A = \widetilde{A} - i\epsilon n \mathcal{G} \left(\widetilde{A}, \widetilde{A}^* \right) + O(\epsilon^2) \tag{6.191}$$

relating the amplitude A to the renormalized one $\widetilde{A}(n)$, we finally obtain the renormalized resonant map

$$\widetilde{A}(n+1) = \widetilde{A}(n) + i\epsilon \widetilde{\mathcal{G}}(n) + \epsilon^2 \left[i\widetilde{\mathcal{C}}(n) + \widetilde{\mathcal{D}}(n) \right], \tag{6.192}$$

where

$$\widetilde{\Lambda}(n) = \Lambda \left[\widetilde{A}(n), \widetilde{A}^*(n) \right], \tag{6.193}$$

and Λ stands for $\mathcal{C}$, $\mathcal{D}$ or $\mathcal{G}$.

It is worthwhile to mention that the resonant shift in the closed orbit gets renormalized automatically, once the renormalization transformation $A \to \widetilde{A}(n)$ has been performed. The result is

$$X_{co}(n) = \frac{\epsilon \mathcal{S}}{3} \left[-\sqrt{3} + 2\epsilon\delta_1 + O(\epsilon^2) \right] \left| \widetilde{A}(n) \right|^2. \tag{6.194}$$

Note that the closed orbit can be corrected up to third order (in the sextupole strength $\mathcal{S}$) by choosing the first order resonance detuning δ_1 to be $\delta_1 = \sqrt{3}/2$. In terms of betatron tune this implies

$$\Delta\nu = \frac{\sqrt{3}}{4\pi}. \tag{6.195}$$

Since the naive RG map (6.192) does not preserve the symplectic structure of the original *Henon* map, an important step at this point consists in building a symplectic map in appropriate variables equivalent to (6.192). Unfortunately, the regularization procedure described in the previous paragraph cannot be applied to the map (6.192). The reason is that in the resonant case $\left| \widetilde{A}(n) \right|$ is no longer an integral of motion. An alternate way to bypass this difficulty is to represent $\widetilde{A}(n)$ as

$$\widetilde{A}(n) = \sqrt{\mathcal{J}_n} e^{i\varphi_n}, \tag{6.196}$$

and try to find an (implicit) map in terms of the new variables $(\varphi_n, \mathcal{J}_n)$ of the form

$$\varphi_{n+1} = \varphi_n + g(\mathcal{J}_{n+1}, \varphi_n), \qquad \mathcal{J}_{n+1} = \mathcal{J}_n + f(\mathcal{J}_{n+1}, \varphi_n), \quad (6.197)$$

Expanding the unknown functions f and g in a perturbation series

$$f = \sum_{k=1}^{\infty} \epsilon^k f_k, \qquad g = \sum_{k=1}^{\infty} \epsilon^k g_k,$$

and substituting the expression (6.196) into equation (6.192), we can determine f and g up to second order. We obtain

$$f_1(\varphi, \mathcal{J}) = \mathcal{S}\mathcal{J}^{3/2} \sin 3\varphi, \qquad g_1(\varphi, \mathcal{J}) = \delta_1 + \frac{\mathcal{S}}{2}\sqrt{\mathcal{J}} \cos 3\varphi, \quad (6.198)$$

$$f_2(\varphi, \mathcal{J}) = 3\delta_1 \mathcal{S}\mathcal{J}^{3/2} \cos 3\varphi - \frac{\delta_1 \mathcal{S}\sqrt{3}}{2}\mathcal{J}^{3/2} \sin 3\varphi + \frac{3\mathcal{S}^2}{4}\mathcal{J}^2 \cos 6\varphi, \quad (6.199)$$

$$g_2(\varphi, \mathcal{J}) = \delta_2 - \frac{5\mathcal{S}^2\sqrt{3}}{12}\mathcal{J} - \frac{3\delta_1 \mathcal{S}}{4}\sqrt{\mathcal{J}} \sin 3\varphi$$

$$- \frac{\delta_1 \mathcal{S}\sqrt{3}}{4}\sqrt{\mathcal{J}} \cos 3\varphi - \frac{\mathcal{S}^2}{4}\mathcal{J} \sin 6\varphi, \quad (6.200)$$

The map (6.197) is symplectic provided the condition

$$\frac{\partial g}{\partial \varphi_n} + \frac{\partial f}{\partial \mathcal{J}_{n+1}} = 0 \quad (6.201)$$

holds. To show that, we evaluate the determinant of its Jacobian

$$\det\left(\widehat{\mathcal{J}}_{\mathcal{M}}\right) = \det\begin{pmatrix} 1 + g_{\varphi_n} + g_{\mathcal{J}_{n+1}} \dfrac{\partial \mathcal{J}_{n+1}}{\partial \varphi_n} & g_{\mathcal{J}_{n+1}} \dfrac{\partial \mathcal{J}_{n+1}}{\partial \mathcal{J}_n} \\ f_{\varphi_n} + f_{\mathcal{J}_{n+1}} \dfrac{\partial \mathcal{J}_{n+1}}{\partial \varphi_n} & 1 + f_{\mathcal{J}_{n+1}} \dfrac{\partial \mathcal{J}_{n+1}}{\partial \varphi_n} \end{pmatrix},$$

where subscripts denote differentiation with respect to the variables indicated. Taking into account

$$\frac{\partial \mathcal{J}_{n+1}}{\partial \varphi_n} = f_{\varphi_n} + f_{\mathcal{J}_{n+1}} \frac{\partial \mathcal{J}_{n+1}}{\partial \varphi_n} \quad \Longrightarrow \quad \frac{\partial \mathcal{J}_{n+1}}{\partial \varphi_n} = \frac{f_{\varphi_n}}{1 - f_{\mathcal{J}_{n+1}}},$$

$$\frac{\partial \mathcal{J}_{n+1}}{\partial \mathcal{J}_n} = 1 + f_{\mathcal{J}_{n+1}} \frac{\partial \mathcal{J}_{n+1}}{\partial \mathcal{J}_n} \quad \Longrightarrow \quad \frac{\partial \mathcal{J}_{n+1}}{\partial \mathcal{J}_n} = \frac{1}{1 - f_{\mathcal{J}_{n+1}}},$$

we obtain

$$\det\left(\widehat{\mathcal{J}}_{\mathcal{M}}\right) = \frac{1 + g_{\varphi_n}}{1 - f_{\mathcal{J}_{n+1}}}.$$

The requirement $\det\left(\widehat{\mathcal{J}}_{\mathcal{M}}\right) = 1$ leads to the condition (6.201). It is straightforward to check that f and g as given by equations (6.198) - (6.200) satisfy equation (6.201).

The representation (6.196) of the renormalized amplitude $\widetilde{A}(n)$ together with (6.197) can be used as an alternate way to obtain the exponential form (6.157) of the RG map (6.156). The expansions

$$f = \sum_{k=1}^{\infty} \epsilon^{2k} f_{2k}, \qquad g = \sum_{k=1}^{\infty} \epsilon^{2k} g_{2k},$$

when substituted into equation (6.156), and simple and straightforward algebra being performed lead to the result

$$f_2(\varphi, \mathcal{J}) \equiv 0, \qquad f_4(\varphi, \mathcal{J}) \equiv 0, \tag{6.202}$$

$$g_2(\varphi, \mathcal{J}) = \mathcal{F}\mathcal{J}, \qquad g_4(\varphi, \mathcal{J}) = \mathcal{D}\mathcal{J}^2. \tag{6.203}$$

Thus, we obtain the symplectic implicit map

$$\varphi_{n+1} = \varphi_n + \epsilon^2 \mathcal{F}\mathcal{J}_{n+1} + \epsilon^4 \mathcal{D}\mathcal{J}_{n+1}^2, \qquad \mathcal{J}_{n+1} = \mathcal{J}_n, \tag{6.204}$$

which is the RG map (6.157) written for the real part and the argument of the amplitude $\widetilde{A}(n)$.

6.6 Renormalization Group Reduction of a Generic Transfer Map

Let us consider the one-turn transfer map in the most general form [Tzenov (2002b)]

$$X_{n+1} = X_n \cos\omega + [P_n - V'(X_n)] \sin\omega, \tag{6.205}$$

$$P_{n+1} = -X_n \sin\omega + [P_n - V'(X_n)] \cos\omega. \tag{6.206}$$

Eliminating the canonical momentum variable P_n in the same manner as was done in paragraph 3, we obtain again a second-order difference equation for X alone

$$\widehat{\mathcal{L}}X_n = X_{n+1} - 2X_n \cos\omega + X_{n-1} = -\epsilon V'(X_n) \sin\omega, \tag{6.207}$$

where ϵ is again a small parameter.

The naive perturbation expansion (6.126) when substituted into equation (6.207), yields the perturbation equations order by order

$$\widehat{\mathcal{L}}X_n^{(0)} = 0, \tag{6.208}$$

$$\widehat{\mathcal{L}}X_n^{(1)} = -V'\left(X_n^{(0)}\right) \sin\omega, \tag{6.209}$$

$$\widehat{\mathcal{L}}X_n^{(2)} = -X_n^{(1)} V''\left(X_n^{(0)}\right) \sin\omega, \tag{6.210}$$

$$\widehat{\mathcal{L}}X_n^{(3)} = -\left[\frac{X_n^{(1)\mathbf{2}}}{2} V'''\left(X_n^{(0)}\right) + X_n^{(2)} V''\left(X_n^{(0)}\right)\right] \sin\omega. \tag{6.211}$$

Solving equation (6.208) for the zeroth-order contribution, we obtain the lowest-order perturbation solution (6.132), where A is again a complex integration constant.

Let us assume that the potential $V(X)$ is an even function of the horizontal displacement X. In the two previous paragraphs it was shown on the example of the *Henon* map that odd multipoles give rise to a shift in the closed orbit. Their contribution in the general case of an arbitrary potential $V(X)$ can be incorporated in a straightforward manner in the calculations presented below. It is easy to check that the *Fourier* image $V(\lambda)$ of the potential V, defined as

$$V(X) = \frac{1}{2\pi} \int\limits_{-\infty}^{\infty} \mathrm{d}\lambda V(\lambda) e^{i\lambda X}, \qquad V(\lambda) = \int\limits_{-\infty}^{\infty} \mathrm{d}X V(X) e^{-i\lambda X} \tag{6.212}$$

retains the symmetry property of $V(X)$, that is

$$V(-\lambda) = V(\lambda).$$

Using the expansion [Abramowitz (1984)], [Gradshteyn (2000)]

$$e^{iz\cos\varphi} = \sum_{m=-\infty}^{\infty} i^m \mathcal{J}_m(z) e^{im\varphi}, \tag{6.213}$$

where $\mathcal{J}_m(z)$ is the *Bessel* function of the first kind of order m, and the explicit form of the zero-order solution (6.132), we find

$$V'\left(X_n^{(0)}\right) = \sum_{M=1}^{\infty} \mathcal{C}_M A^{2M-1} e^{(2M-1)i\omega n} + c.c.. \qquad (6.214)$$

Here the coefficients $\mathcal{C}_M$ are functions of the amplitude $|A|$ and are given by the expression

$$\mathcal{C}_M(|A|) = \frac{1}{\pi} \frac{(-1)^M}{|A|^{2M-1}} \int_0^{\infty} d\lambda \lambda V(\lambda) \mathcal{J}_{2M-1}(2\lambda|A|). \qquad (6.215)$$

Similarly, for the second derivative $V''\left(X_n^{(0)}\right)$ entering the second-order perturbation equation (6.210), we have

$$V''\left(X_n^{(0)}\right) = \mathcal{D}_0 + \sum_{M=1}^{\infty} \mathcal{D}_M A^{2M} e^{2iM\omega n} + c.c., \qquad (6.216)$$

where

$$\mathcal{D}_M(|A|) = \frac{1}{\pi} \frac{(-1)^{M+1}}{|A|^{2M}} \int_0^{\infty} d\lambda \lambda^2 V(\lambda) \mathcal{J}_{2M}(2\lambda|A|). \qquad (6.217)$$

From the recursion property of the *Bessel* functions [Abramowitz (1984)], [Gradshteyn (2000)]

$$\mathcal{J}_{\nu-1}(z) + \mathcal{J}_{\nu+1}(z) = \frac{2\nu}{z} \mathcal{J}_\nu(z),$$

we deduce an important relation to be used later

$$\mathcal{D}_N - \mathcal{D}_{N+1}|A|^2 = (2N+1)\mathcal{C}_{N+1}. \qquad (6.218)$$

Taking into account (6.218), the solution of the perturbation equations (6.209) and (6.210) can be written as

$$X_n^{(1)} = \frac{in\mathcal{C}_1}{2} A e^{i\omega n} + \frac{\sin\omega}{2} \sum_{M=1}^{\infty} \widetilde{\mathcal{C}}_{M+1} A^{2M+1} e^{(2M+1)i\omega n} + c.c., \qquad (6.219)$$

$$X_n^{(2)} = -\frac{\mathcal{C}_1^2}{8}\left(n^2 + in\cot\omega\right) A e^{i\omega n} + \frac{in\sin\omega}{4} \sum_{N=1}^{\infty} \widetilde{\mathcal{C}}_{N+1} \mathcal{D}_N |A|^{4N} A e^{i\omega n}$$

$$+\frac{i\mathcal{C}_1 \sin\omega}{4}\sum_{N=1}^{\infty}(2N+1)\widetilde{\mathcal{C}}_{N+1}\left[n+\frac{i\sin(2N+1)\omega}{\cos\omega-\cos(2N+1)\omega}\right]A^{2N+1}e^{(2N+1)i\omega n}$$

$$+\frac{\mathcal{D}_0 \sin^2\omega}{4}\sum_{N=1}^{\infty}\frac{\widetilde{\mathcal{C}}_{N+1}}{\cos\omega-\cos(2N+1)\omega}A^{2N+1}e^{(2N+1)i\omega n}$$

$$+\frac{\sin^2\omega}{4}\sum_{M=1}^{\infty}\sum_{N=1}^{\infty}\frac{\widetilde{\mathcal{C}}_{M+1}\mathcal{D}_N}{\cos\omega-\cos[2(M+N)+1]\omega}A^{2(M+N)+1}e^{[2(M+N)+1]i\omega n}$$

$$+\frac{\sin^2\omega}{4}\sum_{M=1}^{\infty}{}'\sum_{N=1}^{\infty}{}'\frac{\widetilde{\mathcal{C}}_{M+1}\mathcal{D}_N|A|^{4N}}{\cos\omega-\cos[2(M-N)+1]\omega}$$

$$\times A^{2(M-N)+1}e^{[2(M-N)+1]i\omega n}+c.c., \tag{6.220}$$

where the summation in the last term of equation (6.220) is performed for $M \neq N$, and

$$\widetilde{\mathcal{C}}_{N+1}=\frac{\mathcal{C}_{N+1}}{\cos\omega-\cos(2N+1)\omega}. \tag{6.221}$$

To remove secular terms proportional to n and n^2, we act as before and define the renormalization transformation $A \to \widetilde{A}(n)$ by collecting all terms proportional to the fundamental harmonic $e^{i\omega n}$

$$\widetilde{A}(n)=A+\epsilon\frac{in\mathcal{C}_1}{2}A+\epsilon^2\left[-\frac{\mathcal{C}_1^2}{8}\left(n^2+in\cot\omega\right)\right.$$

$$\left.+\frac{in\sin\omega}{4}\sum_{N=1}^{\infty}\widetilde{\mathcal{C}}_{N+1}\mathcal{D}_N|A|^{4N}\right]A. \tag{6.222}$$

Solving perturbatively equation (6.222) for A in terms of $\widetilde{A}(n)$, we find

$$A=\left[1-\epsilon\frac{in\mathcal{C}_1}{2}+O(\epsilon^2)\right]\widetilde{A}(n). \tag{6.223}$$

Substituting this in the discrete version of the RG equation [recall equation (6.155)], we obtain the RG map

$$\widetilde{A}(n+1)=\left[1+\epsilon\frac{i\mathcal{C}_1}{2}-\epsilon^2\frac{\mathcal{C}_1^2}{8}(1+i\cot\omega)\right.$$

$$+i\epsilon^2 \frac{\sin \omega}{4} \sum_{N=1}^{\infty} \widetilde{\mathcal{C}}_{N+1} \mathcal{D}_N \left| \widetilde{A}(n) \right|^{4N} \Bigg] \widetilde{A}(n). \tag{6.224}$$

To recover the symplectic symmetry of the naive RG map (6.224), we act as before and cast it in an exponentiated form

$$\widetilde{A}(n+1) = \widetilde{A}(n) \exp \left[i\widetilde{\omega}\left(\left| \widetilde{A}(n) \right| \right) \right], \tag{6.225}$$

where

$$\widetilde{\omega}\left(\left| \widetilde{A}(n) \right| \right) = \frac{\epsilon \mathcal{C}_1}{2} + \frac{\epsilon^2}{8} \left(-\mathcal{C}_1^2 \cot \omega + 2 \sin \omega \sum_{N=1}^{\infty} \widetilde{\mathcal{C}}_{N+1} \mathcal{D}_N \left| \widetilde{A}(n) \right|^{4N} \right). \tag{6.226}$$

The integral of motion of the regularized RG map (6.225) is given by expression (6.159).

It is worthwhile to note that the secular coefficients of the $(2N+1)$-st harmonic $e^{(2N+1)i\omega n}$ can be summed up to give a renormalized coefficient expressed in terms of the renormalized amplitude $\widetilde{A}(n)$.

6.7 The Standard Chirikov-Taylor Map

As we mentioned previously, conversion between Hamiltonian description and maps has wide application to the analysis of betatron motion in cyclic accelerators. Numerical calculations over large number of turns of the non-linear betatron oscillations can easily be performed from a transfer map. It is important to answer the question of how the description of the longitudinal motion can be reduced to a map.

We start with the Hamiltonian (1.76) governing the longitudinal motion

$$\widehat{H}_0 = -\frac{R\mathcal{K}(\theta)}{2} \widehat{\eta}^2 + \frac{e\widehat{\mathcal{E}}_0(s)}{\beta_s E_s} \frac{Rc}{\omega} \cos\left(\frac{\omega \sigma}{c\beta_s} - k\theta + \phi_0 \right), \tag{6.227}$$

where

$$\mathcal{K}(\theta) = K(\theta) D(\theta) - \frac{1}{\gamma_s^2}. \tag{6.228}$$

Applying a sequence of transformations comprised of the non canonical transformation

$$\widehat{\eta} \to -\widehat{\eta}, \qquad \widehat{H}_0 \to -\widehat{H}_0, \tag{6.229}$$

and the canonical transformation

$$\widehat{\eta} = \frac{k}{R} J, \qquad \alpha = \frac{\omega \sigma}{c \beta_s} - k\theta + \phi_0 \pm \pi, \tag{6.230}$$

defined by the generating function

$$\mathcal{S}(\sigma, J; \theta) = J\left(\frac{\omega \sigma}{c \beta_s} - k\theta + \phi_0 \pm \pi\right),$$

we cast the Hamiltonian (6.227) in the standard form

$$\widehat{H}_0 = -kJ + \frac{1}{2} \frac{k^2 \mathcal{K}(\theta)}{R} J^2 + \frac{e\widehat{\mathcal{E}}_0(\theta)}{\beta_s E_s} \frac{Rc}{\omega} \cos \alpha. \tag{6.231}$$

Similar to what was done in paragraph 6.2, we assume that the electric field in the cavity is concentrated in a single point θ_1. Thus, the *Hamilton*'s equations of motion can be written as

$$\dot{\alpha} = -k + \frac{k^2 \mathcal{K}(\theta)}{R} J, \qquad \dot{J} = \frac{c \Delta E_0}{\omega \beta_s E_s} \delta(\theta - \theta_1) \sin \alpha. \tag{6.232}$$

Let us define the one-turn map $\widehat{\mathcal{M}}$ governing the longitudinal dynamics according to the expression

$$(J', \alpha') = \widehat{\mathcal{M}}(J, \alpha), \tag{6.233}$$

where

$$J' = J(\theta_1 + 2\pi - 0), \qquad \alpha' = \alpha(\theta_1 + 2\pi - 0),$$

$$J = J(\theta_1 - 0), \qquad \alpha = \alpha(\theta_1 - 0),$$

The $\widehat{\mathcal{M}}$-map is composed of the successive action of the kick map $\widehat{\mathcal{M}}_K$ and the map $\widehat{\mathcal{M}}_R$ describing the free motion between successive kicks

$$\widehat{\mathcal{M}} = \widehat{\mathcal{M}}_R \circ \widehat{\mathcal{M}}_K. \tag{6.234}$$

For $\widehat{\mathcal{M}}_R$, we have

$$\widehat{\mathcal{M}}_R(J, \alpha) = \left[J, \alpha + 2\pi\left(-k + \frac{k^2 \mathcal{K}}{R} J\right)\right], \tag{6.235}$$

where $\mathcal{K}$ is given by expression (1.80). In order to obtain the map $\widehat{\mathcal{M}}_K$, we integrate the *Hamilton*'s equations (6.232) in a small region surrounding the kick location $(\theta_1 - 0, \theta_1 + 0)$. Taking into account the continuity at θ_1 with respect to the variables J and α, we find

$$J(\theta_1 + 0) - J(\theta_1 - 0) = \frac{c\Delta E_0}{\omega\beta_s E_s} \sin\alpha, \qquad \alpha(\theta_1 + 0) - \alpha(\theta_1 - 0) = 0.$$

Therefore, the kick map $\widehat{\mathcal{M}}_K$ can be written as

$$\widehat{\mathcal{M}}_K(J, \alpha) = \left(J + \frac{c\Delta E_0}{\omega\beta_s E_s} \sin\alpha, \alpha\right), \tag{6.236}$$

Combining equations (6.235) and (6.236), we obtain the explicit form of the map (6.234), namely

$$J' = J + \frac{c\Delta E_0}{\omega\beta_s E_s} \sin\alpha, \qquad \alpha' = \alpha + 2\pi\left(-k + \frac{k^2\mathcal{K}}{R} J'\right). \tag{6.237}$$

This is the so-called standard (*Chirikov-Taylor*) map. The simplest analysis of the *Chirikov-Taylor* map consists in determining the eigenvalues of its Jacobian

$$\widehat{\mathcal{J}}_{\mathcal{M}} = \begin{pmatrix} 1 & \dfrac{c\Delta E_0}{\omega\beta_s E_s} \cos\alpha \\ \dfrac{2\pi k^2\mathcal{K}}{R} & 1 + \dfrac{2\pi k^2\mathcal{K}}{R} \dfrac{c\Delta E_0}{\omega\beta_s E_s} \cos\alpha \end{pmatrix}. \tag{6.238}$$

The equation they satisfy reads as follows

$$\lambda^2 - 2(1 + K_s)\lambda + 1 = 0, \tag{6.239}$$

where

$$K_s = 2\pi^2 K_0^2 \cos\alpha, \qquad K_0^2 = \frac{\Delta E_0}{E_s} \frac{k\mathcal{K}}{2\pi\beta_s^2}. \tag{6.240}$$

The solution to the secular equation (6.239) is easy to find, that is

$$\lambda_{1,2} = 1 + K_s \pm \sqrt{(1 + K_s)^2 - 1}. \tag{6.241}$$

From expression (6.241) it follows that the motion is unstable if

$$K_s > 0 \qquad (\lambda_1 > 1), \tag{6.242}$$

$$K_s < -2 \qquad (\lambda_2 < -1). \tag{6.243}$$

Numerical simulations [Chirikov (1979)], [Lichtenberg (1982)] indicate that a large region where the motion is stochastic arises for

$$K_0 \sim \frac{1}{2\pi}. \tag{6.244}$$

Chapter 7

Statistical Description of Charged Particle Beams

7.1 Introduction

A statistical theory is always associated with a particular dynamical model of some kind. In what follows, we will be interested in a model system comprised of a large number of identical structureless microscopic components (the individual particles in a beam). For such a system we may define a Hamiltonian and formally write the appropriate *Hamilton*'s equations of motion. The solution of the *Hamilton*'s equations with specified initial conditions gives us the complete information about the beam.

This task cannot be accomplished even with the aid of the most powerful computers, and what is more, there is no need in doing so because the result will be unreliable one way or another. That is why, we inevitably have to switch to the statistical description of charged particle beams. There are other factors such as dynamic instability of motion of individual particles, or existence of external uncontrollable perturbations and noise, which in addition call for the transition to irreversible equations of statistical mechanics. The transition to irreversible description is based on the concept of the ensemble of identical macroscopic systems (*Gibbs ensemble*), introduced by *Gibbs* to study equilibrium states. In paragraph 7.3 a proper generalization of this concept for non-equilibrium states will be given and used later to derive the corresponding kinetic equations for charged particle beams.

7.2 The Liouville Theorem and the Liouville Equation

Let us consider the beam propagating in an accelerator or a storage ring as consisting of N identical, indistinguishable particles. For brevity, we adopt in what follows the following shorthand notation

$$\mathbf{X} = (\mathbf{x}_1, \mathbf{x}_2, \ldots, \mathbf{x}_N), \qquad \mathbf{P} = (\mathbf{p}_1, \mathbf{p}_2, \ldots, \mathbf{p}_N). \tag{7.1}$$

In order to describe the time evolution of the system of beam particles, we must solve *Hamilton*'s equations of motion, and express $\mathbf{X}(\theta)$ and $\mathbf{P}(\theta)$ in terms of the initial values $\mathbf{X}_0$ and $\mathbf{P}_0$, namely

$$\mathbf{X}(\theta) = \mathbf{X}(\mathbf{X}_0, \mathbf{P}_0, \theta_0; \theta), \qquad \mathbf{P}(\theta) = \mathbf{P}(\mathbf{X}_0, \mathbf{P}_0, \theta_0; \theta). \tag{7.2}$$

In paragraph 2.4 we defined a function of dynamic variables $I(\mathbf{x}, \mathbf{p}; \theta)$ and saw that it is invariant if its total derivative with respect to θ is zero [see equation (2.62)]. Of special importance in statistical physics are those functions of dynamic variables which satisfy the conditions

$$f_N(\mathbf{X}, \mathbf{P}; \theta) \geq 0, \qquad \int \mathrm{d}^{3N}\mathbf{X}\mathrm{d}^{3N}\mathbf{P} f_N(\mathbf{X}, \mathbf{P}; \theta) = 1. \tag{7.3}$$

They are called *distribution functions.* Let us write the obvious condition

$$f_N(\mathbf{X}, \mathbf{P}; \theta)\mathrm{d}\mathbf{X}\mathrm{d}\mathbf{P} = f_N(\mathbf{X}_0, \mathbf{P}_0; \theta_0)\mathrm{d}\mathbf{X}_0\mathrm{d}\mathbf{P}_0, \tag{7.4}$$

stating the fact that the system will occupy a phase space volume $\mathrm{d}\mathbf{X}\mathrm{d}\mathbf{P}$ in the neighborhood of a point $(\mathbf{X}, \mathbf{P})$ at time θ, provided that it occupied a phase space volume $\mathrm{d}\mathbf{X}_0\mathrm{d}\mathbf{P}_0$ in the vicinity of the point $(\mathbf{X}_0, \mathbf{P}_0)$ at some initial time θ_0. The two volumes are linked via the equation

$$\mathrm{d}\mathbf{X}\mathrm{d}\mathbf{P} = \det\left(\widehat{\mathcal{J}}_{\mathcal{M}}\right)\mathrm{d}\mathbf{X}_0\mathrm{d}\mathbf{P}_0, \tag{7.5}$$

where $\widehat{\mathcal{J}}_{\mathcal{M}}$ is the Jacobian of the transformation formally defined by equation (7.2). Since the Jacobian is a symplectic matrix and its determinant is equal to unity, we must have

$$\mathrm{d}\mathbf{X}\mathrm{d}\mathbf{P} = \mathrm{d}\mathbf{X}_0\mathrm{d}\mathbf{P}_0. \tag{7.6}$$

This result is a statement of the *Liouville theorem* in explicit form. On the basis of the *Liouville* theorem, we can rewrite equation (7.4) as

$$f_N(\mathbf{X}, \mathbf{P}; \theta) = f_N(\mathbf{X}_0, \mathbf{P}_0; \theta_0), \tag{7.7}$$

which implies that the distribution function does not change along the path (Hamiltonian flow) defined by equation (7.2). In other words, the distribution function $f_N(\mathbf{X}, \mathbf{P}; \theta)$ is invariant, and its total derivative with respect to θ must be zero. Therefore, we come to the following equation for the distribution function

$$\frac{\partial f_N}{\partial \theta} + [f_N, \mathcal{H}] = 0, \tag{7.8}$$

where $\mathcal{H}$ is the appropriate Hamiltonian for the beam comprised of N particles. This is the well-known *Liouville equation.*

The *Liouville* equation is a first order partial differential equation. The solution of the *Cauchy* problem for first order partial differential equations reduces to the solution of the equations for the characteristics, which in the case of *Liouville* equation coincide with the *Hamilton*'s equations of motion. Therefore, using equation (7.2), we need to express $\mathbf{X}_0$ and $\mathbf{P}_0$ in terms of $\mathbf{X}$ and $\mathbf{P}$

$$\mathbf{X}_0 = \mathbf{X}_0(\mathbf{X}, \mathbf{P}; \theta_0 - \theta), \qquad \mathbf{P}_0 = \mathbf{P}_0(\mathbf{X}, \mathbf{P}, \theta_0 - \theta). \tag{7.9}$$

and substitute the result into $f_N(\mathbf{X}_0, \mathbf{P}_0; \theta_0)$. Thus, we obtain the solution of the inverse problem for the *Liouville* equation

$$f_N(\mathbf{X}, \mathbf{P}; \theta) = f_N(\mathbf{X}_0(\mathbf{X}, \mathbf{P}; \theta_0 - \theta), \mathbf{P}_0(\mathbf{X}, \mathbf{P}; \theta_0 - \theta); \theta_0). \tag{7.10}$$

The above manipulations imply that the solution of the *Liouville* equation is equivalent to the impossible task of solving the *Hamilton*'s equations of motion. So, to this end, we have not gained anything by passing over (formally) to the statistical description of particle beam propagation.

An important feature of *Liouville*'s equation that is transferred from the Hamiltonian dynamics is its *reversibility.* The invariance with respect to the transformation

$$\theta \Rightarrow -\theta, \qquad \mathbf{x}_k \Rightarrow \mathbf{x}_k, \qquad \mathbf{p}_k \Rightarrow -\mathbf{p}_k, \qquad (k = 1, 2, \dots, N) \tag{7.11}$$

is usually regarded as a sign of reversibility of the equations in question. Directionality of time (or the azimuthal angle θ in our case) is a concept more elemental than reversibility. The replacement $\theta \Rightarrow -\theta$ (once the guiding and accelerating fields have been fixed) is impossible in reality for the particular accelerating device considered as a unalterable hardware. For this reason the meaning of reversibility should be perceived only in the sense of "coming back home". To put it another way, suppose that after a "time" lapse $\theta - \theta_0$ we reverse the signs of the momenta of all particles in the beam, then at $2(\theta - \theta_0)$ the particles will restore their initial positions however, with opposite momenta $-\mathbf{P}_0$. The reversibility of motion means that, in principle, such a process of "returning home" is possible. This is a general property of *Hamilton*'s equations of motion and this is the sense reversibility of the *Liouville* equation should be understood.

7.3 Ensemble of Identical Macroscopic Systems

As already indicated in the introduction, the complete description of beam propagation in an accelerator or a storage ring can be achieved by solving the equations of motion for each particle, provided the initial values of the positions and momenta are known. As stated before, solving the equations of motion for all particles in the beam is not only hopeless, but what is more, specification of the initial conditions for every single particle within the beam is impossible. This inevitably calls for a statistical description.

Consider a charged particle beam propagating in an accelerator. What we know about the beam is the total number of particles contained, their mass and electric charge. The electromagnetic forces with which the external confining and accelerating fields act on every single particle, as well as the type of interaction between individual particles comprising the beam are also known. We may even specify the scattering law of beam particles on the residual gas in the vacuum chamber. These factors however, do not uniquely determine the initial values of the positions and momenta of all particles at a particular location θ_0 adopted as initial. Therefore, they do not uniquely determine the evolution of the beam expressed in terms of the functions $\mathbf{X}(\theta)$ and $\mathbf{P}(\theta)$. Under these circumstances the positions $\mathbf{X}(\theta)$ and momenta $\mathbf{P}(\theta)$ of the beam particles are *random variables.*

In order to properly determine the concept of distribution function describing the state of the beam, we consider an ensemble of identical macroscopic beam systems – *Gibbs* ensemble. These systems are macroscopically identical in the sense that they consist of the same sort and number of particles, and all conditions and factors listed above hold the same for all beam systems under consideration. Since these conditions do not determine uniquely the state of the beam, different values of $\mathbf{X}(\theta)$ and $\mathbf{P}(\theta)$ will correspond to different states of the ensemble.

Let us select a volume $\mathrm{d}\mathbf{X}\mathrm{d}\mathbf{P}$ in the phase space surrounding the point $(\mathbf{X}, \mathbf{P})$. Let at certain location θ this volume contains points characterizing the state of $\mathrm{d}\mathcal{N}$ systems of the ensemble out of the total number of systems $\mathcal{N}$. Then the limit

$$\lim_{\mathcal{N}\to\infty} \frac{\mathrm{d}\mathcal{N}}{\mathcal{N}} = f_N(\mathbf{X}, \mathbf{P}; \theta)\mathrm{d}\mathbf{X}\mathrm{d}\mathbf{P} \tag{7.12}$$

determines the distribution function at "time" θ, characterizing the microscopic states of the systems belonging to the *Gibbs* ensemble. From this definition it follows that the distribution function is normalized according to equation (7.3).

7.4 The Method of Microscopic Phase Space Density

Following [Klimontovich (1986)] and [Tzenov (1997b)], we define the microscopic phase space density in the six-dimensional space of coordinates and momenta according to the expression

$$N_M(\mathbf{x}, \mathbf{p}; t) = \sum_{n=1}^{N} \delta(\mathbf{x} - \mathbf{x}_n(t))\delta(\mathbf{p} - \mathbf{p}_n(t)). \tag{7.13}$$

Because the total derivative of the quantity $N_M(\mathbf{x}, \mathbf{p}; t)$ [as defined by equation (7.13)] with respect to the time t must be equal to its partial derivative

$$\frac{\mathrm{d}N_M}{\mathrm{d}t} = \frac{\partial N_M}{\partial t},$$

it is straightforward to verify that the microscopic phase space density $N_M(\mathbf{x}, \mathbf{p}; t)$ satisfies the equation

$$\frac{\partial N_M}{\partial t} + \nabla_{\mathbf{x}} \cdot (\mathbf{v} N_M) + \nabla_{\mathbf{p}} \cdot \left[\mathbf{F}^{(M)}(\mathbf{x}, \mathbf{v}; t) N_M\right] = 0. \tag{7.14}$$

Here $\mathbf{v}$ is the particle's velocity. In deriving equation (7.14) the following property

$$\frac{\mathrm{d}}{\mathrm{d}t}\delta(\mathbf{x} - \mathbf{y}(t)) = -\nabla_{\mathbf{x}} \cdot \left[\frac{\mathrm{d}\mathbf{y}}{\mathrm{d}t}\delta(\mathbf{x} - \mathbf{y}(t))\right] \tag{7.15}$$

of the delta-function has been used[1]. The microscopic force $\mathbf{F}^{(M)}(\mathbf{x}, \mathbf{v}; t)$ entering equation (7.14) is the *Lorentz* force acting on each individual par-

[1]To prove this property it suffices to multiply both sides of equation (7.15) by an arbitrary function $f(\mathbf{x})$ and integrate over $\mathbf{x}$. For the left-hand-side, we have

$$\frac{\mathrm{d}}{\mathrm{d}t}\int \mathrm{d}^3\mathbf{x} f(\mathbf{x})\delta(\mathbf{x} - \mathbf{y}(t)) = \frac{\mathrm{d}}{\mathrm{d}t} f(\mathbf{y}(t)) = \frac{\mathrm{d}\mathbf{y}}{\mathrm{d}t} \cdot \nabla_{\mathbf{y}} f(\mathbf{y}(t)).$$

For the right-hand-side, we obtain

$$-\int \mathrm{d}^3\mathbf{x} f(\mathbf{x})\nabla_{\mathbf{x}} \cdot \left[\frac{\mathrm{d}\mathbf{y}}{\mathrm{d}t}\delta(\mathbf{x} - \mathbf{y}(t))\right] = \int \mathrm{d}^3\mathbf{x}\delta(\mathbf{x} - \mathbf{y}(t))\frac{\mathrm{d}\mathbf{y}}{\mathrm{d}t} \cdot \nabla_{\mathbf{x}} f(\mathbf{x}) = \frac{\mathrm{d}\mathbf{y}}{\mathrm{d}t} \cdot \nabla_{\mathbf{y}} f(\mathbf{y}(t)).$$

Comparison of the above two results completes the proof of equation (7.15).

ticle, which is given by the expression

$$\mathbf{F}^{(M)}(\mathbf{x}, \mathbf{v}; t) = e\mathbf{E}^{(M)}(\mathbf{x}; t) + e\mathbf{v} \times \mathbf{B}^{(M)}(\mathbf{x}; t) + \mathbf{F}_0(\mathbf{x}, \mathbf{v}; t), \qquad (7.16)$$

where $\mathbf{F}_0(\mathbf{x}, \mathbf{v}; t)$ is the external force specified by the confining and accelerating electromagnetic fields. The microscopic electromagnetic fields $\mathbf{E}^{(M)}(\mathbf{x}; t)$ and $\mathbf{B}^{(M)}(\mathbf{x}; t)$ are determined as solution to the *Maxwell* equations

$$\nabla_{\mathbf{x}} \times \mathbf{B}^{(M)} = \frac{1}{c^2} \frac{\partial \mathbf{E}^{(M)}}{\partial t} + \mu_0 e \int \mathrm{d}^3\mathbf{p} \mathbf{v} N_M(\mathbf{x}, \mathbf{p}; t), \qquad (7.17)$$

$$\nabla_{\mathbf{x}} \times \mathbf{E}^{(M)} = -\frac{\partial \mathbf{B}^{(M)}}{\partial t}, \qquad \nabla_{\mathbf{x}} \cdot \mathbf{B}^{(M)} = 0, \qquad (7.18)$$

$$\nabla_{\mathbf{x}} \cdot \mathbf{E}^{(M)} = \frac{e}{\epsilon_0} \int \mathrm{d}^3\mathbf{p} N_M(\mathbf{x}, \mathbf{p}; t), \qquad (7.19)$$

where ϵ_0 and μ_0 are the electric and the magnetic susceptibility of vacuum, respectively. The electric field $\mathbf{E}^{(M)}(\mathbf{x}; t)$ and the magnetic field $\mathbf{B}^{(M)}(\mathbf{x}; t)$ can be defined in terms of the electromagnetic potentials $\varphi^{(M)}(\mathbf{x}; t)$ and $\mathbf{A}^{(M)}(\mathbf{x}; t)$ according to the relations

$$\mathbf{E}^{(M)} = -\nabla_{\mathbf{x}} \varphi^{(M)} - \frac{\partial \mathbf{A}^{(M)}}{\partial t}, \qquad \mathbf{B}^{(M)} = \nabla_{\mathbf{x}} \times \mathbf{A}^{(M)}. \qquad (7.20)$$

Using the *Lorentz* gauge

$$\nabla_{\mathbf{x}} \cdot \mathbf{A}^{(M)} + \frac{1}{c^2} \frac{\partial \varphi^{(M)}}{\partial t} = 0, \qquad (7.21)$$

we obtain the wave equations for the electromagnetic potentials

$$\nabla_{\mathbf{x}}^2 \mathbf{A}^{(M)} - \frac{1}{c^2} \frac{\partial^2 \mathbf{A}^{(M)}}{\partial t^2} = -\mu_0 e \int \mathrm{d}^3\mathbf{p} \mathbf{v} N_M(\mathbf{x}, \mathbf{p}; t), \qquad (7.22)$$

$$\nabla_{\mathbf{x}}^2 \varphi^{(M)} - \frac{1}{c^2} \frac{\partial^2 \varphi^{(M)}}{\partial t^2} = -\frac{e}{\epsilon_0} \int \mathrm{d}^3\mathbf{p} N_M(\mathbf{x}, \mathbf{p}; t). \qquad (7.23)$$

The microscopic phase space density $N_M(\mathbf{x}, \mathbf{p}; t)$ can be regarded as a dynamic variable that is an invariant

$$\frac{\mathrm{d}N_M}{\mathrm{d}t} = \frac{\partial N_M}{\partial t} + [N_M, \mathcal{H}] = 0, \qquad (7.24)$$

where the Hamiltonian $\mathcal{H}$ is specified by equation (1.8). Moreover, as mentioned in paragraph 1.5, the electromagnetic potentials consist of two parts

$$\mathbf{A} = \mathbf{A}^{(M)} + \mathbf{A}_0, \qquad \varphi = \varphi^{(M)} + \varphi_0. \tag{7.25}$$

The part $\mathbf{A}_0$ of the vector potential represents the external electromagnetic fields giving rise to the force $\mathbf{F}_0(\mathbf{x}, \mathbf{v}; t)$, responsible for the beam confinement and acceleration. Recall that we have taken an account of $\mathbf{A}_0$ in the treatment of paragraph 1.5, while the scalar potential φ_0 has been set to zero. In what follows some of the calculations of paragraph 1.5 will be repeated with the microscopic fields (or equivalently the microscopic electromagnetic potentials) taken into account.

Recall that the components of the kinetic momenta are related to the ones of the canonical momenta according to the expressions

$$p_{x,z}^{(k)} = p_{x,z} - eA_{x,z}, \qquad p_s^{(k)} = \frac{1}{1 + xK}\left(\frac{p_s}{1 + xK} - eA_s\right), \tag{7.26}$$

$$h^{(k)} = h - \frac{e\varphi}{\beta_s^2 E_s}. \tag{7.27}$$

Obviously, the square root in equation (1.54) can be written in the form

$$\sqrt{\left(\beta_s h - \frac{e\varphi}{\beta_s E_s}\right)^2 - \frac{1}{\beta_s^2\gamma_s^2} - \left(\widetilde{p}_x - \frac{eA_x}{p_{0s}}\right)^2 - \left(\widetilde{p}_z - \frac{eA_z}{p_{0s}}\right)^2}$$

$$= (1 + xK)\frac{p_s^{(k)}}{p_{0s}}. \tag{7.28}$$

With equations (7.26), (7.27) and (7.28) in hand, we can write the equations of motion following from the Hamiltonian (1.54). They are

$$\frac{\mathrm{d}\tau}{\mathrm{d}\theta} = -R\frac{\beta_s^2 h^{(k)}}{\widetilde{p}_s^{(k)}} \approx -R\beta_s(1 + xK)\frac{\widetilde{p}_s^{(k)}}{\widetilde{p}_s^{(k)}}, \tag{7.29}$$

$$\frac{\mathrm{d}h^{(k)}}{\mathrm{d}\theta} = \frac{eR\widehat{\mathcal{E}}_0(s)}{\beta_s^2 E_s}\sin\left(\frac{\omega\tau}{c\beta_s} + \phi_0\right) + \frac{eR}{\widetilde{p}_s^{(k)}\beta_s^2 E_s}\left(\widetilde{\mathbf{p}}^{(k)} \cdot \mathbf{E}^{(M)}\right), \tag{7.30}$$

$$\frac{\mathrm{d}x}{\mathrm{d}\theta} = R\frac{\widetilde{p}_x^{(k)}}{\widetilde{p}_s^{(k)}}, \qquad \frac{\mathrm{d}z}{\mathrm{d}\theta} = R\frac{\widetilde{p}_z^{(k)}}{\widetilde{p}_s^{(k)}}, \tag{7.31}$$

$$\frac{\mathrm{d}\widetilde{p}_x^{(k)}}{\mathrm{d}\theta} = -\frac{\partial \mathcal{V}}{\partial x} + RK(1+xK)\widetilde{p}_s^{(k)} + \frac{eR\beta_s h^{(k)}}{cp_s^{(k)}} E_x^{(M)} + \frac{eR}{p_s^{(k)}} \left(\widetilde{\mathbf{p}}^{(k)} \times \mathbf{B}^{(M)} \right)_x, \tag{7.32}$$

$$\frac{\mathrm{d}\widetilde{p}_z^{(k)}}{\mathrm{d}\theta} = -\frac{\partial \mathcal{V}}{\partial z} + \frac{eR\beta_s h^{(k)}}{cp_s^{(k)}} E_z^{(M)} + \frac{eR}{p_s^{(k)}} \left(\widetilde{\mathbf{p}}^{(k)} \times \mathbf{B}^{(M)} \right)_z, \tag{7.33}$$

where the potential function $\mathcal{V}(x,z;\theta)$ entering equations (7.32) and (7.33) is defined as

$$\mathcal{V}(x,z;\theta) = RKx + \frac{1}{2R}\left(G_x x^2 + G_z z^2\right) + \frac{\lambda_0}{6R^2}\left(x^3 - 3xz^2\right)$$
$$+ \frac{\mu_0}{24R^3}\left(x^4 - 6x^2z^2 + z^4\right) + \dots. \tag{7.34}$$

Similar to paragraph 1.5, we perform the sequence of two canonical transformations defined by the generating functions displayed in equations (1.65) and (1.72). We must find the equations of motion in terms of the new variables $\widehat{\sigma}$, $\widehat{x}$, $\widehat{z}$ and

$$\widehat{\eta}^{(k)} = h^{(k)} - \frac{1}{\beta_s^2}, \qquad \widehat{p}_x^{(k)} = \widetilde{p}_x^{(k)} - \frac{\widehat{\eta}}{R}\frac{\mathrm{d}D}{\mathrm{d}\theta}, \qquad \widehat{p}_z^{(k)} = \widetilde{p}_z^{(k)}, \tag{7.35}$$

starting from the equations we have just derived. Retaining only linear terms in the new hat-variables, linear terms in the microscopic fields $\mathbf{E}^{(M)}$ and $\mathbf{B}^{(M)}$, and assuming as usual that there is no coupling between the longitudinal and the transverse motion, we obtain

$$\frac{\mathrm{d}\widehat{\sigma}}{\mathrm{d}\theta} = -RK\widehat{\eta}^{(k)}, \tag{7.36}$$

$$\frac{\mathrm{d}\widehat{\eta}^{(k)}}{\mathrm{d}\theta} = \frac{\Delta E_0}{2\pi\beta_s^2 E_s} \sin\left(\frac{\omega\widehat{\sigma}}{c\beta_s} + \Phi_0\right)$$
$$+ \frac{eR}{\beta_s^2 E_s}\left[(1+xK)E_s^{(M)} + \left(\widehat{\mathbf{p}}^{(k)} \cdot \mathbf{E}^{(M)}\right)\right], \tag{7.37}$$

$$\frac{\mathrm{d}\widehat{x}}{\mathrm{d}\theta} = R\widehat{p}_x^{(k)}, \qquad \frac{\mathrm{d}\widehat{z}}{\mathrm{d}\theta} = R\widehat{p}_z^{(k)}, \tag{7.38}$$

$$\frac{\mathrm{d}\widehat{p}_x^{(k)}}{\mathrm{d}\theta} = -\frac{\partial \mathcal{U}}{\partial \widehat{x}} + \frac{eR}{\beta_s^2 E_s}(1+\widehat{x}K)\left(E_x^{(M)} - v_{0s}B_z^{(M)}\right) + \frac{eR}{p_{0s}}\left(\widehat{\mathbf{p}}^{(k)} \times \mathbf{B}^{(M)}\right)_x, \tag{7.39}$$

$$\frac{\mathrm{d}\widehat{p}_z^{(k)}}{\mathrm{d}\theta} = -\frac{\partial \mathcal{U}}{\partial \widehat{z}} + \frac{eR}{\beta_s^2 E_s}(1 + \widehat{x}K)\Big(E_z^{(M)} + v_{0s} B_x^{(M)}\Big) + \frac{eR}{p_{0s}}\Big(\widehat{\mathbf{p}}^{(k)} \times \mathbf{B}^{(M)}\Big)_z, \tag{7.40}$$

where

$$\widehat{\mathbf{p}}^{(k)} = \Big(\widehat{p}_x^{(k)}, \widehat{p}_z^{(k)}, \alpha_M \widehat{\eta}^{(k)}\Big), \qquad \mathcal{U}(\widehat{x}, \widehat{z}; \theta) = \mathcal{V}(\widehat{x}, \widehat{z}; \theta) - \widehat{x}RK. \tag{7.41}$$

The equation for the microscopic phase space density (7.14) can be written in terms of the new variables (7.35) according to

$$\frac{\partial N_M}{\partial \theta} + \Big(\widehat{\mathbf{v}} \cdot \widehat{\nabla}_{\mathbf{x}}\Big) N_M + \widehat{\nabla}_{\mathbf{p}} \cdot \Big[\Big(\widehat{\mathbf{F}}_0 + \widehat{\mathbf{F}}^{(M)}\Big) N_M\Big] = 0, \tag{7.42}$$

where

$$\widehat{\mathbf{v}} = \Big(R\widehat{p}_x^{(k)}, R\widehat{p}_z^{(k)}, -R\mathcal{K}\widehat{\eta}^{(k)}\Big), \qquad \widehat{\nabla}_{\mathbf{x}} = \left(\frac{\partial}{\partial \widehat{x}}, \frac{\partial}{\partial \widehat{z}}, \frac{\partial}{\partial \widehat{\sigma}}\right), \tag{7.43}$$

$$\widehat{\nabla}_{\mathbf{p}} = \left(\frac{\partial}{\partial \widehat{p}_x^{(k)}}, \frac{\partial}{\partial \widehat{p}_z^{(k)}}, \frac{\partial}{\partial \widehat{\eta}^{(k)}}\right), \tag{7.44}$$

$$\widehat{\mathbf{F}}_0 = \left(-\frac{\partial \mathcal{U}}{\partial \widehat{x}}, -\frac{\partial \mathcal{U}}{\partial \widehat{z}}, \frac{\Delta E_0}{2\pi \beta_s^2 E_s} \sin\left(\frac{\omega \widehat{\sigma}}{c\beta_s} + \Phi_0\right)\right), \tag{7.45}$$

$$\widehat{\mathbf{F}}^{(M)} = \frac{eR}{\beta_s^2 E_s}\Big\{(1 + \widehat{x}K)\Big[\mathbf{E}^{(M)} + v_{0s}\Big(\mathbf{e}_s \times \mathbf{B}^{(M)}\Big)\Big] + \mathbf{e}_s\Big(\widehat{\mathbf{p}}^{(k)} \cdot \mathbf{E}^{(M)}\Big)\Big\}$$
$$+ \frac{eR}{p_{0s}}\Big(\widehat{\mathbf{p}}^{(k)} \times \mathbf{B}^{(M)}\Big)_{\mathbf{u}}, \tag{7.46}$$

$$\mathbf{e}_s = (0, 0, 1). \tag{7.47}$$

The main task of the further exposition in this chapter will be to implement the transition from the reversible equations of mechanics describing the motion of individual particles, to the irreversible equations of statistical theory. The most widely used for this purpose tool is the approach based on the so-called *BBGKY hierarchy* [Bogoliubov (1965)] of equations for the reduced (multi-)particle distribution functions. Here, we follow the method of the microscopic phase space density, which works on the basis of equations (7.42)–(7.47).

Since complete information about the evolution of the particles in the beam is not available, the microscopic phase space density $N_M\big(\widehat{\mathbf{x}}, \widehat{\mathbf{p}}^{(k)}; \theta\big)$

and the microscopic force $\widehat{\mathbf{F}}^{(M)}\left(\widehat{\mathbf{x}}, \widehat{\mathbf{p}}^{(k)}; \theta\right)$ must be treated as random functions. That is why the transition to statistical description of charged particle beam propagation requires averaging the equations involving these functions over the *Gibbs* ensemble.

First of all, let us define the s-particle ($s < N$) distribution function as

$$f_s(\mathbf{x}_1, \mathbf{p}_1, \dots, \mathbf{x}_s, \mathbf{p}_s; t) = V^s \int \mathrm{d}^3\mathbf{x}_{s+1}\mathrm{d}^3\mathbf{p}_{s+1} \dots \mathrm{d}^3\mathbf{x}_N\mathrm{d}^3\mathbf{p}_N f_N(\mathbf{X}, \mathbf{P}; t), \tag{7.48}$$

with the normalization condition

$$\frac{1}{V^s} \int \mathrm{d}^3\mathbf{x}_1\mathrm{d}^3\mathbf{p}_1 \dots \mathrm{d}^3\mathbf{x}_s\mathrm{d}^3\mathbf{p}_s f_s(\mathbf{x}_1, \mathbf{p}_1, \dots, \mathbf{x}_s, \mathbf{p}_s; t) = 1. \tag{7.49}$$

The average of the microscopic phase space density over the *Gibbs* ensemble is related to the one-particle distribution function $f\left(\widehat{\mathbf{x}}, \widehat{\mathbf{p}}^{(k)}; \theta\right)$. We have

$$\left\langle N_M\left(\widehat{X}; \theta\right)\right\rangle = \int \mathrm{d}^{3N}\widehat{\mathbf{X}}\mathrm{d}^{3N}\widehat{\mathbf{P}}^{(k)} f_N\left(\widehat{\mathbf{X}}, \widehat{\mathbf{P}}^{(k)}; \theta\right) \sum_{n=1}^{N} \delta(\widehat{\mathbf{x}} - \widehat{\mathbf{x}}_n)\delta\left(\widehat{\mathbf{p}}^{(k)} - \widehat{\mathbf{p}}_n^{(k)}\right)$$

$$= N \int \mathrm{d}^{3N}\widehat{\mathbf{X}}\mathrm{d}^{3N}\widehat{\mathbf{P}}^{(k)} f_N\left(\widehat{\mathbf{X}}, \widehat{\mathbf{P}}^{(k)}; \theta\right) \delta(\widehat{\mathbf{x}} - \widehat{\mathbf{x}}_1)\delta\left(\widehat{\mathbf{p}}^{(k)} - \widehat{\mathbf{p}}_1^{(k)}\right),$$

where the following cumulative notation

$$\widehat{X} = \left(\widehat{\mathbf{x}}, \widehat{\mathbf{p}}^{(k)}\right), \tag{7.50}$$

has been introduced. Using the definition of one-particle distribution function following from equation (7.48), we obtain

$$\left\langle N_M\left(\widehat{X}; \theta\right)\right\rangle = \frac{N}{V} \int \mathrm{d}^6\widehat{X}_1 f\left(\widehat{X}_1; \theta\right) \delta(\widehat{\mathbf{x}} - \widehat{\mathbf{x}}_1)\delta\left(\widehat{\mathbf{p}}^{(k)} - \widehat{\mathbf{p}}_1^{(k)}\right)$$

$$= \frac{N}{V} f\left(\widehat{X}; \theta\right) = n f\left(\widehat{X}; \theta\right). \tag{7.51}$$

In the second equality of equation (7.51) the thermodynamic limit has been taken. The latter implies that if both the number of particles N as well as the volume V they occupy tend to infinity, their ratio N/V remains a finite number n. The second moment of the microscopic phase space density is related to the two-particle distribution function. Indeed, according to the definition of the second moment, we have

$$\left\langle N_M\left(\widehat{X}; \theta\right) N_M\left(\widehat{X}_a; \theta\right)\right\rangle = \int \mathrm{d}^{3N}\widehat{\mathbf{X}}\mathrm{d}^{3N}\widehat{\mathbf{P}}^{(k)} f_N\left(\widehat{\mathbf{X}}, \widehat{\mathbf{P}}^{(k)}; \theta\right)$$

$$\times \sum_{m,n=1}^{N} \delta(\widehat{\mathbf{x}} - \widehat{\mathbf{x}}_m)\delta\left(\widehat{\mathbf{p}}^{(k)} - \widehat{\mathbf{p}}_m^{(k)}\right)\delta(\widehat{\mathbf{x}}_a - \widehat{\mathbf{x}}_n)\delta\left(\widehat{\mathbf{p}}_a^{(k)} - \widehat{\mathbf{p}}_n^{(k)}\right). \tag{7.52}$$

Performing the integral with a due account of the definition of the two-particle distribution function, we obtain

$$\left\langle N_M\left(\widehat{X};\theta\right)N_M\left(\widehat{X}_a;\theta\right)\right\rangle = \frac{N}{V}\delta(\widehat{\mathbf{x}} - \widehat{\mathbf{x}}_a)\delta\left(\widehat{\mathbf{p}}^{(k)} - \widehat{\mathbf{p}}_a^{(k)}\right)f\left(\widehat{X};\theta\right)$$

$$+\frac{N(N-1)}{V^2}f_2\left(\widehat{X},\widehat{X}_a;\theta\right). \tag{7.53}$$

Similar expression in terms of f, f_2 and f_3 can be obtained for the third moment.

Let us introduce the *fluctuation* of the microscopic phase space density according to the relation

$$\delta N_M\left(\widehat{X};\theta\right) = N_M\left(\widehat{X};\theta\right) - \left\langle N_M\left(\widehat{X};\theta\right)\right\rangle. \tag{7.54}$$

Clearly, $\left\langle \delta N_M\left(\widehat{X};\theta\right)\right\rangle = 0$. Thus, the second central moment is defined as

$$\langle \delta N_M \delta N_M \rangle_{\widehat{X},\widehat{X}_1;\theta} = \left\langle N_M\left(\widehat{X};\theta\right)N_M\left(\widehat{X}_1;\theta\right)\right\rangle$$

$$-\left\langle N_M\left(\widehat{X};\theta\right)\right\rangle\left\langle N_M\left(\widehat{X}_1;\theta\right)\right\rangle. \tag{7.55}$$

Furthermore,

$$\langle \delta N_M \delta N_M \rangle_{\widehat{X},\widehat{X}_1;\theta} = \frac{N(N-1)}{V^2}g_2\left(\widehat{X},\widehat{X}_1;\theta\right) + \frac{N}{V}\delta\left(\widehat{X} - \widehat{X}_1\right)f\left(\widehat{X};\theta\right)$$

$$-\frac{N}{V^2}f\left(\widehat{X};\theta\right)f\left(\widehat{X}_1;\theta\right), \tag{7.56}$$

where

$$g_2\left(\widehat{X},\widehat{X}_1;\theta\right) = f_2\left(\widehat{X},\widehat{X}_1;\theta\right) - f\left(\widehat{X};\theta\right)f\left(\widehat{X}_1;\theta\right), \tag{7.57}$$

is the so-called *correlation function.* Taking the thermodynamic limit, we can further simplify equation (7.56) and write it in the form

$$\langle \delta N_M \delta N_M \rangle_{\widehat{X},\widehat{X}_1;\theta} = n^2 g_2\left(\widehat{X},\widehat{X}_1;\theta\right) + n\delta\left(\widehat{X} - \widehat{X}_1\right)f\left(\widehat{X};\theta\right). \tag{7.58}$$

The method of microscopic phase space density is rather simple and straightforward to use. It is very effective (as compared to the *BBGKY* method) in all cases in which it is necessary to take into account the collective effects due to interactions between particles, as well as inter-particle interactions through electromagnetic field.

7.5 The Equation for the Microscopic Phase Space Density with a Small Source

Suppose that at some initial "time" θ_0 the microscopic phase space density is known to be $N_{M0}\left(\widehat{X};\theta_0\right)$. Then, the formal solution of equation (7.42) for arbitrary "time" θ can be found in principle, and may be written as

$$N_M\left(\widehat{X};\theta\right) = \widehat{\mathcal{S}}(\theta;\theta_0)N_{M0}\left(\widehat{X};\theta_0\right), \qquad (7.59)$$

where $\widehat{\mathcal{S}}(\theta;\theta_0)$ is the evolution operator (S-matrix). The choice of the initial $N_{M0}\left(\widehat{X};\theta_0\right)$ should be based on our knowledge of the microscopic characteristics of the system. This knowledge, as was mentioned a number of times before, is incomplete. Due to the extremely complex dynamics of all particles within the beam entire dynamic description is not feasible. Two fundamental "time" scales typical for a charged particle beam considered as a microscopic system (and not only for it [Klimontovich (1986)]) can be defined, namely: the minimum *correlation time* θ_{cor}, and the *relaxation time* θ_{rel} ($\theta_{cor} \ll \theta_{rel}$) towards a local equilibrium state. An intermediate "time" scale θ_{ph} should exist that corresponds to a coarse-grained description. Consequently, the conforming physically infinitesimal length scale l_{ph} confines the "point" of continuous medium itself, which depends on the level of description adopted. The physically infinitesimal time mentioned above is estimated as a characteristic time for the development of a dynamic instability of motion. The effect of dynamic instability of motion may be taken into account by smoothing the reversible dynamics over the physically infinitesimal scales. Thus, we assume

$$N_{M0}\left(\widehat{X};\theta_0\right) = \widetilde{N}_M\left(\widehat{X};\theta_0\right) = \int \mathrm{d}^3\mathbf{z}G(\widehat{\mathbf{x}}|\mathbf{z})N_M\left(\widehat{\mathbf{z}},\widehat{\mathbf{p}}^{(k)};\theta_0\right), \qquad (7.60)$$

where $G(\widehat{\mathbf{x}}|\mathbf{z})$ is a smoothing function to be specified later. It is expedient to displace θ_0 at the infinitely far past, such that in doing so, we remove the unphysical dependence on the initial "time". The limit $\theta_0 \to -\infty$ should not be performed directly in equation (7.60) because the dynamics of the system (i.e. its evolution in phase space) would drop off. Applying the evolution operator $\widehat{\mathcal{S}}(\theta; \theta_0)$ on both sides of equation (7.60) by virtue of equation (7.59), we find [Zubarev (1993)]

$$N_M\left(\widehat{X}; \theta\right) = \lim_{\tau \to -\infty} \widehat{\mathcal{S}}(\theta; \theta + \tau) \widetilde{N}_M\left(\widehat{X}; \theta + \tau\right), \tag{7.61}$$

where $\tau = \theta_0 - \theta$. In order to take an account of the initial preparation of the system, we must average equation (7.61) over some interval T in the past with smoothing function[2]

$$f(\tau) = \frac{\epsilon e^{\epsilon \tau}}{1 - e^{-\epsilon T}}, \qquad (T \to \infty). \tag{7.62}$$

This yields

$$N_M\left(\widehat{X}; \theta\right) = \epsilon \lim_{T \to \infty} \frac{1}{1 - e^{-\epsilon T}} \int\limits_{-T}^{0} \mathrm{d}\tau e^{\epsilon \tau} \widehat{\mathcal{S}}(\theta; \theta + \tau) \widetilde{N}_M\left(\widehat{X}; \theta + \tau\right), \tag{7.63}$$

or in equivalent form

$$N_M\left(\widehat{X}; \theta\right) = \epsilon \int\limits_{-\infty}^{0} \mathrm{d}\tau e^{\epsilon \tau} \widehat{\mathcal{S}}(\theta; \theta + \tau) \widetilde{N}_M\left(\widehat{X}; \theta + \tau\right), \tag{7.64}$$

where the limit $\epsilon \to +0$ is implied. It is important to note that the ϵ-limit must be performed after taking the thermodynamic limit. As a matter of fact, the boundary condition (7.64) is a direct consequence of equation (7.61) by virtue of *Abel's* theorem (see e.g. arbitrary course of complex analysis) with $\epsilon \to +0$ being equivalent to $\theta_0 \to -\infty$. Note that if we had performed the limit $\epsilon \to +0$ in equation (7.63), we would have obtained the boundary condition

$$N_M\left(\widehat{X}; \theta\right) = \lim_{T \to \infty} \frac{1}{T} \int\limits_{-T}^{0} \mathrm{d}\tau \widehat{\mathcal{S}}(\theta; \theta + \tau) \widetilde{N}_M\left(\widehat{X}; \theta + \tau\right), \tag{7.65}$$

---

[2] Note that the smoothing function defined by equation (7.62) satisfies the obvious condition

$$\int\limits_{-T}^{0} \mathrm{d}\tau f(\tau) = 1.$$

with a clearer physical meaning as compared to (7.64), implying an average over an infinite time interval in the past. It is not suited for our purposes and we will concentrate our attention on the boundary condition in the form of (7.64).

Equation (7.64) can be written alternatively as

$$N_M\left(\widehat{X};\theta\right) = \epsilon \int\limits_{-\infty}^{\theta} \mathrm{d}\lambda e^{\epsilon(\lambda-\theta)} \widehat{\mathcal{S}}(\theta;\lambda) \widetilde{N}_M\left(\widehat{X};\lambda\right). \tag{7.66}$$

Differentiating equation (7.66) with respect to θ, we readily obtain

$$\frac{\partial N_M}{\partial\theta} + \left(\widehat{\mathbf{v}}\cdot\widehat{\nabla}_{\mathbf{x}}\right)N_M + \widehat{\nabla}_{\mathbf{p}}\cdot\left[\left(\widehat{\mathbf{F}}_0 + \widehat{\mathbf{F}}^{(M)}\right)N_M\right] = \epsilon\left(\widetilde{N}_M - N_M\right). \tag{7.67}$$

Since the physically infinitesimal time interval θ_{ph} is long compared to the correlation time θ_{cor} ($\theta_{ph} \gg \theta_{cor}$), we can set $\epsilon \sim 1/\theta_{ph}$. As a result, we arrive at the equation for the microscopic phase space density with a small source

$$\frac{\partial N_M}{\partial\theta} + \left(\widehat{\mathbf{v}}\cdot\widehat{\nabla}_{\mathbf{x}}\right)N_M + \widehat{\nabla}_{\mathbf{p}}\cdot\left[\left(\widehat{\mathbf{F}}_0 + \widehat{\mathbf{F}}^{(M)}\right)N_M\right] = \frac{1}{\theta_{ph}}\left(\widetilde{N}_M - N_M\right). \tag{7.68}$$

Nevertheless the physically infinitesimal time θ_{ph} is infinitely small quantity, the right-hand-side of equation (7.68) is small at the kinetic level of description, because the left-hand-side contains intrinsically the quantity $1/\theta_{cor}$. Equation (7.68) differs from equation (7.42) by the infinitely small source that violates the "time" invariance. Its right-hand-side describes the attuning of the microscopic distribution of beam particles to the smoothed distribution over the time θ_{ph}. The first term in the brackets may be regarded as a quasi-equilibrium density distribution.

7.6 The Generalized Kinetic Equation

To obtain the desired kinetic equation for the one-particle distribution function, we perform averaging in equation (7.68) over the *Gibbs* ensemble. Using equations (7.51) and (7.54) with the microscopic phase space density N_M taken in the form

$$N_M = nf + \delta N_M, \tag{7.69}$$

as well as the representation

$$\widehat{\mathbf{F}}^{(M)} = \left\langle \widehat{\mathbf{F}} \right\rangle + \delta \widehat{\mathbf{F}}, \qquad \left\langle N_M \widehat{\mathbf{F}}^{(M)} \right\rangle = nf \left\langle \widehat{\mathbf{F}} \right\rangle + \left\langle \delta N_M \delta \widehat{\mathbf{F}} \right\rangle, \tag{7.70}$$

we obtain the generalized kinetic equation

$$\frac{\partial f}{\partial \theta} + \left(\widehat{\mathbf{v}} \cdot \widehat{\nabla}_{\mathbf{x}} \right) f + \widehat{\nabla}_{\mathbf{p}} \cdot \left[\left(\widehat{\mathbf{F}}_0 + \left\langle \widehat{\mathbf{F}} \right\rangle \right) f \right] = \mathcal{J}_{col} \left(\widehat{X}; \theta \right) + \widetilde{\mathcal{J}} \left(\widehat{X}; \theta \right), \tag{7.71}$$

where

$$\mathcal{J}_{col} \left(\widehat{X}; \theta \right) = -\frac{1}{n} \widehat{\nabla}_{\mathbf{p}} \cdot \left\langle \delta \widehat{\mathbf{F}} \delta N_M \right\rangle, \qquad \widetilde{\mathcal{J}} \left(\widehat{X}; \theta \right) = \frac{1}{\theta_{ph}} \left(\widetilde{f} - f \right), \tag{7.72}$$

are the collision integrals. The first one depends on the correlator of the fluctuation of the microscopic force and the fluctuation of the microscopic phase density. It has the form of the well-known *Balescu-Lenard* or *Landau* collision integral, which we will derive in detail in the next paragraph. In the case of collisionless beam $\mathcal{J}_{col}$ vanishes, since it is of second order in the fluctuations of the microscopic phase space density. This is the so-called *Vlasov* limit, which will be also considered in the next paragraph.

For the time being, let us concentrate our attention on the collision term $\widetilde{\mathcal{J}}$. Consider the quantity

$$\mathcal{A} \left(\widehat{\mathbf{p}}^{(k)}; \theta \right) = \frac{1}{\theta_{ph}} \int \mathrm{d}^3 \mathbf{x} g(\mathbf{x}) \left[\widetilde{f}(\mathbf{x}) - f(\mathbf{x}) \right], \tag{7.73}$$

where $g(\mathbf{x})$ is a generic function and in order to simplify notations, we have dropped the rest of the arguments in the one-particle distribution function. Further, we manipulate the first term in equation (7.73) as follows

$$\frac{1}{\theta_{ph}} \int \mathrm{d}^3 \mathbf{x} g(\mathbf{x}) \widetilde{f}(\mathbf{x}) = \frac{1}{\theta_{ph}} \int \mathrm{d}^3 \mathbf{x} g(\mathbf{x}) \int \mathrm{d}^3 \mathbf{z} G(\mathbf{x}|\mathbf{z}) f(\mathbf{z})$$

$$= \frac{1}{\theta_{ph}} \int \mathrm{d}^3 \mathbf{z} f(\mathbf{z}) \int \mathrm{d}^3 \mathbf{x} G(\mathbf{x}|\mathbf{z})$$

$$\times \left[g(\mathbf{z}) + (x_k - z_k) \nabla_k g(\mathbf{z}) + \frac{1}{2} (x_k - z_k)(x_l - z_l) \nabla_k \nabla_l g(\mathbf{z}) + \dots \right]$$

$$= \frac{1}{\theta_{ph}} \int \mathrm{d}^3 \mathbf{x} g(\mathbf{x}) f(\mathbf{x}) + \int \mathrm{d}^3 \mathbf{x} f(\mathbf{x}) \left[A_k(\mathbf{x}) \nabla_k g(\mathbf{x}) + \frac{B_{kl}(\mathbf{x})}{2} \nabla_k \nabla_l g(\mathbf{x}) \right], \tag{7.74}$$

where

$$A_k(\mathbf{x}) = \frac{1}{\theta_{ph}} \int \mathrm{d}^3\mathbf{z}(z_k - x_k)G(\mathbf{z}|\mathbf{x}), \tag{7.75}$$

$$B_{kl}(\mathbf{x}) = \frac{1}{\theta_{ph}} \int \mathrm{d}^3\mathbf{z}(z_k - x_k)(z_l - x_l)G(\mathbf{z}|\mathbf{x}), \tag{7.76}$$

and summation over repeated indices is implied. Thus, we obtain

$$\mathcal{A}\left(\widehat{\mathbf{p}}^{(k)};\theta\right) = \int \mathrm{d}^3\mathbf{x} g(\mathbf{x}) \left\{ -\nabla_k [A_k(\mathbf{x}) f(\mathbf{x})] + \frac{1}{2} \nabla_k \nabla_l [B_{kl}(\mathbf{x}) f(\mathbf{x})] \right\}. \tag{7.77}$$

Comparing equations (7.77) and (7.73) and bearing in mind that $g(\mathbf{x})$ is arbitrary, we finally arrive at

$$\widetilde{\mathcal{J}}\left(\widehat{X};\theta\right) = -\widehat{\nabla}_k [A_k(\widehat{\mathbf{x}}) f(\widehat{\mathbf{x}})] + \frac{1}{2} \widehat{\nabla}_k \widehat{\nabla}_l [B_{kl}(\widehat{\mathbf{x}}) f(\widehat{\mathbf{x}})]. \tag{7.78}$$

The most important question to answer is how to choose the particular form of the smoothing function $G(\mathbf{z}|\mathbf{x})$. This choice depends on a number of factors that influence the onset of dynamic instability. At the kinetic stage of relaxation the characteristic length playing the role of an external parameter becomes the mean free path (of the order of the relaxation length), while a local quasi-equilibrium within the elementary volume unit of continuous medium (the "point" of the physical continuous medium) is determined by the physically infinitesimal length l_{ph}. Therefore, $G(\mathbf{z}|\mathbf{x})$ can be expected to describe some local equilibrium distribution of particles contained in the "point" that corresponds to a coarse-grained hydrodynamic picture.

To find the particular form of the smoothing function $G(\mathbf{z}|\mathbf{x})$, we consider the conditional probability $G(\mathbf{z}(\theta + \tau)|\mathbf{x}(\theta))$ for two successive instants of time θ and $\theta + \tau$. By definition, the entropy can be written as

$$S = -k_B \int \mathrm{d}^3\mathbf{z} G(\mathbf{z}|\mathbf{x}) \ln G(\mathbf{z}|\mathbf{x}), \tag{7.79}$$

where k_B is the Boltzmann constant. Let us further introduce the following restrictions

$$\langle z_k \rangle_{\mathbf{x}} = \int \mathrm{d}^3\mathbf{z} z_k G(\mathbf{z}|\mathbf{x}), \qquad \langle z_k z_l \rangle_{\mathbf{x}} = \int \mathrm{d}^3\mathbf{z} z_k z_l G(\mathbf{z}|\mathbf{x}), \tag{7.80}$$

that is, we consider the first and the second moment of $\mathbf{z}$ at the instant $\theta + \tau$, provided that $\mathbf{z}$ measured at the instant θ equals $\mathbf{x}$ [i.e. $\mathbf{z}(\theta) = \mathbf{x}$].

A third obvious restriction follows from the normalizability condition

$$\int \mathrm{d}^3\mathbf{z} G(\mathbf{z}|\mathbf{x}) = 1. \tag{7.81}$$

Taking into account the restrictions (7.80) and (7.81), we construct the functional of information entropy

$$\mathcal{I} = \frac{S}{k_B} - (\lambda - 1) \int \mathrm{d}^3\mathbf{z} G(\mathbf{z}) - \lambda_k \int \mathrm{d}^3\mathbf{z} z_k G(\mathbf{z}) - \lambda_{kl} \int \mathrm{d}^3\mathbf{z} z_k z_l G(\mathbf{z}), \tag{7.82}$$

where summation over repeated indices is implied and λ, λ_k and λ_{kl} are yet unknown *Lagrange* multipliers. The principle of maximum of the information entropy first formulated by *Jaynes* [Jaynes (1957a)], [Jaynes (1957b)] (see also [Haken (1990)]) states that an equilibrium distribution $G(\mathbf{z}|\mathbf{x})$ is reached for an extremum of the information entropy

$$\frac{\delta \mathcal{I}}{\delta G(\mathbf{z}|\mathbf{x})} = 0. \tag{7.83}$$

This gives

$$G(\mathbf{z}|\mathbf{x}) = \exp\left[-\lambda(\mathbf{x}) - \lambda_k(\mathbf{x}) z_k - \lambda_{kl}(\mathbf{x}) z_k z_l\right].$$

The *Lagrange* multipliers λ, λ_k and λ_{kl} can be determined from the restrictions (7.80) and (7.81), so that, we obtain finally

$$G(\mathbf{z}|\mathbf{x}) = \frac{1}{\pi^{3/2} \sqrt{\left|\det \widehat{\mathcal{G}}\right|}} \exp\left[-\left(\mathbf{z} - \langle \mathbf{z} \rangle_{\mathbf{x}}\right)^T \widehat{\mathcal{G}}^{-1}(\mathbf{x}) \left(\mathbf{z} - \langle \mathbf{z} \rangle_{\mathbf{x}}\right)\right], \tag{7.84}$$

where $\widehat{\mathcal{G}}(\mathbf{x})$ is the covariance matrix

$$\mathcal{G}_{kl}(\mathbf{x}) = 2\left[\langle z_k z_l \rangle_{\mathbf{x}} - \langle z_k \rangle_{\mathbf{x}} \langle z_l \rangle_{\mathbf{x}}\right]. \tag{7.85}$$

Using the explicit form (7.84) of the smoothing function, as well as equations (7.80) to be substituted in the expressions (7.75) and (7.76) for the drift vector $A_k(\mathbf{x})$ and the diffusion matrix $B_{kl}(\mathbf{x})$ respectively, we obtain

$$A_k(\mathbf{x}) = \frac{1}{\theta_{ph}} \langle \Delta z_k \rangle_{\mathbf{x}}, \qquad B_{kl}(\mathbf{x}) = \frac{1}{\theta_{ph}} \langle \Delta z_k \Delta z_l \rangle_{\mathbf{x}}, \qquad (\Delta \mathbf{z} = \mathbf{z} - \mathbf{x}). \tag{7.86}$$

For Hamiltonian systems the well-known relation [Lichtenberg (1982)]

$$A_k(\mathbf{x}) = \frac{1}{2} \nabla_l B_{kl}(\mathbf{x}) \tag{7.87}$$

holds, so that, we can write the collision integral (7.78) in the form

$$\widetilde{\mathcal{J}}\left(\widehat{X};\theta\right) = \frac{1}{2}\widehat{\nabla}_k\left[B_{kl}(\widehat{\mathbf{x}})\widehat{\nabla}_l f(\widehat{\mathbf{x}})\right]. \tag{7.88}$$

It is worthwhile to note that a careful inspection of equations (7.87) shows that the diffusion matrix $B_{kl}(\widehat{\mathbf{x}})$ has the meaning of an action averaged over the quasi-equilibrium distribution (7.84). This can be regarded as a stochastic generalization of the classical action invariant. An important feature that needs to be particularly emphasized is the self-diffusion process intrinsically contained in the collision integral (7.88). It is specified by means of a proper physical definition of continuous medium with an elementary cell ("point") determined by physically infinitesimal scales. Note that knowledge of the initial microscopic characteristics (initial conditions) of the beam consists in the microscopic phase space density smoothed over the quasi-equilibrium distribution. These initial correlations due to mixing of particles contained in the "point" are then transferred to the entire evolution of the system thus giving rise to a relaxation term in the equation for the microscopic phase space density.

7.7 The Balescu-Lenard Kinetic Equation

Averaging the *Maxwell* equations (7.17) - (7.19) over the *Gibbs* ensemble, we obtain

$$\nabla_{\mathbf{x}} \times \langle\mathbf{B}\rangle = \frac{1}{c^2}\frac{\partial\langle\mathbf{E}\rangle}{\partial t} + \mu_0 en\mathbf{j}(\mathbf{x};t), \qquad \nabla_{\mathbf{x}} \times \langle\mathbf{E}\rangle = -\frac{\partial\langle\mathbf{B}\rangle}{\partial t}, \tag{7.89}$$

$$\nabla_{\mathbf{x}} \cdot \langle\mathbf{B}\rangle = 0, \qquad \nabla_{\mathbf{x}} \cdot \langle\mathbf{E}\rangle = \frac{en}{\epsilon_0}\rho(\mathbf{x};t), \tag{7.90}$$

where

$$\rho(\mathbf{x};t) = \int \mathrm{d}^3\mathbf{p}^{(k)} f\left(\mathbf{x},\mathbf{p}^{(k)};t\right), \qquad \mathbf{j}(\mathbf{x};t) = \int \mathrm{d}^3\mathbf{p}^{(k)}\mathbf{v} f\left(\mathbf{x},\mathbf{p}^{(k)};t\right). \tag{7.91}$$

The equations for the fluctuating fields are similar to (7.89) and (7.90) and read as

$$\nabla_{\mathbf{x}} \times \delta\mathbf{B} = \frac{1}{c^2}\frac{\partial\delta\mathbf{E}}{\partial t} + \mu_0 e\delta\mathbf{j}(\mathbf{x};t), \qquad \nabla_{\mathbf{x}} \times \delta\mathbf{E} = -\frac{\partial\delta\mathbf{B}}{\partial t}, \tag{7.92}$$

$$\nabla_{\mathbf{x}} \cdot \delta\mathbf{B} = 0, \qquad \nabla_{\mathbf{x}} \cdot \delta\mathbf{E} = \frac{e}{\epsilon_0}\delta\rho(\mathbf{x};t), \tag{7.93}$$

where

$$\delta\rho(\mathbf{x};t) = \int \mathrm{d}^3\mathbf{p}^{(k)} \delta N_M \Big(\mathbf{x}, \mathbf{p}^{(k)}; t\Big), \tag{7.94}$$

$$\delta\mathbf{j}(\mathbf{x};t) = \int \mathrm{d}^3\mathbf{p}^{(k)} \mathbf{v} \delta N_M \Big(\mathbf{x}, \mathbf{p}^{(k)}; t\Big). \tag{7.95}$$

Taking the divergence of the first of equations (7.92) and utilizing the last of equations (7.93), it can be easily verified that the continuity equation for fluctuating quantities holds

$$\frac{\partial}{\partial t}\delta\rho(\mathbf{x};t) + \nabla_{\mathbf{x}} \cdot \delta\mathbf{j}(\mathbf{x};t) = 0. \tag{7.96}$$

It should be pointed out that the microscopic electromagnetic fields depend on the coordinates $\mathbf{x} = (x, z, \sigma)$ through the microscopic phase space density N_M written in these coordinates. Before proceeding further, we derive some useful relations that will be needed for the subsequent exposition. Consider the simple change of variables

$$\mathrm{d}^3\mathbf{x}\mathrm{d}^3\mathbf{p}^{(k)} = (1 + xK)^2 \mathrm{d}x\mathrm{d}z\mathrm{d}s\mathrm{d}p_x^{(k)}\mathrm{d}p_z^{(k)}\mathrm{d}p_s^{(k)}$$

$$= (1 + xK)^2 |\det \mathcal{J}_1| \mathrm{d}x\mathrm{d}z\mathrm{d}\sigma\mathrm{d}\widetilde{p}_x^{(k)}\mathrm{d}\widetilde{p}_z^{(k)}\mathrm{d}h^{(k)}.$$

Noting that

$$x = \widetilde{x}, \qquad z = \widetilde{z}, \qquad s = \sigma + v_{0s}t,$$

$$p_x^{(k)} = p_{0s}\widetilde{p}_x^{(k)}, \qquad p_z^{(k)} = p_{0s}\widetilde{p}_z^{(k)}, \qquad p_s^{(k)} = \frac{p_{0s}\mathcal{S}}{1 + xK},$$

$$\mathcal{S} = \sqrt{\beta_s^2 h^{(k)2} - \frac{1}{\beta_s^2\gamma_s^2} - \widetilde{p}_x^{(k)2} - \widetilde{p}_z^{(k)2}},$$

we easily find

$$|\det \mathcal{J}_1| = p_{0s}^3 \frac{\beta_s^2 h^{(k)}}{\mathcal{S}(1 + xK)}.$$

Hence

$$\mathrm{d}^3\mathbf{x}\mathrm{d}^3\mathbf{p}^{(k)} = p_{0s}^3(1 + xK)\frac{\beta_s^2 h^{(k)}}{\mathcal{S}} \mathrm{d}x\mathrm{d}z\mathrm{d}\sigma\mathrm{d}\widetilde{p}_x^{(k)}\mathrm{d}\widetilde{p}_z^{(k)}\mathrm{d}h^{(k)}.$$

Continuing further, we use the relations

$$x = \widehat{x} + \widehat{\eta} D, \qquad \widetilde{p}_x^{(k)} = \widehat{p}_x^{(k)} + \frac{\widehat{\eta}}{R} \frac{\mathrm{d}D}{\mathrm{d}\theta}, \qquad z = \widehat{z}, \qquad \widetilde{p}_z^{(k)} = \widehat{p}_z^{(k)},$$

$$\sigma = \widehat{\sigma} + \widehat{p}_x D - \frac{\widehat{x}}{R} \frac{\mathrm{d}D}{\mathrm{d}\theta}, \qquad h^{(k)} = \widehat{\eta}^{(k)} + \frac{1}{\beta_s^2}.$$

similar to the derived in chapter 1, and obtain finally

$$\mathrm{d}^3\mathbf{x}\mathrm{d}^3\mathbf{p}^{(k)} = p_{0s}^3(1 + xK)\frac{1 + \beta_s^2 \widehat{\eta}^{(k)}}{\mathcal{S}} \mathrm{d}\widehat{x}\mathrm{d}\widehat{z}\mathrm{d}\widehat{\sigma}\mathrm{d}\widehat{p}_x^{(k)}\mathrm{d}\widehat{p}_z^{(k)}\mathrm{d}\widehat{\eta}^{(k)}.$$

Since

$$\mathcal{S} \approx \sqrt{1 + 2\widehat{\eta}^{(k)} + \beta_s^2 \widehat{\eta}^{(k)2}} \approx 1 + \widehat{\eta}^{(k)},$$

for $\beta_s \approx 1$, we can write

$$\mathrm{d}^3\mathbf{x}\mathrm{d}^3\mathbf{p}^{(k)} = p_{0s}^3(1 + xK)\mathrm{d}^3\widehat{\mathbf{x}}\mathrm{d}^3\widehat{\mathbf{p}}^{(k)}. \tag{7.97}$$

Thus, integration in the expressions for the charge and current density

$$\delta\rho(\mathbf{x}; t) = \int \mathrm{d}^3\widehat{\mathbf{p}}^{(k)} \delta N_M\left(\mathbf{x}, \widehat{\mathbf{p}}^{(k)}; \theta\right), \tag{7.98}$$

$$\delta\mathbf{j}(\mathbf{x}; t) = \int \mathrm{d}^3\widehat{\mathbf{p}}^{(k)} \mathbf{v} \delta N_M\left(\mathbf{x}, \widehat{\mathbf{p}}^{(k)}; \theta\right), \tag{7.99}$$

goes approximately over the new kinetic momenta $\widehat{\mathbf{p}}^{(k)}$.

In order to determine the first of the collision integrals (7.72), we have to solve the equation

$$\left[\frac{\partial}{\partial\theta} + \widehat{\mathbf{v}} \cdot \widehat{\nabla}_{\mathbf{x}} + \widehat{\nabla}_{\mathbf{p}} \cdot \left(\widehat{\mathbf{F}}_0 + \left\langle \widehat{\mathbf{F}} \right\rangle\right)\right] \delta N_M$$

$$= -n\widehat{\nabla}_{\mathbf{p}} \cdot \left(f\delta\widehat{\mathbf{F}}\right) + \widehat{\nabla}_{\mathbf{p}} \cdot \left[\left\langle \delta\widehat{\mathbf{F}}\delta N_M \right\rangle - \delta\widehat{\mathbf{F}}\delta N_M\right], \tag{7.100}$$

governing the evolution of the fluctuating part δN_M of the microscopic phase space density. Under the assumption that fluctuations are small, the second term on the right-hand-side of equation (7.100) can be neglected, so that the equation to be analyzed in what follows is

$$\left[\frac{\partial}{\partial\theta} + \widehat{\mathbf{v}} \cdot \widehat{\nabla}_{\mathbf{x}} + \widehat{\nabla}_{\mathbf{p}} \cdot \left(\widehat{\mathbf{F}}_0 + \left\langle \widehat{\mathbf{F}} \right\rangle\right)\right] \delta N_M(\widehat{\mathbf{x}}; \theta) = -n\widehat{\nabla}_{\mathbf{p}} \cdot \left[f(\widehat{\mathbf{x}}; \theta)\delta\widehat{\mathbf{F}}(\widehat{\mathbf{x}}; \theta)\right]. \tag{7.101}$$

Note that in general the right-hand-side of equation (7.100) contains additional small source term

$$\frac{1}{\theta_{ph}}\left(\delta\widetilde{N}_M - \delta N_M\right), \tag{7.102}$$

which has been dropped off as non relevant for the dynamics of small-scale fluctuations.

The term containing the mean force in equation (7.101) can be neglected. This is justified when calculating the small-scale fluctuations if

$$\omega_p \gg \nu_{bs}\omega_s, \tag{7.103}$$

where ν_{bs} stands for the betatron tunes in the two transverse directions, as well as for the synchrotron tune, and ω_s is the angular frequency of the synchronous particle. Furthermore, ω_p is the beam plasma frequency defined as

$$\omega_p^2 = \frac{e^2 n}{\epsilon_0 m_0 \gamma_s}. \tag{7.104}$$

Another important characteristic is the *Debye* radius

$$r_D^2 = \frac{\epsilon_0 k_B T}{e^2 n}, \tag{7.105}$$

where T is the temperature of the beam. It is worth to note that the physical meaning of the *Debye* radius for particle beams is somewhat different from the one commonly used in plasma physics. In fact, the *Debye* radius is an equilibrium characteristic of the beam indicating the exponential decay of the self-field, needed to maintain this equilibrium state.

The contribution of small-scale fluctuations can be better extracted if a small source proportional to Δ is introduced into the left-hand-side of equation (7.101) as follows

$$\left(\frac{\partial}{\partial\theta} + \widetilde{\mathbf{v}}\cdot\nabla_{\mathbf{x}} + \Delta\right)\delta N_M(\mathbf{x};\theta) = -n\widehat{\nabla}_{\mathbf{p}}\cdot\left[f(\mathbf{x};\theta)\delta\widehat{\mathbf{F}}(\mathbf{x};\theta)\right], \tag{7.106}$$

$$\widetilde{\mathbf{v}} = \left(R\widetilde{p}_x^{(k)}, R\widetilde{p}_z^{(k)}, -R\mathcal{K}\widehat{\eta}^{(k)}\right).$$

In going over from equation (7.101) to equation (7.106) the left-hand-side has been represented in terms of the variables $\mathbf{x} = (x, z, \sigma)$. The general solution of the above equation can be written as

$$\delta N_M\left(\mathbf{x}, \widehat{\mathbf{p}}^{(k)}; \theta\right) = \delta N_M^s\left(\mathbf{x}, \widehat{\mathbf{p}}^{(k)}; \theta\right) + \delta N_M^{ind}\left(\mathbf{x}, \widehat{\mathbf{p}}^{(k)}; \theta\right), \tag{7.107}$$

where δN_M^{ind} is a generic solution of equation (7.106), while δN_M^s accounts for the discrete structure of the beam as a collection of particles. The latter can be determined from [Klimontovich (1986)]

$$\left(\frac{\partial}{\partial\theta} + \widetilde{\mathbf{v}}\cdot\nabla_{\mathbf{x}} + \Delta\right)\langle\delta N_M^s(\mathbf{X};\theta)\delta N_M^s(\mathbf{X}_1;\theta_1)\rangle = 0, \qquad \mathbf{X} = \left(\mathbf{x}, \widehat{\mathbf{p}}^{(k)}\right) \tag{7.108}$$

with the initial condition

$$\langle\delta N_M^s(\mathbf{X};\theta)\delta N_M^s(\mathbf{X}_1;\theta)\rangle = n\delta(\mathbf{x}-\mathbf{x}_1)\delta\left(\widehat{\mathbf{p}}^{(k)} - \widehat{\mathbf{p}}_1^{(k)}\right)f\left(\mathbf{x}, \widehat{\mathbf{p}}^{(k)};\theta\right). \tag{7.109}$$

When small-scale fluctuations are computed $f\left(\mathbf{x}, \widehat{\mathbf{p}}^{(k)};\theta\right)$ can be considered a smooth enough function (not varying considerably), so that, we can assume that the correlator $\langle\delta N_M^s(\mathbf{X};\theta)\delta N_M^s(\mathbf{X}_1;\theta_1)\rangle$ depends on $\theta-\theta_1$ and $\mathbf{x}-\mathbf{x}_1$ only. Introducing the *Fourier* transform

$$\langle\delta N_M(\mathbf{X};\theta)\delta N_M(\mathbf{X}_1;\theta_1)\rangle = \langle\delta N_M\delta N_M\rangle\left(\theta-\theta_1, \mathbf{x}-\mathbf{x}_1; \widehat{\mathbf{p}}^{(k)}, \widehat{\mathbf{p}}_1^{(k)}\right)$$

$$= \frac{1}{(2\pi)^3}\int \mathrm{d}^3\mathbf{k}\left(\widetilde{\delta N_M\delta N_M}\right)\left(\theta-\theta_1, \mathbf{k}; \widehat{\mathbf{p}}^{(k)}, \widehat{\mathbf{p}}_1^{(k)}\right)\exp\left[i\mathbf{k}\cdot(\mathbf{x}-\mathbf{x}_1)\right], \tag{7.110}$$

we cast equation (7.108) into the form

$$\left(\frac{\partial}{\partial\theta} + i\mathbf{k}\cdot\widetilde{\mathbf{v}} + \Delta\right)\left(\widetilde{\delta N_M\delta N_M}\right)^s\left(\tau, \mathbf{k}; \widehat{\mathbf{p}}^{(k)}, \widehat{\mathbf{p}}_1^{(k)}\right) = 0, \qquad (\tau = \theta-\theta_1), \tag{7.111}$$

$$\left(\widetilde{\delta N_M\delta N_M}\right)^s\left(\tau, \mathbf{k}; \widehat{\mathbf{p}}^{(k)}, \widehat{\mathbf{p}}_1^{(k)}\right)\Big|_{\tau=0} = n\delta\left(\widehat{\mathbf{p}}^{(k)} - \widehat{\mathbf{p}}_1^{(k)}\right)f\left(\mathbf{x}, \widehat{\mathbf{p}}^{(k)};\theta\right). \tag{7.112}$$

Further, we introduce the one-sided *Fourier* transform in the time domain

$$\left(\widetilde{\delta N_M\delta N_M}\right)^\dagger\left(\omega, \mathbf{k}; \widehat{\mathbf{p}}^{(k)}, \widehat{\mathbf{p}}_1^{(k)}\right) = \int_0^\infty \mathrm{d}\tau\left(\widetilde{\delta N_M\delta N_M}\right)\left(\tau, \mathbf{k}; \widehat{\mathbf{p}}^{(k)}, \widehat{\mathbf{p}}_1^{(k)}\right)e^{i\omega\tau}. \tag{7.113}$$

Multiplication of equation (7.111) by $e^{i\omega\tau}$ and subsequent integration on τ yields

$$(-i\omega + i\mathbf{k}\cdot\widetilde{\mathbf{v}} + \Delta)\left(\widetilde{\delta N_M\delta N_M}\right)^\dagger\left(\omega, \mathbf{k}; \widehat{\mathbf{p}}^{(k)}, \widehat{\mathbf{p}}_1^{(k)}\right)$$

$$= \left(\widetilde{\delta N_M\delta N_M}\right)^s\left(0, \mathbf{k}; \widehat{\mathbf{p}}^{(k)}, \widehat{\mathbf{p}}_1^{(k)}\right),$$

or

$$\left(\delta \widetilde{N_M \delta N_M}\right)^{\dagger}\left(\omega, \mathbf{k}; \widehat{\mathbf{p}}^{(k)}, \widehat{\mathbf{p}}_1^{(k)}\right) = \frac{inf\left(\mathbf{x}, \widehat{\mathbf{p}}^{(k)}; \theta\right)}{\omega - \mathbf{k}\cdot\widetilde{\mathbf{v}} + i\Delta}\delta\left(\widehat{\mathbf{p}}^{(k)} - \widehat{\mathbf{p}}_1^{(k)}\right). \quad (7.114)$$

Using the equation

$$\left(\delta \widetilde{N_M \delta N_M}\right)\left(\omega, \mathbf{k}; \widehat{\mathbf{p}}^{(k)}, \widehat{\mathbf{p}}_1^{(k)}\right) = \left(\delta \widetilde{N_M \delta N_M}\right)^{\dagger}\left(\omega, \mathbf{k}; \widehat{\mathbf{p}}^{(k)}, \widehat{\mathbf{p}}_1^{(k)}\right)$$

$$+\left[\left(\delta \widetilde{N_M \delta N_M}\right)^{\dagger}\left(\omega, \mathbf{k}; \widehat{\mathbf{p}}^{(k)}, \widehat{\mathbf{p}}_1^{(k)}\right)\right]^{*}_{\widehat{\mathbf{p}}\leftrightarrow\widehat{\mathbf{p}}_1},$$

relating the one-sided and the two-sided *Fourier* transform, we find

$$\left(\delta \widetilde{N_M \delta N_M}\right)^{s}\left(\omega, \mathbf{k}; \widehat{\mathbf{p}}^{(k)}, \widehat{\mathbf{p}}_1^{(k)}\right)$$

$$= \frac{2\Delta}{\left(\omega - \mathbf{k}\cdot\widetilde{\mathbf{v}}\right)^2 + \Delta^2} nf\left(\mathbf{x}, \widehat{\mathbf{p}}^{(k)}; \theta\right)\delta\left(\widehat{\mathbf{p}}^{(k)} - \widehat{\mathbf{p}}_1^{(k)}\right). \quad (7.115)$$

The definition of the *Dirac*'s δ-function

$$\lim_{\Delta\to 0}\frac{\Delta}{\left(\omega - \mathbf{k}\cdot\widetilde{\mathbf{v}}\right)^2 + \Delta^2} = \pi\delta(\omega - \mathbf{k}\cdot\widetilde{\mathbf{v}}),$$

gives finally

$$\left(\delta \widetilde{N_M \delta N_M}\right)^{s}\left(\omega, \mathbf{k}; \widehat{\mathbf{p}}^{(k)}, \widehat{\mathbf{p}}_1^{(k)}\right)$$

$$= 2\pi nf\left(\mathbf{x}, \widehat{\mathbf{p}}^{(k)}; \theta\right)\delta\left(\widehat{\mathbf{p}}^{(k)} - \widehat{\mathbf{p}}_1^{(k)}\right)\delta(\omega - \mathbf{k}\cdot\widetilde{\mathbf{v}}). \quad (7.116)$$

To obtain an arbitrary solution of equation (7.106), we perform the *Fourier* transform

$$\delta N_M\left(\mathbf{x}, \widehat{\mathbf{p}}^{(k)}; \theta\right) = \frac{1}{(2\pi)^4}\int \mathrm{d}\omega \mathrm{d}^3\mathbf{k}\,\delta\widetilde{N}_M\left(\omega, \mathbf{k}, \widehat{\mathbf{p}}^{(k)}\right)e^{i(\mathbf{k}\cdot\mathbf{x} - \omega\theta)}, \quad (7.117)$$

$$\delta\widetilde{N}_M\left(\omega, \mathbf{k}, \widehat{\mathbf{p}}^{(k)}\right) = \int \mathrm{d}\theta \mathrm{d}^3\mathbf{x}\,\delta N_M\left(\mathbf{x}, \widehat{\mathbf{p}}^{(k)}; \theta\right)e^{i(\omega\theta - \mathbf{k}\cdot\mathbf{x})}, \quad (7.118)$$

and find

$$\delta\widetilde{N}_M\left(\omega, \mathbf{k}, \widehat{\mathbf{p}}^{(k)}\right) = \delta\widetilde{N}_M^{s}\left(\omega, \mathbf{k}, \widehat{\mathbf{p}}^{(k)}\right)$$

$$-\frac{in}{\omega - \mathbf{k}\cdot\widetilde{\mathbf{v}} + i\Delta}\widehat{\nabla}_{\mathbf{p}}\cdot\left[f(\widehat{\mathbf{x}};\theta)\delta\widetilde{\widehat{\mathbf{F}}}(\omega,\mathbf{k})\right]. \tag{7.119}$$

What remains now is to compute the spectral density of the fluctuating force $\delta\widetilde{\widehat{\mathbf{F}}}$. In doing so, we consider a generic function $F(\mathbf{x};\theta)$. Let the same function, written in the variables $\mathbf{r} = (x, z, s = R\theta)$ and t be $F_r(\mathbf{r};t)$. Further, we have

$$F_r(\mathbf{r};t) = \frac{1}{(2\pi)^4}\int \mathrm{d}\nu \mathrm{d}^3\mathbf{m}\widetilde{F}_r(\nu,\mathbf{m})e^{i(\mathbf{m}\cdot\mathbf{r}-\nu t)}$$

$$= \frac{1}{(2\pi)^4}\int \mathrm{d}\nu \mathrm{d}^3\mathbf{m}\widetilde{F}_r(\nu,\mathbf{m})\exp\left\{i\left[m_x x + m_z z + m_s R\theta - \frac{\nu(R\theta-\sigma)}{v_{0s}}\right]\right\}$$

$$= \frac{\omega_s}{(2\pi)^4}\int \mathrm{d}\omega \mathrm{d}^3\mathbf{k}\widetilde{F}_r\left(v_{0s}k_\sigma; k_x, k_z, k_\sigma - \frac{\omega}{R}\right)e^{i(\mathbf{k}\cdot\mathbf{x}-\omega\theta)},$$

where the following change of variables

$$\mathbf{m} = \left(k_x, k_z, k_\sigma - \frac{\omega}{R}\right), \qquad \nu = v_{0s}k_\sigma \tag{7.120}$$

has been introduced. Therefore, the relation sought-for reads as

$$\widetilde{F}(\omega;\mathbf{k}) = \omega_s\widetilde{F}_r\left(v_{0s}k_\sigma; k_x, k_z, k_\sigma - \frac{\omega}{R}\right). \tag{7.121}$$

Fourier analyzing equations (7.92) and (7.93), we obtain

$$i\mathbf{m}\times\delta\widetilde{\mathbf{B}}_r = -\frac{i\nu}{c^2}\delta\widetilde{\mathbf{E}}_r + \mu_0 e\delta\widetilde{\mathbf{j}}_r, \qquad \delta\widetilde{\mathbf{B}}_r = \frac{1}{\nu}\mathbf{m}\times\delta\widetilde{\mathbf{E}}_r, \tag{7.122}$$

$$\mathbf{m}\cdot\delta\widetilde{\mathbf{B}}_r = 0, \qquad i\mathbf{m}\cdot\delta\widetilde{\mathbf{E}}_r = \frac{e}{\epsilon_0}\delta\widetilde{\rho}_r. \tag{7.123}$$

Next, we represent the electromagnetic fields as a sum of longitudinal and transversal components

$$\delta\widetilde{\mathbf{E}}_r = \delta\widetilde{\mathbf{E}}_r^{\parallel} + \delta\widetilde{\mathbf{E}}_r^{\perp}, \qquad \left(\mathbf{m}\times\delta\widetilde{\mathbf{E}}_r^{\parallel} = 0, \qquad \mathbf{m}\cdot\delta\widetilde{\mathbf{E}}_r^{\perp} = 0\right), \tag{7.124}$$

and further simplify the problem by considering

$$\delta\widetilde{\mathbf{j}}_r = v_{0s}\mathbf{e}_s\delta\widetilde{\rho}_r, \tag{7.125}$$

From the continuity equation (7.96), we deduce

$$\delta\widetilde{\rho}_r = \frac{1}{\nu}\mathbf{m}\cdot\delta\widetilde{\mathbf{j}}_r, \tag{7.126}$$

and using equations (7.121) and (7.125), we conclude that $\mathbf{m} = \mathbf{k}$. Thus, we obtain

$$\delta\widetilde{\mathbf{E}}^{\parallel}(\omega, \mathbf{k}) = -\frac{ie\mathbf{k}}{\epsilon_0 k^2}\delta\widetilde{\rho}(\omega, \mathbf{k}), \tag{7.127}$$

$$\delta\widetilde{\mathbf{E}}^{\perp}(\omega, \mathbf{k}) = \frac{ie\beta_s^2 k_\sigma}{\epsilon_0 k^2(k^2 - \beta_s^2 k_\sigma^2)}[\mathbf{k} \times (\mathbf{e}_s \times \mathbf{k})]\delta\widetilde{\rho}(\omega, \mathbf{k}), \tag{7.128}$$

$$\delta\widetilde{\mathbf{B}}(\omega, \mathbf{k}) = \frac{1}{v_{0s}k_\sigma}\mathbf{k} \times \delta\widetilde{\mathbf{E}}^{\perp}(\omega, \mathbf{k}) = \frac{ie\beta_s^2}{\epsilon_0 v_{0s}(k^2 - \beta_s^2 k_\sigma^2)}(\mathbf{k} \times \mathbf{e}_s)\delta\widetilde{\rho}(\omega, \mathbf{k}). \tag{7.129}$$

Retaining the leading terms only, we can write the fluctuating force $\delta\widetilde{\widehat{\mathbf{F}}}$ as

$$\delta\widetilde{\widehat{\mathbf{F}}}(\omega, \mathbf{k}) = \frac{e}{\beta_s^2 E_s}\left[\delta\widetilde{\mathbf{E}} + v_{0s}\left(\mathbf{e}_s \times \delta\widetilde{\mathbf{B}}\right)\right]$$

$$= -\frac{ie^2\mathbf{k}}{\epsilon_0\beta_s^2\gamma_s^2 E_s(k^2 - \beta_s^2 k_\sigma^2)}\delta\widetilde{\rho}(\omega, \mathbf{k}). \tag{7.130}$$

Integrating equation (7.119) on $\widehat{\mathbf{p}}^{(k)}$, we obtain

$$\delta\widetilde{\rho}(\omega, \mathbf{k}) = \delta\widetilde{\rho}^s(\omega, \mathbf{k}) - in\int d^3\widehat{\mathbf{p}}^{(k)}\frac{\widehat{\nabla}_{\mathbf{p}}f(\widehat{\mathbf{x}}; \theta)}{\omega - \mathbf{k}\cdot\widetilde{\mathbf{v}} + i\Delta}\cdot\delta\widetilde{\widehat{\mathbf{F}}}(\omega, \mathbf{k}).$$

With equation (7.130) in hand, the fluctuating force $\delta\widetilde{\widehat{\mathbf{F}}}(\omega, \mathbf{k})$ can be eliminated yielding the final result

$$\widetilde{\varepsilon}_b(\omega, \mathbf{k})\delta\widetilde{\rho}(\omega, \mathbf{k}) = \delta\widetilde{\rho}^s(\omega, \mathbf{k}), \tag{7.131}$$

where

$$\widetilde{\varepsilon}_b(\omega, \mathbf{k}) = 1 + \frac{e^2 n}{\epsilon_0\beta_s^2\gamma_s^2 E_s(k^2 - \beta_s^2 k_\sigma^2)}\int d^3\widehat{\mathbf{p}}^{(k)}\frac{\mathbf{k}\cdot\widehat{\nabla}_{\mathbf{p}}f(\widehat{\mathbf{x}}; \theta)}{\omega - \mathbf{k}\cdot\widetilde{\mathbf{v}} + i\Delta} \tag{7.132}$$

is the dielectric susceptibility of the beam. Thus, for the spectral density of the fluctuating force, we have the following expression

$$\delta\widetilde{\widehat{\mathbf{F}}}(\omega, \mathbf{k}) = -\frac{ie^2\mathbf{k}}{\epsilon_0\widetilde{\varepsilon}_b(\omega, \mathbf{k})\beta_s^2\gamma_s^2 E_s(k^2 - \beta_s^2 k_\sigma^2)}\delta\widetilde{\rho}^s(\omega, \mathbf{k}). \tag{7.133}$$

According to equation (7.72) the collision integral is given by the expression

$$\mathcal{J}_{col}\left(\widehat{\mathbf{x}}, \widehat{\mathbf{p}}^{(k)}; \theta\right) = -\frac{1}{n}\widehat{\nabla}_{\mathbf{p}} \cdot \left\langle \delta\widehat{\mathbf{F}}\delta N_M \right\rangle \left(\widehat{\mathbf{x}}, \widehat{\mathbf{p}}^{(k)}, \theta; \widehat{\mathbf{x}}, \widehat{\mathbf{p}}^{(k)}, \theta\right). \qquad (7.134)$$

We shall express the right-hand-side of equation (7.134) in terms of the spectral density of fluctuations $\widetilde{\delta\widehat{\mathbf{F}}}$ and $\delta\widetilde{N}_M$. Let $\mathcal{F}(\mathbf{x};\theta)$ and $\mathcal{G}(\mathbf{x}_1;\theta_1)$ be two random functions. The second moment in the variables $\mathbf{x} - \mathbf{x}_1$ and $\theta - \theta_1$ can be written as

$$\langle \mathcal{F}\mathcal{G} \rangle(\mathbf{x}, \theta; \mathbf{x}_1, \theta_1) = \langle \mathcal{F}\mathcal{G} \rangle(\mathbf{x}, \theta; \mathbf{x} - \mathbf{x}_1, \theta - \theta_1)$$

$$= \frac{1}{(2\pi)^4} \int \mathrm{d}\omega \mathrm{d}^3\mathbf{k} \left(\widetilde{\mathcal{F}\mathcal{G}}\right)(\omega, \mathbf{k}; \mathbf{x}, \theta) \exp\left\{i[\mathbf{k}\cdot(\mathbf{x} - \mathbf{x}_1) - \omega(\theta - \theta_1)]\right\}. \qquad (7.135)$$

Since the second moment is a real function, the spectral density obeys the identity

$$\left(\widetilde{\mathcal{F}\mathcal{G}}\right)(\omega, \mathbf{k}; \mathbf{x}, \theta) = \left(\widetilde{\mathcal{F}\mathcal{G}}\right)^*(-\omega, -\mathbf{k}; \mathbf{x}, \theta). \qquad (7.136)$$

Letting now $\mathbf{x}_1 = \mathbf{x}$ and $\theta_1 = \theta$ in equation (7.135) and using the above identity (7.136), we find

$$\langle \mathcal{F}\mathcal{G} \rangle(\mathbf{x}, \theta; \mathbf{x}, \theta) = \frac{1}{(2\pi)^4} \int \mathrm{d}\omega \mathrm{d}^3\mathbf{k} \mathrm{Re}\left(\widetilde{\mathcal{F}\mathcal{G}}\right)(\omega, \mathbf{k}; \mathbf{x}, \theta). \qquad (7.137)$$

Making use of equation (7.137) and taking into account only leading terms in $\widetilde{\delta\widehat{\mathbf{F}}}$, we rewrite the expression for the collision integral as

$$\mathcal{J}_{col}\left(\widehat{\mathbf{x}}, \widehat{\mathbf{p}}^{(k)}; \theta\right) = -\frac{1}{n}\widehat{\nabla}_{\mathbf{p}} \cdot \int \frac{\mathrm{d}\omega \mathrm{d}^3\mathbf{k}}{(2\pi)^4} \mathrm{Re}\left(\widetilde{\delta\widehat{\mathbf{F}}\delta N_M}\right)\left(\omega, \mathbf{k}; \widehat{\mathbf{x}}, \widehat{\mathbf{p}}^{(k)}, \theta\right). \qquad (7.138)$$

Utilizing equations (7.119) and (7.133), we obtain

$$\left(\widetilde{\delta\widehat{\mathbf{F}}\delta N_M}\right)\left(\omega, \mathbf{k}; \widehat{\mathbf{x}}, \widehat{\mathbf{p}}^{(k)}, \theta\right) = -\frac{in\mathbf{k}\cdot\widehat{\nabla}_{\mathbf{p}} f(\widehat{\mathbf{x}};\theta)}{\omega - \mathbf{k}\cdot\widetilde{\mathbf{v}} + i\Delta}\mathbf{k}\left(\widetilde{\delta\widehat{F}\delta\widehat{F}}\right)_{\omega,\mathbf{k}}$$

$$-\frac{ie^2 n\mathbf{k}}{\epsilon_0\widetilde{\varepsilon}_b\beta_s^2\gamma_s^2 E_s(k^2 - \beta_s^2 k_\sigma^2)} f(\widehat{\mathbf{x}};\theta) 2\pi\delta(\omega - \mathbf{k}\cdot\widetilde{\mathbf{v}}),$$

where

$$\left(\widetilde{\delta\widehat{F}\delta\widehat{F}}\right)_{\omega,\mathbf{k}} = \frac{e^4 n}{\epsilon_0^2 |\widetilde{\varepsilon}_b|^2 \beta_s^4 \gamma_s^4 E_s^2 (k^2 - \beta_s^2 k_\sigma^2)^2} \int \mathrm{d}^3\widehat{\mathbf{p}}^{(k)} f(\widehat{\mathbf{x}};\theta) 2\pi\delta(\omega - \mathbf{k}\cdot\widetilde{\mathbf{v}}). \tag{7.139}$$

In equation (7.138) representing the collision integral, only the real part of $\left(\widetilde{\delta\widehat{\mathbf{F}}\delta N_M}\right)(\omega,\mathbf{k};\widehat{\mathbf{x}},\widehat{\mathbf{p}}^{(k)},\theta)$ enters. Therefore, the expression to be substituted back into (7.138) reads as

$$\begin{aligned}
&\mathrm{Re}\left(\widetilde{\delta\widehat{\mathbf{F}}\delta N_M}\right)\left(\omega,\mathbf{k};\widehat{\mathbf{x}},\widehat{\mathbf{p}}^{(k)},\theta\right) \\
&= -\pi n\delta(\omega - \mathbf{k}\cdot\widetilde{\mathbf{v}})\mathbf{k}\left(\widetilde{\delta\widehat{F}\delta\widehat{F}}\right)_{\omega,\mathbf{k}} \mathbf{k}\cdot\widehat{\nabla}_{\mathbf{p}} f(\widehat{\mathbf{x}};\theta) \\
&- \frac{2\pi e^2 n\mathbf{k}}{\epsilon_0 \beta_s^2\gamma_s^2 E_s (k^2 - \beta_s^2 k_\sigma^2)} \delta(\omega - \mathbf{k}\cdot\widetilde{\mathbf{v}}) \frac{\mathrm{Im}\widetilde{\varepsilon}_b(\omega,\mathbf{k})}{|\widetilde{\varepsilon}_b(\omega,\mathbf{k})|^2} f(\widehat{\mathbf{x}};\theta),
\end{aligned} \tag{7.140}$$

where

$$\mathrm{Im}\widetilde{\varepsilon}_b(\omega,\mathbf{k}) = -\frac{\pi e^2 n}{\epsilon_0 \beta_s^2 \gamma_s^2 E_s (k^2 - \beta_s^2 k_\sigma^2)} \int \mathrm{d}^3\widehat{\mathbf{p}}^{(k)} \delta(\omega - \mathbf{k}\cdot\widetilde{\mathbf{v}})\mathbf{k}\cdot\widehat{\nabla}_{\mathbf{p}} f(\widehat{\mathbf{x}};\theta). \tag{7.141}$$

Finally, the collision integral (7.138) can be written in the form of *Balescu-Lenard* as

$$\begin{aligned}
\mathcal{J}_{col}^{(BL)}\left(\widehat{\mathbf{x}},\widehat{\mathbf{p}}^{(k)};\theta\right) &= \frac{\pi e^4 n}{\epsilon_0^2 \beta_s^4 \gamma_s^4 E_s^2} \widehat{\nabla}_{\mathbf{p}} \cdot \int \frac{\mathrm{d}^3\mathbf{k}\mathrm{d}^3\widehat{\mathbf{p}}_1^{(k)}}{(2\pi)^3} \delta(\mathbf{k}\cdot\widetilde{\mathbf{v}} - \mathbf{k}\cdot\widetilde{\mathbf{v}}_1) \\
&\times \frac{\mathbf{k}\mathbf{k}}{|\widetilde{\varepsilon}_b(\mathbf{k}\cdot\widetilde{\mathbf{v}},\mathbf{k})|^2 (k^2 - \beta_s^2 k_\sigma^2)^2} \\
&\cdot \left[f\left(\widehat{\mathbf{x}},\widehat{\mathbf{p}}_1^{(k)};\theta\right) \widehat{\nabla}_{\mathbf{p}} f\left(\widehat{\mathbf{x}},\widehat{\mathbf{p}}^{(k)};\theta\right) - f\left(\widehat{\mathbf{x}},\widehat{\mathbf{p}}^{(k)};\theta\right) \widehat{\nabla}_{\mathbf{p}_1} f\left(\widehat{\mathbf{x}},\widehat{\mathbf{p}}_1^{(k)};\theta\right) \right].
\end{aligned} \tag{7.142}$$

The collision integral specified by equation (7.138) can be expressed in an equivalent form of a nonlinear *Fokker-Planck* operator according to

$$\begin{aligned}
\mathcal{J}_{col}^{(BL)}\left(\widehat{\mathbf{x}},\widehat{\mathbf{p}}^{(k)};\theta\right) &= \widehat{\nabla}_{\mathbf{p}} \cdot \left[\widehat{\mathcal{D}}^{(BL)} \cdot \widehat{\nabla}_{\mathbf{p}} f\left(\widehat{\mathbf{x}},\widehat{\mathbf{p}}^{(k)};\theta\right) \right] \\
&+ \widehat{\nabla}_{\mathbf{p}} \cdot \left[\mathbf{A}^{(BL)} f\left(\widehat{\mathbf{x}},\widehat{\mathbf{p}}^{(k)};\theta\right) \right],
\end{aligned} \tag{7.143}$$

where the drift and the diffusion coefficients

$$\widehat{\mathcal{D}}^{(BL)} = \pi \int \frac{\mathrm{d}\omega \mathrm{d}^3\mathbf{k}}{(2\pi)^4} \delta(\omega - \mathbf{k} \cdot \widetilde{\mathbf{v}}) \mathbf{k} \left(\widetilde{\delta \widehat{F} \delta \widehat{F}} \right)_{\omega, \mathbf{k}} \mathbf{k}, \tag{7.144}$$

$$\mathbf{A}^{(BL)} = \frac{e^2}{\epsilon_0 \beta_s^2 \gamma_s^2 E_s} \int \frac{\mathrm{d}\omega \mathrm{d}^3\mathbf{k}}{(2\pi)^4} \delta(\omega - \mathbf{k} \cdot \widetilde{\mathbf{v}}) \frac{\mathbf{k}}{k^2 - \beta_s^2 k_\sigma^2} \frac{\mathrm{Im}\widetilde{\varepsilon}_b(\omega, \mathbf{k})}{|\widetilde{\varepsilon}_b(\omega, \mathbf{k})|^2}, \tag{7.145}$$

depend on the distribution function itself.

7.8 The Landau Kinetic Equation

The dielectric function (7.132) depends on the distribution function and consequently the corresponding kinetic equation with the collision term in the form of *Balescu-Lenard* is extremely complicated to solve. Therefore, we must seek for reasonable ways to further simplify the kinetic description of charged particle beams. A considerable simplification can be achieved if the collision integral is transformed in the form of *Landau*.

In order to obtain the collision integral in the form of *Landau*, we set $\widetilde{\varepsilon}_b(\mathbf{k} \cdot \widetilde{\mathbf{v}}, \mathbf{k}) = 1$ in equation (7.142) and simultaneously take into account the effect of polarization by altering the domain of integration on k for small k. As far as the large values of k are concerned, the upper limit of integration can be obtained from the condition that perturbation treatment is amenable. To proceed further, it is convenient to change variables in the *Balescu-Lenard* kinetic equation according to

$$\widehat{\eta}^{(k)} \longrightarrow \mathrm{sign}(\mathcal{K}) \frac{\widehat{\eta}^{(k)}}{\sqrt{|\mathcal{K}|}}, \qquad k_\sigma \longrightarrow \frac{k_\sigma}{\sqrt{|\mathcal{K}|}}.$$

This means that the canonical coordinate σ has been transformed according to $\sigma \longrightarrow \sigma\sqrt{|\mathcal{K}|}$, and in order to retain the Hamiltonian structure of the microscopic equations of motion the σ-component of the force should also be transformed as $\widehat{F}_\sigma \longrightarrow -\mathrm{sign}(\mathcal{K})\widehat{F}_\sigma$. Taking into account the fact that the *Balescu-Lenard* collision integral is proportional to the square of the fluctuating force, we can write

$$\mathcal{J}_{col}^{(BL)}\left(\widehat{\mathbf{x}}, \widehat{\mathbf{p}}^{(k)}; \theta\right) = \frac{\pi e^4 n}{\epsilon_0^2 \beta_s^4 \gamma_s^4 E_s^2} \widehat{\nabla}_{\mathbf{p}} \cdot \int \frac{\mathrm{d}^3\mathbf{k} \mathrm{d}^3 \widehat{\mathbf{p}}_1^{(k)}}{(2\pi)^3} \delta\left(\mathbf{k} \cdot \widetilde{\mathbf{p}}^{(k)} - \mathbf{k} \cdot \widetilde{\mathbf{p}}_1^{(k)}\right)$$

$$\times \frac{\mathbf{k}\mathbf{k}}{\left|\widetilde{\varepsilon}_b\left(\mathbf{k}\cdot\widetilde{\mathbf{p}}^{(k)},\mathbf{k}\right)\right|^2\left(k_x^2+k_z^2+k_\sigma^2/\gamma_s^2|\mathcal{K}|\right)^2}$$

$$\cdot\left[f\left(\widehat{\mathbf{x}},\widehat{\mathbf{p}}_1^{(k)};\theta\right)\widehat{\nabla}_{\mathbf{p}}f\left(\widehat{\mathbf{x}},\widehat{\mathbf{p}}^{(k)};\theta\right)-f\left(\widehat{\mathbf{x}},\widehat{\mathbf{p}}^{(k)};\theta\right)\widehat{\nabla}_{\mathbf{p}_1}f\left(\widehat{\mathbf{x}},\widehat{\mathbf{p}}_1^{(k)};\theta\right)\right]. \quad (7.146)$$

where

$$\widetilde{\mathbf{p}}^{(k)}=\left(R\widetilde{p}_x^{(k)},R\widetilde{p}_z^{(k)},R\widetilde{\eta}^{(k)}\right).$$

Handling the integral

$$\widehat{\mathbf{I}}_L(\mathbf{g})=\int \mathrm{d}^3\mathbf{k}\frac{\mathbf{k}\mathbf{k}}{\left(k_x^2+k_z^2+k_\sigma^2/\gamma_s^2|\mathcal{K}|\right)^2}\delta(\mathbf{k}\cdot\mathbf{g}), \qquad \mathbf{g}=\widetilde{\mathbf{p}}^{(k)}-\widetilde{\mathbf{p}}_1^{(k)}, \quad (7.147)$$

by choosing a reference frame in which the vector $\mathbf{g}$ points along the σ-axis, and by using cylindrical coordinates in this frame, we find

$$\widehat{\mathbf{I}}_L(\mathbf{g})=\int_0^\infty \mathrm{d}k_\perp k_\perp\int_0^{2\pi}\mathrm{d}\Phi\int_{-\infty}^{\infty}\mathrm{d}k_\sigma\delta(k_\sigma g)\frac{1}{\left(k_\perp^2+k_\sigma^2/\gamma_s^2|\mathcal{K}|\right)^2}$$

$$\times\begin{pmatrix}k_\perp\cos\Phi\\ k_\perp\sin\Phi\\ k_\sigma\end{pmatrix}\begin{pmatrix}k_\perp\cos\Phi\\ k_\perp\sin\Phi\\ k_\sigma\end{pmatrix}=\frac{\pi}{g}\left(\widehat{\mathbf{I}}-\mathbf{e}_s\mathbf{e}_s\right)\int_{k_D}^{k_L}\frac{\mathrm{d}k_\perp}{k_\perp}=\frac{\pi\mathcal{L}}{g}\left(\widehat{\mathbf{I}}-\mathbf{e}_s\mathbf{e}_s\right). \quad (7.148)$$

As was mentioned above, in order to avoid logarithmic divergences at both limits of integration on $k_\perp$ in equation (7.148), we have altered these limits according to

$$k_D=\frac{1}{\gamma_s r_D}, \qquad k_L=\frac{4\pi\epsilon_0 k_B T}{e^2\gamma_s}. \quad (7.149)$$

Therefore, for the *Coulomb logarithm* $\mathcal{L}$ entering the *Landau* integral in equation (7.148), we obtain

$$\mathcal{L}=\ln\left[\frac{4\pi}{e^3\sqrt{n}}\left(\epsilon_0 k_B T\right)^{3/2}\right]. \quad (7.150)$$

The tensor $\widehat{\mathbf{I}}_L(\mathbf{g})$ can be evaluated in an arbitrary frame. This gives

$$\widehat{\mathbf{I}}_L(\mathbf{g})=\frac{\pi\mathcal{L}}{g}\left(\widehat{\mathbf{I}}-\frac{\mathbf{g}\mathbf{g}}{g^2}\right). \quad (7.151)$$

Finally, the collision integral (7.146) can be represented in the form of *Landau* as

$$\mathcal{J}_{col}^{(L)}\left(\widehat{\mathbf{x}}, \widehat{\mathbf{p}}^{(k)}; \theta\right) = \frac{e^4 n \mathcal{L}}{8\pi\epsilon_0^2 \beta_s^4 \gamma_s^4 E_s^2} \widehat{\nabla}_{\mathbf{p}} \cdot \int \mathrm{d}^3 \widehat{\mathbf{p}}_1^{(k)} \widehat{\mathbf{G}}_L(\mathbf{g})$$

$$\cdot \left[f\left(\widehat{\mathbf{x}}, \widehat{\mathbf{p}}_1^{(k)}; \theta\right) \widehat{\nabla}_{\mathbf{p}} f\left(\widehat{\mathbf{x}}, \widehat{\mathbf{p}}^{(k)}; \theta\right) - f\left(\widehat{\mathbf{x}}, \widehat{\mathbf{p}}^{(k)}; \theta\right) \widehat{\nabla}_{\mathbf{p}_1} f\left(\widehat{\mathbf{x}}, \widehat{\mathbf{p}}_1^{(k)}; \theta\right) \right], \quad (7.152)$$

where

$$\widehat{\mathbf{G}}_L(\mathbf{g}) = \frac{1}{g} \left(\widehat{\mathbf{I}} - \frac{\mathbf{g}\mathbf{g}}{g^2} \right). \quad (7.153)$$

is the *Landau* tensor [Balescu (1988)].

7.9 The Approximate Collision Integral and the Generalized Kinetic Equation

The local equilibrium state is defined as a solution to the equation

$$\mathcal{J}_{col}\left(\widehat{\mathbf{x}}, \widehat{\mathbf{p}}^{(k)}; \theta\right) = 0, \quad (7.154)$$

where the collision integral is taken either in the *Balescu-Lenard* or *Landau* form. This solution is well known to be the Maxwellian distribution

$$f_q\left(\widehat{\mathbf{x}}, \widehat{\mathbf{p}}^{(k)}; \theta\right) = \rho \left(\frac{1}{2\pi\epsilon} \right)^{3/2} \exp\left[-\frac{1}{2\epsilon} \left(\widehat{\mathbf{p}}^{(k)} - \mathbf{u} \right)^2 \right], \quad (7.155)$$

$$\int \mathrm{d}^3 \widehat{\mathbf{x}} \rho(\widehat{\mathbf{x}}; \theta) = V,$$

where $\rho(\widehat{\mathbf{x}}; \theta)$, $\epsilon(\widehat{\mathbf{x}}; \theta)$ and $u(\widehat{\mathbf{x}}; \theta)$ are functions of $\widehat{\mathbf{x}}$ and θ. It should be clear that the local equilibrium state is not a true thermodynamic equilibrium state, since the latter must be homogeneous and stationary. To prove that the distribution (7.155) is a solution of equation (7.154) when the collision integral is taken in the *Landau* form, it suffices to take into account the obvious identity

$$\widehat{\mathbf{G}}_L(\mathbf{a}) \cdot \mathbf{a} = \mathbf{a}^T \cdot \widehat{\mathbf{G}}_L(\mathbf{a}) = 0. \quad (7.156)$$

Next , we note that the *Landau* collision integral can be written as a non-linear *Fokker-Planck* operator

$$\mathcal{J}_{col}^{(L)}\left(\widehat{\mathbf{x}},\widehat{\mathbf{p}}^{(k)};\theta\right)=\mathcal{B}\left[\widehat{\nabla}_{\mathbf{p}}\cdot\left(\widehat{\mathcal{D}}\cdot\widehat{\nabla}_{\mathbf{p}}\right)-\widehat{\nabla}_{\mathbf{p}}\cdot\mathbf{A}\right]f\left(\widehat{\mathbf{x}},\widehat{\mathbf{p}}^{(k)};\theta\right), \tag{7.157}$$

where

$$\mathcal{B}=\frac{\epsilon^2\mathcal{L}}{8\pi n r_D^4\gamma_s^4}, \tag{7.158}$$

$$\widehat{\mathcal{D}}=\int \mathrm{d}^3\widehat{\mathbf{p}}_1^{(k)}\widehat{\mathbf{G}}_L(\mathbf{g})f\left(\widehat{\mathbf{p}}_1^{(k)}\right), \qquad \mathbf{A}=\int \mathrm{d}^3\widehat{\mathbf{p}}_1^{(k)}\widehat{\mathbf{G}}_L(\mathbf{g})\cdot\widehat{\nabla}_{\mathbf{p}_1}f\left(\widehat{\mathbf{p}}_1^{(k)}\right). \tag{7.159}$$

Our goal in what follows is to match the transition to the unified kinetic equation. To approach this, it is sufficient to compute the drift and diffusion coefficients (7.159) using the local equilibrium distribution (7.155). The interested reader is referred to [Balescu (1988)], where more systematic approximation methods using the linearized *Landau* collision integral can be found. Going over to the new variable

$$\mathbf{C}=\sqrt{\frac{1}{2\epsilon}}\left(\widehat{\mathbf{p}}^{(k)}-\mathbf{u}\right)=\sqrt{\frac{1}{2\epsilon}}\delta\mathbf{p}, \tag{7.160}$$

we write

$$\widehat{\mathcal{D}}(\mathbf{C})=2\epsilon\int \mathrm{d}^3\mathbf{C}_1\widehat{\mathbf{G}}_L(\mathbf{g})f_q(\mathbf{C}_1)=\mathcal{A}_0(C)\widehat{\mathbf{G}}_L(\delta\widehat{\mathbf{p}})+\mathcal{A}_1(C)\frac{\delta\widehat{\mathbf{p}}\delta\widehat{\mathbf{p}}}{(\delta\widehat{p})^4}, \tag{7.161}$$

where $\mathbf{g}=\mathbf{C}-\mathbf{C}_1$ and $\mathcal{A}_0(C)$, $\mathcal{A}_1(C)$ are functions of the modulus of the vector $\mathbf{C}$, and are represented by the following expressions

$$\mathcal{A}_1(C)=\delta\widehat{\mathbf{p}}\cdot\widehat{\mathcal{D}}\cdot\delta\widehat{\mathbf{p}}=2\epsilon\int \mathrm{d}^3\mathbf{C}_1\delta\widehat{\mathbf{p}}\cdot\widehat{\mathbf{G}}_L(\mathbf{g})\cdot\delta\widehat{\mathbf{p}}f_q(\mathbf{C}_1), \tag{7.162}$$

$$\mathcal{A}_0(C)=\frac{\delta\widehat{p}}{2}\mathrm{Sp}\left(\widehat{\mathcal{D}}\right)-\frac{1}{2\delta\widehat{p}}\mathcal{A}_1(C), \tag{7.163}$$

$$\mathrm{Sp}\left(\widehat{\mathcal{D}}\right)=4\epsilon\int \mathrm{d}^3\mathbf{C}_1\frac{f_q(\mathbf{C}_1)}{|\mathbf{C}-\mathbf{C}_1|}. \tag{7.164}$$

In order to calculate the integrals (7.162) and (7.164), we use spherical coordinates in a reference frame in which the vector $\mathbf{C}$ point along the

σ-axis. We find

$$\mathrm{Sp}\left(\widehat{\mathcal{D}}\right) = \frac{2\rho}{\pi^{3/2}}\sqrt{\frac{1}{2\epsilon}}\int\limits_0^\infty \mathrm{d}C_1 C_1^2 \int\limits_0^{2\pi} \mathrm{d}\Phi \int\limits_{-1}^{1} \mathrm{d}\cos\Theta \frac{e^{-C_1^2}}{g},$$

$$\mathcal{A}_1(C) = \frac{\rho\sqrt{2\epsilon}}{\pi^{3/2}}\int\limits_0^\infty \mathrm{d}C_1 C_1^2 \int\limits_0^{2\pi} \mathrm{d}\Phi \int\limits_{-1}^{1} \mathrm{d}\cos\Theta \frac{e^{-C_1^2}}{g}\left[C^2 - \frac{\left(C^2 - CC_1\cos\Theta\right)^2}{g^2}\right],$$

where we have used $\mathbf{g}\cdot\mathbf{C} = C^2 - CC_1\cos\Theta$. The change of variables in the above integrals according to

$$g^2 = C^2 + C_1^2 - 2CC_1\cos\Theta, \qquad \mathrm{d}\cos\Theta = -\frac{g}{CC_1}\mathrm{d}g,$$

yields the result

$$\mathrm{Sp}\left(\widehat{\mathcal{D}}\right) = \frac{4\rho}{C}\sqrt{\frac{1}{2\pi\epsilon}}\int\limits_0^\infty \mathrm{d}C_1 C_1 e^{-C_1^2}\int\limits_{|C-C_1|}^{C+C_1} \mathrm{d}g$$

$$= \frac{8\rho}{C}\sqrt{\frac{1}{2\pi\epsilon}}\left[\int\limits_0^C \mathrm{d}C_1 C_1^2 e^{-C_1^2} + C\int\limits_C^\infty \mathrm{d}C_1 C_1 e^{-C_1^2}\right] = \frac{2\rho}{C}\sqrt{\frac{1}{2\epsilon}}\mathrm{erf}(C),$$

and similarly

$$\mathcal{A}_1(C) = \frac{2\rho}{C}\sqrt{\frac{2\epsilon}{\pi}}\int\limits_0^\infty \mathrm{d}C_1 C_1 e^{-C_1^2}\int\limits_{|C-C_1|}^{C+C_1} \mathrm{d}g\left[C^2 - \frac{\left(g^2 - C_1^2 + C^2\right)^2}{4g^2}\right]$$

$$= \frac{8\rho}{3C}\sqrt{\frac{2\epsilon}{\pi}}\left[\int\limits_0^C \mathrm{d}C_1 C_1^4 e^{-C_1^2} + C^3\int\limits_C^\infty \mathrm{d}C_1 C_1 e^{-C_1^2}\right]$$

$$= \frac{\rho}{C}\sqrt{2\epsilon}\left(1 - C\frac{\mathrm{d}}{\mathrm{d}C}\right)\mathrm{erf}(C),$$

where $\mathrm{erf}(z)$ is the error function. Thus, for the coefficients $\mathcal{A}_0$, and $\mathcal{A}_1$ in equations (7.162) and (7.163), we have

$$\mathcal{A}_0(C) = \rho\mathcal{C}(C), \qquad \mathcal{A}_1(C) = \frac{\rho}{C}\sqrt{2\epsilon}\left(1 - C\frac{\mathrm{d}}{\mathrm{d}C}\right)\mathrm{erf}(C), \tag{7.165}$$

where

$$\mathcal{C}(C) = \frac{1}{2C^2}\left(2C^2 - 1 + C\frac{\mathrm{d}}{\mathrm{d}C}\right)\mathrm{erf}(C) \tag{7.166}$$

is the *Chandrasekhar* function. The drift vector can be written as

$$\mathbf{A}(\delta\widehat{\mathbf{p}}) = -\frac{1}{\epsilon}\widehat{\mathcal{D}} \cdot \delta\widehat{\mathbf{p}} = -\frac{1}{\epsilon}\mathcal{A}_1(C)\frac{\delta\widehat{\mathbf{p}}}{(\delta\widehat{p})^2}. \tag{7.167}$$

The drift and the diffusion coefficients can be further evaluated by replacing $\delta\widehat{\mathbf{p}}$ with the r.m.s. value

$$(\delta\widehat{p}_i)_{rms} = \sqrt{\epsilon(\widehat{\mathbf{x}};\theta)}, \qquad (C_i)_{rms} = \frac{1}{\sqrt{2}}. \tag{7.168}$$

Thus, we obtain

$$\widehat{\mathcal{D}} = D\widehat{\mathbf{I}}, \qquad \mathbf{A} = -\frac{D}{\epsilon}\delta\widehat{\mathbf{p}}, \tag{7.169}$$

where

$$D = \frac{1}{3}\mathrm{Sp}\left(\widehat{\mathcal{D}}\right) = \frac{2\mathrm{erf}\left(\sqrt{3/2}\right)}{(3\epsilon)^{3/2}}\epsilon(\widehat{\mathbf{x}};\theta), \tag{7.170}$$

for $\rho(\widehat{\mathbf{x}};\theta) \sim 1$. This enables us to cast the collision integral (7.157) in the form

$$\mathcal{J}_{col}^{(L)}\left(\widehat{\mathbf{x}},\widehat{\mathbf{p}}^{(k)};\theta\right) = \frac{1}{\theta_{rel}}\left\{\epsilon(\widehat{\mathbf{x}};\theta)\widehat{\nabla}_{\mathbf{p}}^2 + \widehat{\nabla}_{\mathbf{p}} \cdot \left[\widehat{\mathbf{p}}^{(k)} - \mathbf{u}(\widehat{\mathbf{x}};\theta)\right]\right\}f\left(\widehat{\mathbf{x}},\widehat{\mathbf{p}}^{(k)};\theta\right), \tag{7.171}$$

where

$$\theta_{rel} = \frac{12\pi\sqrt{3}}{\mathrm{erf}\left(\sqrt{3/2}\right)}\frac{nr_D^4\gamma_s^4}{\mathcal{L}\sqrt{\epsilon}} \tag{7.172}$$

is the relaxation "time".

The transition to local equilibrium, that is the kinetic stage of relaxation, is described by the *Balescu-Lenard* or the *Landau* kinetic equation. The latter with due account of the approximate collision integral (7.171) can be written as

$$\frac{\partial f}{\partial\theta} + \left(\widehat{\mathbf{p}}^{(k)} \cdot \widehat{\nabla}_{\mathbf{x}}\right)f + \left(\widehat{\mathbf{F}} \cdot \widehat{\nabla}_{\mathbf{p}}\right)f = \frac{1}{\theta_{rel}}\left[\epsilon\widehat{\nabla}_{\mathbf{p}}^2 + \widehat{\nabla}_{\mathbf{p}} \cdot \left(\widehat{\mathbf{p}}^{(k)} - \mathbf{u}\right)\right]f, \tag{7.173}$$

where

$$\widehat{\mathbf{F}} = \widehat{\mathbf{F}}_0 + \left\langle\widehat{\mathbf{F}}\right\rangle.$$

It is well-known [Gardiner (1983)] that the kinetic equation (7.173) is equivalent to the system of *Langevin* equations

$$\frac{\mathrm{d}\widehat{\mathbf{x}}}{\mathrm{d}\theta} = \widehat{\mathbf{p}}^{(k)}, \tag{7.174}$$

$$\frac{\mathrm{d}\widehat{\mathbf{p}}^{(k)}}{\mathrm{d}\theta} = -\frac{1}{\theta_{rel}}\left(\widehat{\mathbf{p}}^{(k)} - \mathbf{u}\right) + \widehat{\mathbf{F}} + \sqrt{\frac{\epsilon}{\theta_{rel}}}\mathbf{W}(\theta). \tag{7.175}$$

Here $\mathbf{W}(\theta)$ is a white-noise random variable with formal correlation properties

$$\langle \mathbf{W}(\theta) \rangle \equiv 0, \qquad \langle W_m(\theta) W_n(\theta_1) \rangle = 2\delta_{mn}\delta(\theta - \theta_1). \tag{7.176}$$

The generalized kinetic equation (7.71) describes the evolution of the beam for time scales greater than the relaxation time θ_{rel}. In order to determine the additional collision integral $\widetilde{\mathcal{J}}\left(\widehat{\mathbf{x}}, \widehat{\mathbf{p}}^{(k)}; \theta\right)$, we use the method of adiabatic elimination of fast variables, which in our case are the kinetic momenta $\widehat{\mathbf{p}}^{(k)}$. In the limit of small times θ_{rel} (compared to the time scale of physical interest), equation (7.175) describes the sufficiently fast relaxation towards the quasi-stationary (local equilibrium) state for which $\mathrm{d}\widehat{\mathbf{p}}^{(k)}/\mathrm{d}\theta \to 0$. Thus, we find

$$\widehat{\mathbf{p}}^{(k)} = \mathbf{u} + \theta_{rel}\widehat{\mathbf{F}} + \sqrt{\epsilon\theta_{rel}}\mathbf{W}(\theta). \tag{7.177}$$

Substituting this into equation (7.174), we arrive at

$$\frac{\mathrm{d}\widehat{\mathbf{x}}}{\mathrm{d}\theta} = \mathbf{u} + \theta_{rel}\widehat{\mathbf{F}} + \sqrt{\epsilon\theta_{rel}}\mathbf{W}(\theta). \tag{7.178}$$

The above equation (7.178) governs the evolution of particles within the elementary cell of continuous medium, where local equilibrium state is established. Such a coarse-graining procedure gives rise to the additional collision integral in the generalized kinetic equation (7.71). The latter follows in a straightforward manner from equation (7.178), and can be written in the form

$$\widetilde{\mathcal{J}}\left(\widehat{\mathbf{x}}, \widehat{\mathbf{p}}^{(k)}; \theta\right) = \theta_{rel}\left\{\widehat{\nabla}_{\mathbf{x}} \cdot \left[\epsilon(\widehat{\mathbf{x}}; \theta)\widehat{\nabla}_{\mathbf{x}}\right] - \left(\widehat{\nabla}_{\mathbf{x}} \cdot \widehat{\mathbf{F}}\right)\right\} f\left(\widehat{\mathbf{x}}, \widehat{\mathbf{p}}^{(k)}; \theta\right). \tag{7.179}$$

We have studied the role of electromagnetic interactions between particles on the evolution of a high energy beam. The interparticle forces that have been taken into account are due to space charge alone. Starting from the reversible dynamics of individual particles and applying a smoothing procedure over the physically infinitesimal spatial scales, we derived a generalized kinetic equation capable to describe in a unified manner the kinetic, hydrodynamic and diffusion processes.

It should be particularly emphasized that the irreversibility of the beam evolution has been introduced at the very beginning in the initial equation (7.68) for the microscopic phase space density with a small source. Smoothing destroys information about the motion of individual particles within the unit cell of continuous medium, hence the reversible description becomes no longer feasible. Details of particle dynamics is being lost and motion smears out due to dynamic instability and to the resulting mixing of trajectories in phase space.

Chapter 8

Statistical Description of Non Integrable Hamiltonian Systems

8.1 Introduction

It is well-known that dynamical systems may exhibit irregular motion in certain regions of phase space [Chirikov (1979)], [Lichtenberg (1982)]. These regions differ in size, from being considerably small, to occupying large parts of the phase space. This depends mostly on the strength of the perturbation, as well as on the intrinsic characteristics of the system. It was already demonstrated in previous chapters that for comparatively small perturbations the regularity of the motion is expressed in the existence of adiabatic action invariants. In the course of nonlinear interaction the action invariants vary within a certain range, prescribed by the integrals of motion (if such exist). For chaotic systems some (or all) of the integrals of motion are destroyed, causing specific trajectories to become extremely complicated. These trajectories look random in their behavior, therefore it is natural to explore the statistical properties of chaotic dynamical systems.

Much experimental and theoretical evidence [see e.g. the contributions in the proceedings [Tzenov (1995)] and [Tzenov (1997b)]] of nonlinear effects characterizing the dynamics of particles in accelerators and storage rings is available at present. An individual particle propagating in an accelerator experiences growth of amplitude of betatron oscillations in a plane transverse to the particle orbit, whenever a perturbing force acts on it. This force may be of various origin, for instance high order multipole magnetic field errors, space charge forces, beam-beam interaction force, power supply ripple or other external and collective forces. Therefore, the Hamiltonian governing the motion of a beam particle is far from being integrable, and an irregular behavior of the beam is clearly observed, especially for a large number of revolutions.

The idea to treat the evolution of chaotic dynamical systems in a statistical sense is not new; many rigorous results related to the statistical properties of such systems can be found in the book by *Arnold* and *Avez* [Arnold (1968)]. Many details concerning the transport phenomena in the space of adiabatic action invariants only can be found also in the book by *Lichtenberg* and *Lieberman* [Lichtenberg (1982)].

8.2 Projection Operator Method

In chapter 2, we elaborated the details [see equations (2.64) and (2.70)] and in the subsequent chapters, we used the fact that single particle dynamics in cyclic accelerators and storage rings is most properly described by the adiabatic action invariants (Courant-Snyder invariants [Courant (1958)]) and the canonically conjugate to them angle variables. However, to be more general, we consider again (as was done in chapter 3 and later) a dynamical system with N degrees of freedom, governed by the Hamiltonian written in action-angle variables $(\mathbf{J}, \boldsymbol{\alpha})$ as

$$H(\boldsymbol{\alpha}, \mathbf{J}; \theta) = H_0(\mathbf{J}) + \epsilon V(\boldsymbol{\alpha}, \mathbf{J}; \theta), \tag{8.1}$$

where θ is the independent azimuthal variable, playing the role of time and $\mathbf{J}$ and $\boldsymbol{\alpha}$ are N-dimensional vectors

$$\mathbf{J} = (J_1, J_2, \ldots, J_N), \qquad \boldsymbol{\alpha} = (\alpha_1, \alpha_2, \ldots, \alpha_N). \tag{8.2}$$

Moreover $H_0(\mathbf{J})$ is the integrable part of the Hamiltonian, ϵ is a formal small parameter, while $V(\boldsymbol{\alpha}, \mathbf{J}; \theta)$ is the perturbation periodic in the angle variables

$$V(\boldsymbol{\alpha}, \mathbf{J}; \theta) = {\sum_{\mathbf{m}}}' V_{\mathbf{m}}(\mathbf{J}; \theta) \exp{(i\mathbf{m} \cdot \boldsymbol{\alpha})}, \tag{8.3}$$

where $\sum'$ denotes exclusion of the harmonic $\mathbf{m} = (0, 0, \ldots, 0)$ from the above sum. The *Hamilton*'s equations of motion are

$$\frac{\mathrm{d}\alpha_k}{\mathrm{d}\theta} = \omega_{0k}(\mathbf{J}) + \epsilon \frac{\partial V}{\partial J_k}, \qquad \frac{\mathrm{d}J_k}{\mathrm{d}\theta} = -\epsilon \frac{\partial V}{\partial \alpha_k}, \tag{8.4}$$

where

$$\omega_{0k}(\mathbf{J}) = \frac{\partial H_0}{\partial J_k}. \tag{8.5}$$

In what follows, we will assume that the nonlinearity coefficients

$$\gamma_{kl}(\mathbf{J}) = \frac{\partial^2 H_0}{\partial J_k \partial J_l} \tag{8.6}$$

are small and can be neglected. The *Liouville* equation governing the evolution of the phase space density $P(\boldsymbol{\alpha}, \mathbf{J}; \theta)$ can be written as

$$\frac{\partial}{\partial \theta} P(\boldsymbol{\alpha}, \mathbf{J}; \theta) = \left[\widehat{\mathcal{L}}_0 + \epsilon \widehat{\mathcal{L}}_v(\theta) \right] P(\boldsymbol{\alpha}, \mathbf{J}; \theta). \tag{8.7}$$

Here the operators $\widehat{\mathcal{L}}_0$ and $\widehat{\mathcal{L}}_v$ are given by the expressions

$$\widehat{\mathcal{L}}_0 = -\omega_{0k}(\mathbf{J}) \frac{\partial}{\partial \alpha_k}, \qquad \widehat{\mathcal{L}}_v = \frac{\partial V}{\partial \alpha_k} \frac{\partial}{\partial J_k} - \frac{\partial V}{\partial J_k} \frac{\partial}{\partial \alpha_k}, \tag{8.8}$$

where summation over repeated indices is implied. It is convenient and effective to apply the *projection operator method* first introduced by *Zwanzig* [Zwanzig (1961)]. Let us define the projection operator onto the subspace of action variables by the following integral

$$\widehat{\mathcal{P}} f(\mathbf{J}; \theta) = \frac{1}{(2\pi)^N} \int_0^{2\pi} \mathrm{d}\alpha_1 \ldots \int_0^{2\pi} \mathrm{d}\alpha_N f(\boldsymbol{\alpha}, \mathbf{J}; \theta), \tag{8.9}$$

where $f(\boldsymbol{\alpha}, \mathbf{J}; \theta)$ is a generic function of its arguments. We introduce also the functions

$$F = \widehat{\mathcal{P}} P, \qquad G = \left(1 - \widehat{\mathcal{P}}\right) P = \widehat{\mathcal{C}} P, \qquad (P = F + G). \tag{8.10}$$

From equation (8.7) with the obvious relations

$$\widehat{\mathcal{P}} \widehat{\mathcal{L}}_0 = \widehat{\mathcal{L}}_0 \widehat{\mathcal{P}} \equiv 0, \qquad \widehat{\mathcal{P}} \widehat{\mathcal{L}}_v \widehat{\mathcal{P}} \equiv 0 \tag{8.11}$$

in hand, it is straightforward to obtain the equations

$$\frac{\partial F}{\partial \theta} = \epsilon \widehat{\mathcal{P}} \widehat{\mathcal{L}}_v G = \epsilon \frac{\partial}{\partial J_k} \widehat{\mathcal{P}} \left(\frac{\partial V}{\partial \alpha_k} G \right), \tag{8.12}$$

$$\frac{\partial G}{\partial \theta} = \widehat{\mathcal{L}}_0 G + \epsilon \widehat{\mathcal{C}} \widehat{\mathcal{L}}_v G + \epsilon \widehat{\mathcal{L}}_v F. \tag{8.13}$$

Our goal in the subsequent exposition is to analyze equations (8.12) and (8.13) using the RG method. It will prove efficient to eliminate the dependence on the angle variables in G and V by noting that the eigenfunctions of the operator $\widehat{\mathcal{L}}_0$ form a complete set, so that every function periodic in the angle variables can be expanded in this basis. Using Dirac's "bra-ket" notation, we write

$$|\mathbf{n}\rangle = \frac{1}{(2\pi)^{N/2}} \exp{(i\mathbf{n} \cdot \boldsymbol{\alpha})}, \qquad \langle\mathbf{n}| = \frac{1}{(2\pi)^{N/2}} \exp{(-i\mathbf{n} \cdot \boldsymbol{\alpha})}. \quad (8.14)$$

The projection operator $\widehat{\mathcal{P}}$ can be represented in the form [Sudbery (1986)]

$$\widehat{\mathcal{P}} = \widehat{\mathcal{P}}_{\mathbf{0}} = |\mathbf{0}\rangle\langle\mathbf{0}|. \quad (8.15)$$

One can also define a set of projection operators $\widehat{\mathcal{P}}_{\mathbf{n}}$ according to the expression [Sudbery (1986)]

$$\widehat{\mathcal{P}}_{\mathbf{n}} = |\mathbf{n}\rangle\langle\mathbf{n}|. \quad (8.16)$$

It is easy to check the completeness relation

$$\sum_{\mathbf{n}} \widehat{\mathcal{P}}_{\mathbf{n}} = 1, \quad (8.17)$$

from which and from equation (8.15) it follows that

$$\widehat{\mathcal{C}} = \sum_{\mathbf{n}\neq\mathbf{0}}{}' |\mathbf{n}\rangle\langle\mathbf{n}|. \quad (8.18)$$

Decomposing the quantities F, G and V in the basis (8.14) as

$$F = F(\mathbf{J};\theta)|\mathbf{0}\rangle, \quad (8.19)$$

$$G(\boldsymbol{\alpha}, \mathbf{J}; \theta) = \sum_{\mathbf{m}\neq\mathbf{0}}{}' G_{\mathbf{m}}(\mathbf{J};\theta)|\mathbf{m}\rangle, \quad (8.20)$$

$$V(\boldsymbol{\alpha}, \mathbf{J}; \theta) = \sum_{\mathbf{n}\neq\mathbf{0}}{}' V_{\mathbf{n}}(\mathbf{J};\theta)|\mathbf{n}\rangle, \quad (8.21)$$

from equations (8.12) and (8.13), we obtain

$$\frac{\partial F}{\partial \theta} = i\epsilon \frac{\partial}{\partial J_k}\left(\sum_{\mathbf{n}}{}' n_k V_{\mathbf{n}} G_{-\mathbf{n}}\right), \quad (8.22)$$

$$\frac{\partial G_{\mathbf{n}}}{\partial \theta} = -in_k\omega_{0k}G_{\mathbf{n}}$$

$$+i\epsilon\sum_{\mathbf{m}}{}' \left[n_k V_{\mathbf{n}-\mathbf{m}} \frac{\partial G_{\mathbf{m}}}{\partial J_k} - m_k \frac{\partial}{\partial J_k}(V_{\mathbf{n}-\mathbf{m}} G_{\mathbf{m}})\right] + i\epsilon n_k V_{\mathbf{n}} \frac{\partial F}{\partial J_k}. \quad (8.23)$$

Equations (8.22) and (8.23) comprise the starting point for the renormalization group analysis outlined in the next section. We are primarily interested in the long-time evolution of the original system governed by certain amplitude equations. These will turn out to be precisely the RG equations.

8.3 Renormalization Group Reduction of the Liouville Equation

In paragraph 5.7, we performed the renormalization group reduction of the *Hamilton*'s equations of motion. In the present paragraph, we consider the solution of equations (8.22) and (8.23) for small ϵ by means of the RG method. For that purpose, we perform again the naive perturbation expansion

$$F = F^{(0)} + \epsilon F^{(1)} + \epsilon^2 F^{(2)} + \cdots, \qquad G_{\mathbf{n}} = G_{\mathbf{n}}^{(0)} + \epsilon G_{\mathbf{n}}^{(1)} + \epsilon^2 G_{\mathbf{n}}^{(2)} + \ldots, \tag{8.24}$$

and substitute it into equations (8.22) and (8.23). The lowest order perturbation equations have the obvious solution

$$F^{(0)} = F_0(\mathbf{J}), \qquad G_{\mathbf{n}}^{(0)} = W_{\mathbf{n}}(\mathbf{J}) \exp\left(-i n_k \omega_{0k} \theta\right). \tag{8.25}$$

The first order perturbation equations read as:

$$\frac{\partial F^{(1)}}{\partial \theta} = i \frac{\partial}{\partial J_k} \left[\sum_{\mathbf{n}}{}' n_k V_{\mathbf{n}} W_{-\mathbf{n}} \exp\left(i n_l \omega_{0l} \theta\right) \right], \tag{8.26}$$

$$\frac{\partial G_{\mathbf{n}}^{(1)}}{\partial \theta} = -i n_k \omega_{0k} G_{\mathbf{n}}^{(1)} + i n_k V_{\mathbf{n}} \frac{\partial F_0}{\partial J_k}$$

$$+ i \sum_{\mathbf{m}}{}' \left[n_k V_{\mathbf{n}-\mathbf{m}} \frac{\partial W_{\mathbf{m}}}{\partial J_k} - m_k \frac{\partial}{\partial J_k} \left(V_{\mathbf{n}-\mathbf{m}} W_{\mathbf{m}} \right) \right] \exp\left(-i m_l \omega_{0l} \theta\right). \tag{8.27}$$

We again assume that the modes $V_{\mathbf{n}}(\mathbf{J}; \theta)$ are periodic in θ, so that they can be expanded in a Fourier series (5.127). If the original system (8.1) exhibits primary resonances of the form (5.128) in the case when ω_{0k} does not depend on the action variables, we can solve the first order perturbation equations (8.26) and (8.27). The result is as follows

$$F^{(1)} = i(\theta - \theta_0)\sum_{\mathbf{n}^{(R)}}'' n_k^{(R)} \frac{\partial}{\partial J_k}\left[V_{\mathbf{n}^{(R)}}\left(-\frac{n_l^{(R)}\omega_{0l}}{\nu_R}\right)W_{-\mathbf{n}^{(R)}}\right]$$

$$+\sum_{\mathbf{n}}'\sum_{\mu}' n_k \frac{\partial}{\partial J_k}\left[V_{\mathbf{n}}(\mu)W_{-\mathbf{n}}\right]\frac{\exp\left[i(n_l\omega_{0l} + \mu\nu_{\mathbf{n}})\theta\right]}{n_l\omega_{0l} + \mu\nu_{\mathbf{n}}}, \tag{8.28}$$

$$G_{\mathbf{n}}^{(1)} = \mathcal{G}_{\mathbf{n}} \exp\left(-in_k\omega_{0k}\theta\right), \tag{8.29}$$

where

$$\mathcal{G}_{\mathbf{n}} = i(\theta - \theta_0)\delta_{\mathbf{n}\mathbf{n}^{(R)}} n_k \frac{\partial F_0}{\partial J_k} V_{\mathbf{n}}\left(-\frac{n_l^{(R)}\omega_{0l}}{\nu_R}\right)$$

$$+i(\theta - \theta_0)\sum_{\mathbf{n}-\mathbf{n}^{(R)}}''\left\{n_k V_{\mathbf{n}^{(R)}}\left(-\frac{n_l^{(R)}\omega_{0l}}{\nu_R}\right)\frac{\partial W_{\mathbf{n}-\mathbf{n}^{(R)}}}{\partial J_k}\right.$$

$$\left.-\left(n_k - n_k^{(R)}\right)\frac{\partial}{\partial J_k}\left[V_{\mathbf{n}^{(R)}}\left(-\frac{n_l^{(R)}\omega_{0l}}{\nu_R}\right)W_{\mathbf{n}-\mathbf{n}^{(R)}}\right]\right\}$$

$$+n_k \frac{\partial F_0}{\partial J_k}\sum_{\mu}' V_{\mathbf{n}}(\mu)\frac{\exp\left[i(n_l\omega_{0l} + \mu\nu_{\mathbf{n}})\theta\right]}{n_l\omega_{0l} + \mu\nu_{\mathbf{n}}}$$

$$+\sum_{\mathbf{m}}'\sum_{\mu}'\left\{n_k V_{\mathbf{n}-\mathbf{m}}(\mu)\frac{\partial W_{\mathbf{m}}}{\partial J_k} - m_k \frac{\partial}{\partial J_k}\left[V_{\mathbf{n}-\mathbf{m}}(\mu)W_{\mathbf{m}}\right]\right\}$$

$$\times\frac{\exp\left\{i\left[(n_l - m_l)\omega_{0l} + \mu\nu_{\mathbf{n}-\mathbf{m}}\right]\theta\right\}}{(n_l - m_l)\omega_{0l} + \mu\nu_{\mathbf{n}-\mathbf{m}}}. \tag{8.30}$$

In the above expressions $\sum''$ denotes summation over all primary resonances (5.128). To obtain the desired RG equations, we proceed in the same way as in paragraph 5.7. The first order RG equations are

$$\frac{\partial F_0}{\partial \theta} = i\epsilon\sum_{\mathbf{n}^{(R)}}'' n_k^{(R)}\frac{\partial}{\partial J_k}\left[V_{\mathbf{n}^{(R)}}(\mathbf{J};\mu_R)W_{-\mathbf{n}^{(R)}}(\mathbf{J})\right], \tag{8.31}$$

$$\frac{\partial W_{\mathbf{n}}}{\partial \theta} = i\epsilon\delta_{\mathbf{n}\mathbf{n}^{(R)}} n_k \frac{\partial F_0}{\partial J_k} V_{\mathbf{n}}(\mathbf{J};\mu_R)$$

$$+i\epsilon \sum_{\mathbf{n}-\mathbf{n}^{(R)}}{}'' \Bigg\{ n_k V_{\mathbf{n}^{(R)}}(\mathbf{J}; \mu_R) \frac{\partial W_{\mathbf{n}-\mathbf{n}^{(R)}}}{\partial J_k}$$

$$- \left(n_k - n_k^{(R)} \right) \frac{\partial}{\partial J_k} \left[V_{\mathbf{n}^{(R)}}(\mathbf{J}; \mu_R) W_{\mathbf{n}-\mathbf{n}^{(R)}}(\mathbf{J}) \right] \Bigg\}, \tag{8.32}$$

where

$$\mu_R = -\frac{n_k^{(R)} \omega_{0k}}{\nu_R}. \tag{8.33}$$

Equations (8.31) and (8.32) describe the resonant mode coupling when strong primary resonances are present in the original system.

Let us now assume that the original system is far from resonances. Solving the second order perturbation equation for $F^{(2)}$ and $G_{\mathbf{n}}^{(2)}$

$$\frac{\partial F^{(2)}}{\partial \theta} = i \frac{\partial}{\partial J_k} \left(\sum_{\mathbf{n}}{}' n_k V_{\mathbf{n}} G_{-\mathbf{n}}^{(1)} \right), \tag{8.34}$$

$$\frac{\partial G_{\mathbf{n}}^{(2)}}{\partial \theta} + i n_k \omega_{0k} G_{\mathbf{n}}^{(2)} = i n_k V_{\mathbf{n}} \frac{\partial F^{(1)}}{\partial J_k}$$

$$+i \sum_{\mathbf{m}}{}' \left[n_k V_{\mathbf{n}-\mathbf{m}} \frac{\partial G_{\mathbf{m}}^{(1)}}{\partial J_k} - m_k \frac{\partial}{\partial J_k} \left(V_{\mathbf{n}-\mathbf{m}} G_{\mathbf{m}}^{(1)} \right) \right], \tag{8.35}$$

we obtain

$$F^{(2)} = 2\pi(\theta - \theta_0) \frac{\partial}{\partial J_k} \left[\sum_{\mathbf{n}>0}{}' \sum_{\lambda} n_k n_l |V_{\mathbf{n}}(\lambda)|^2 \frac{\partial F_0}{\partial J_l} \Re(\gamma;\ n_s \omega_{0s} + \lambda \nu_{\mathbf{n}}) \right]$$

$$+\text{oscillating terms}, \tag{8.36}$$

$$G_{\mathbf{n}}^{(2)} = \mathcal{F}_{\mathbf{n}} \exp\left(-i n_s \omega_{0s} \theta\right), \tag{8.37}$$

where

$$\mathcal{F}_{\mathbf{n}} = i(\theta - \theta_0) n_k n_l \sum_{\mu} \frac{V_{\mathbf{n}}(\mu)}{n_s \omega_{0s} + \mu \nu_{\mathbf{n}}} \frac{\partial^2}{\partial J_k \partial J_l} [V_{\mathbf{n}}^*(\mu) W_{\mathbf{n}}]$$

$$+i(\theta - \theta_0) \sum_{\mathbf{m}}{}' \sum_{\mu} \frac{1}{(n_s - m_s)\omega_{0s} + \mu \nu_{\mathbf{n}-\mathbf{m}}}$$

$$\times\left\{-n_k m_l V_{\mathbf{n}-\mathbf{m}}(\mu)\frac{\partial}{\partial J_k}\left(V^*_{\mathbf{n}-\mathbf{m}}(\mu)\frac{\partial W_\mathbf{n}}{\partial J_l}\right)\right.$$

$$+n_k n_l V_{\mathbf{n}-\mathbf{m}}(\mu)\frac{\partial^2}{\partial J_k \partial J_l}\left(V^*_{\mathbf{n}-\mathbf{m}}(\mu)W_\mathbf{n}\right)+m_k m_l\frac{\partial}{\partial J_k}\left(|V_{\mathbf{n}-\mathbf{m}}(\mu)|^2\frac{\partial W_\mathbf{n}}{\partial J_l}\right)$$

$$\left.-m_k n_l\frac{\partial}{\partial J_k}\left[V_{\mathbf{n}-\mathbf{m}}(\mu)\frac{\partial}{\partial J_l}\left(V^*_{\mathbf{n}-\mathbf{m}}(\mu)W_\mathbf{n}\right)\right]\right\}+\text{oscillating terms}, \qquad (8.38)$$

and the function $\Re(x;\ y)$ is given by equation (5.134). It is now straightforward to write the second order RG equations. They are:

$$\frac{\partial F_0}{\partial \theta}=2\pi\epsilon^2\frac{\partial}{\partial J_k}\left[\sum_{\mathbf{n}>0}{}'\sum_{\lambda} n_k n_l |V_\mathbf{n}(\lambda)|^2\frac{\partial F_0}{\partial J_l}\Re(\gamma;\ n_s\omega_{0s}+\lambda\nu_\mathbf{n})\right], \qquad (8.39)$$

$$\frac{\partial W_\mathbf{n}}{\partial \theta}=i\epsilon^2 n_k n_l\sum_\mu\frac{V_\mathbf{n}(\mu)}{n_s\omega_{0s}+\mu\nu_\mathbf{n}}\frac{\partial^2}{\partial J_k\partial J_l}\left[V^*_\mathbf{n}(\mu)W_\mathbf{n}\right]$$

$$+i\epsilon^2\sum_{\mathbf{m}}{}'\sum_{\mu}\frac{1}{(n_s-m_s)\omega_{0s}+\mu\nu_{\mathbf{n}-\mathbf{m}}}$$

$$\times\left\{-n_k m_l V_{\mathbf{n}-\mathbf{m}}(\mu)\frac{\partial}{\partial J_k}\left(V^*_{\mathbf{n}-\mathbf{m}}(\mu)\frac{\partial W_\mathbf{n}}{\partial J_l}\right)\right.$$

$$+n_k n_l V_{\mathbf{n}-\mathbf{m}}(\mu)\frac{\partial^2}{\partial J_k \partial J_l}\left(V^*_{\mathbf{n}-\mathbf{m}}(\mu)W_\mathbf{n}\right)+m_k m_l\frac{\partial}{\partial J_k}\left(|V_{\mathbf{n}-\mathbf{m}}(\mu)|^2\frac{\partial W_\mathbf{n}}{\partial J_l}\right)$$

$$\left.-m_k n_l\frac{\partial}{\partial J_k}\left[V_{\mathbf{n}-\mathbf{m}}(\mu)\frac{\partial}{\partial J_l}\left(V^*_{\mathbf{n}-\mathbf{m}}(\mu)W_\mathbf{n}\right)\right]\right\}. \qquad (8.40)$$

The RG equation (8.39) is a *Fokker-Planck* equation describing the diffusion of the adiabatic action invariant. It has been derived previously by many authors (see e.g. the book by *Lichtenberg* and *Lieberman* [Lichtenberg (1982)] and the references therein). It is important to note that our derivation does not require the initial assumption concerning the fast stochastization of the angle variable. The fact that the latter is indeed a stochastic variable is clearly visible from the second RG equation (8.40), governing the slow amplitude evolution of the angle-dependent part of the phase space density. Nevertheless it looks complicated, its most important feature is that equations for the amplitudes of different modes are decoupled. In the case of isolated nonlinear resonance equation (8.40) acquires a very simple form as will be shown in the next paragraph.

8.4 Modulational Diffusion

As an example to demonstrate the theory developed in the previous paragraph , we consider again the example of paragraph 5.7 [see equation (5.138)]

$$H_0(J) = \lambda J + H_s(J), \qquad V(\alpha, J; \theta) = V(J)\cos(\alpha + \xi \sin \nu_m \theta). \quad (8.41)$$

The Hamiltonian (8.41), written in resonant canonical variables describes an isolated nonlinear resonance of one-dimensional betatron motion of particles in an accelerator with modulated resonant phase (or modulated linear betatron tune). It has been already mentioned that the modulation may come from various sources: ripple in the power supply of quadrupole magnets, synchro-betatron coupling or ground motion. The resonance detuning λ defines the distance from the resonance, Ξ is the amplitude of modulation of the linear betatron tune and ν_m is the modulation frequency, where $\xi = \Xi/\nu_m$. Without loss of generality, we consider ξ positive.

Since

$$\omega_0 = \lambda + \frac{\mathrm{d}H_s}{\mathrm{d}J}, \qquad V_1(J;\mu) = \frac{1}{2}V(J)\mathcal{J}_\mu(\xi), \quad (8.42)$$

where $\mathcal{J}_n(z)$ is the *Bessel* function of order n, the RG equation for the amplitude A can be rewritten as [see equation (5.140)]

$$\frac{\mathrm{d}A}{\mathrm{d}\theta} = \frac{\pi\epsilon^2}{2\nu_m}\left\{\frac{\partial}{\partial A}\left[V^2(A)\mathcal{J}^2_{[\frac{\omega_0}{\nu_m}]}(\xi)\right]\right.$$

$$\left. -\gamma(A)V^2(A)\frac{\partial}{\partial a}\left[\mathcal{J}^2_a(\xi)\right]\right|_{a=-\frac{\omega_0}{\nu_m}}\Bigg\}. \quad (8.43)$$

Here the square brackets $[z]$ encountered in the index of the Bessel function imply integer part of z. Moreover, in deriving the expression for $V_1(J;\mu)$ in equation (8.42) use has been made of the identity

$$\exp(iq\sin z) == \sum_{n=-\infty}^{\infty} \mathcal{J}_n(|q|)\exp[inz\mathrm{sgn}(q)], \quad (8.44)$$

and finally, the appropriate limit $\gamma \to 0$ in the corresponding RG equation has been taken. For small value of ξ utilizing the approximate expression for the derivative of *Bessel* functions with respect to the order, we obtain

$$\frac{\mathrm{d}A}{\mathrm{d}\theta} = \frac{\pi\epsilon^2}{2\nu_m}\left\{\frac{\partial}{\partial A}\left[V^2(A)\mathcal{J}^2_{\left[\frac{\omega_0}{\nu_m}\right]}(\xi)\right]\right.$$

$$\left. -2\mathcal{J}^2_{\left[\frac{\omega_0}{\nu_m}\right]}(\xi)\ln\left(\frac{\xi}{2}\right)\gamma(A)V^2(A)\right\}. \tag{8.45}$$

Let us now turn to the RG equations (8.39) and (8.40). They can be rewritten in the form:

$$\frac{\partial F_0}{\partial\theta} = \frac{\pi\epsilon^2}{2\nu_m}\frac{\partial}{\partial J}\left[V^2(J)\mathcal{J}^2_{\left[\frac{\omega_0}{\nu_m}\right]}(\xi)\frac{\partial F_0}{\partial J}\right], \tag{8.46}$$

$$\frac{1}{W_n}\frac{\partial W_n}{\partial\theta} = \frac{i\pi\epsilon^2 n}{2\nu_m\sin(\pi\omega_0/\nu_m)}\mathcal{J}^2_{-\frac{\omega_0}{\nu_m}}(\xi)\frac{\mathrm{d}}{\mathrm{d}J}\left(V\frac{\mathrm{d}V}{\mathrm{d}J}\right)$$

$$+\frac{\pi\epsilon^2 n}{2\nu_m}\mathcal{J}^2_{\left[\frac{\omega_0}{\nu_m}\right]}(\xi)\frac{\mathrm{d}}{\mathrm{d}J}\left(V\frac{\mathrm{d}V}{\mathrm{d}J}\right). \tag{8.47}$$

Equation (8.47) suggests that the amplitudes W_n of the angular modes G_n exhibit exponential growth with an increment

$$\Gamma = \frac{\pi\epsilon^2}{2\nu_m}\mathcal{J}^2_{\left[\frac{\omega_0}{\nu_m}\right]}(\xi)\frac{\mathrm{d}}{\mathrm{d}J}\left(V\frac{\mathrm{d}V}{\mathrm{d}J}\right). \tag{8.48}$$

Equation (8.46) is a *Fokker-Planck* equation for the angle-independent part of the phase space density with a diffusion coefficient

$$\mathcal{D}(J) = \frac{\pi\epsilon^2}{2\nu_m}V^2(J)\mathcal{J}^2_{\left[\frac{\omega_0}{\nu_m}\right]}(\xi). \tag{8.49}$$

One can also define the reduced diffusion coefficient

$$\mathcal{D}^{(R)}\left(J,\frac{\Xi}{\nu_m}\right) = \frac{2\Xi}{\pi\epsilon^2 V^2(J)}\mathcal{D}(J). \tag{8.50}$$

Three typical regimes corresponding to different values of λ/Ξ used as a control parameter can be observed. In the first one the resonance detuning can be chosen twice as large as the amplitude of the modulation ($\lambda = 2\Xi$). In this case there is no crossing of the main resonance described by the Hamiltonian (8.41) and the diffusion coefficient decreases very rapidly after reaching its maximum value at $\xi = 0.25$. The cases of periodic resonance crossings for $\lambda = \Xi$ and $\lambda = \Xi/2$ are characterized by strong modulational diffusion as shown in [Tzenov (2002a)].

8.5 The Liouville Operator and the Frobenius-Perron Operator

Complex systems giving rise to chaos or bifurcations exhibit remarkable sensitivity to small errors, external disturbances, or small internal fluctuations since initial states which are experimentally indistinguishable may evolve in quite different way in the course of time. It was pointed out in the introduction that the statistical approach to deterministic chaos appears to be a natural technique to study non integrable dynamical systems. A powerful tool to study the behaviour of complex dynamical systems, which we shall introduce in this paragraph is the method of the *Liouville operator* together with its discrete version - the so-called *Frobenius-Perron operator.*

In general, we consider a continuous multi-dimensional finite degree-of-freedom dynamical system defined by the state vector $\mathbf{x}(t)$, where t denotes the independent time variable in a general sense. The evolution of the system is described by the set of coupled first-order differential equations

$$\frac{\mathrm{d}\mathbf{x}}{\mathrm{d}t} = \mathbf{F}(\mathbf{x}, \lambda; t), \tag{8.51}$$

where $\mathbf{F}(\mathbf{x}, \lambda; t)$ is a vector field and λ is a set of parameters. In order to be even more general, let us suppose that an external random force of arbitrary origin acts on our system. Instead of the dynamical equations (8.51), we consider the *Langevin* equations

$$\frac{\mathrm{d}\mathbf{x}}{\mathrm{d}t} = \mathbf{F}(\mathbf{x}, \lambda; t) + \widehat{\mathcal{B}}(\mathbf{x}, \lambda; t) \cdot \mathbf{W}(t), \tag{8.52}$$

where $\widehat{\mathcal{B}}(\mathbf{x}, \lambda; t)$ is a tensor field accounting for the intensity of the external noise $\mathbf{W}(t)$. The latter has the formal correlation properties

$$\langle \mathbf{W}(t) \rangle = 0, \qquad \langle W_i(t) W_j(t') \rangle = 2\delta_{ij}\delta(t - t'), \tag{8.53}$$

where the average is performed over a concrete realization of the stochastic process $\mathbf{W}(t)$. Note that the presence of the *Kronecker*'s delta-symbol δ_{ij} implies that various components are uncorrelated, that is statistically independent, while the delta-function $\delta(t - t')$ shows that the correlation time of the external noise is negligibly small.

We further define the dynamic distribution

$$N(\mathbf{x};t) = \delta(\mathbf{x} - \mathbf{x}(t)), \tag{8.54}$$

analogous to the microscopic phase space density (7.13). In analogy with equation (7.14), we write the equation for the dynamic distribution $N(\mathbf{x};t)$ as

$$\frac{\partial N}{\partial t} + \nabla \cdot (\mathbf{F}N) = -\nabla \cdot \left(\widehat{\mathcal{B}} \cdot \mathbf{W}N\right). \tag{8.55}$$

Let us define the distribution function $f(\mathbf{x};t)$ and the fluctuation $\delta N(\mathbf{x};t)$ according to the relations

$$f(\mathbf{x};t) = \langle N(\mathbf{x};t)\rangle \quad ; \quad \delta N(\mathbf{x};t) = N(\mathbf{x};t) - f(\mathbf{x};t), \tag{8.56}$$

and write down the equations for f and δN

$$\frac{\partial f}{\partial t} + \nabla \cdot (\mathbf{F}f) = -\nabla \cdot \left(\widehat{\mathcal{B}} \cdot \langle \mathbf{W}\delta N\rangle\right), \tag{8.57}$$

$$\frac{\partial \delta N}{\partial t} + \nabla \cdot (\mathbf{F}\delta N) = -\nabla \cdot \left(\widehat{\mathcal{B}} \cdot \mathbf{W}f\right) - \nabla \cdot \left(\widehat{\mathcal{B}} \cdot \mathbf{W}\delta N\right) + \nabla \cdot \left(\widehat{\mathcal{B}} \cdot \langle \mathbf{W}\delta N\rangle\right). \tag{8.58}$$

Equation (8.58) can be further simplified. First of all, fluctuations can be considered small and therefore nonlinear terms (in $\mathbf{W}$ and δN) on the right-hand-side of equation (8.58) can be neglected. Secondly, since the stochastic process $\mathbf{W}(t)$ is delta-correlated, we can drop the second term on the left-hand-side of equation (8.58) as it does not manage to contribute for infinitely small correlation times. As a result equation (8.58) acquires the form

$$\frac{\partial \delta N}{\partial t} = -\nabla \cdot \left(\widehat{\mathcal{B}} \cdot \mathbf{W}f\right). \tag{8.59}$$

The equation for the fluctuation (8.59) can be solved in a straightforward manner. We obtain the obvious result

$$\delta N(t) = -\nabla \cdot \int\limits_0^\infty d\tau \widehat{\mathcal{B}}(t-\tau) \cdot \mathbf{W}(t-\tau) f(t-\tau), \tag{8.60}$$

which substituted back into equation (8.57) yields the kinetic form of the

Fokker-Planck equation

$$\frac{\partial f}{\partial t} = -\sum_{k} \nabla_k \left[\left(F_k - \sum_{m,l} \widehat{\mathcal{B}}_{km} \nabla_l \widehat{\mathcal{B}}_{lm} \right) f \right] + \sum_{k,l} \nabla_k \left(\widehat{\mathcal{D}}_{kl} \nabla_l f \right). \quad (8.61)$$

The diffusion tensor $\widehat{\mathcal{D}}_{km}$ is given by the expression:

$$\widehat{\mathcal{D}}_{kl} = \widehat{\mathcal{D}}_{lk} = \sum_{m} \widehat{\mathcal{B}}_{km} \widehat{\mathcal{B}}_{lm}. \quad (8.62)$$

Suppose now that the external noise is zero $\left(\widehat{\mathcal{B}} \equiv 0\right)$. Then equation (8.61) reduces to the *Liouville* equation

$$\frac{\partial f(\mathbf{x};t)}{\partial t} = -\nabla \cdot [\mathbf{F}(\mathbf{x}, \lambda; t) f(\mathbf{x};t)] = \widehat{\mathcal{L}} f(\mathbf{x};t), \quad (8.63)$$

induced by the dynamical equations (8.51) of motion for the state vector $\mathbf{x}$. Here $\widehat{\mathcal{L}}$ is the *Liouville* operator. Note that if the dynamical system is symplectic, in other words, if the dynamical equations (8.51) are the *Hamilton* equations of motion, then equation (8.63) coincides with the well-known *Liouville* equation of classical statistical mechanics.

Let us write the formal solution of equations (8.51) as

$$\mathbf{x}(t) = \mathbf{X}(\mathbf{x}_0, \lambda; t), \quad (8.64)$$

where $\mathbf{x}_0$ is the initial condition. One can then verify straightforwardly that the solution of the *Liouville* equation (8.63) reads

$$f(\mathbf{x};t) = \int d\mathbf{z} \delta(\mathbf{x} - \mathbf{X}(\mathbf{z}, \lambda; t)) f_0(\mathbf{z}). \quad (8.65)$$

Assuming the invertibility of $\mathbf{X}(\mathbf{x}_0, \lambda; t)$, in agreement with the uniqueness of the solution of equations (8.51), we can write equation (8.65) in alternative form

$$f(\mathbf{x};t) = f_0\left[\mathbf{X}^{-1}(\mathbf{x}, \lambda; t)\right] \left| \frac{\partial \mathbf{X}^{-1}(\mathbf{x}, \lambda; t)}{\partial \mathbf{x}} \right|, \quad (8.66)$$

where the second factor on the right-hand-side is the absolute value of the Jacobian determinant of the inverse transformation $\mathbf{X}^{-1}(\mathbf{x}, \lambda; t)$.

For one-dimensional maps of the form

$$x_{n+1} = F(x_n, \lambda), \quad (8.67)$$

equation (8.65) can be written as

$$f_{n+1}(x) = \widehat{U} f_n(x) = \int \mathrm{d}z \delta(x - F(z, \lambda)) f_n(z), \tag{8.68}$$

where $\widehat{U}$ is the *Frobenius-Perron operator*. We can also write

$$\widehat{U} f_n(x) = \sum_b \frac{f_n\big(F_b^{-1}(x, \lambda)\big)}{\big|F'\big(F_b^{-1}(x, \lambda)\big)\big|}, \tag{8.69}$$

where the index b runs over all the various branches of the inverse map F^{-1}. The generalization to multidimensional maps is straightforward.

For chaotic systems the probability density approaches an equilibrium density [Gaspard (1998)], which depends only on the system's characteristics and not on the initial density. For hyperbolic systems (the so-called A systems), the relaxation towards an equilibrium state is exponential, and the relaxation rates are the *Ruelle resonances* [Ruelle (1978)]. These rates are usually studied in the framework of the *Frobenius-Perron* operator and are known to be the poles of the matrix elements of the resolvent to the *Frobenius-Perron* operator [Hasegawa (1992)].

In the present paragraph, we will study the properties of the *Frobenius-Perron* operator directly. As an application, we consider the standard *Chirikov-Taylor* map (6.237) introduced in paragraph 6.7. The standard map given by equation (6.237) can be written in a simple form as

$$a_{n+1} = a_n + I_{n+1}, \qquad I_{n+1} = I_n - K \sin a_n, \tag{8.70}$$

where

$$a = \alpha \pm \pi, \qquad I = 2\pi\left(-k + \frac{k^2 \mathcal{K}}{R} J\right), \tag{8.71}$$

and

$$K = \frac{2\pi c k^2 \mathcal{K} \Delta E_0}{\omega R \beta_s E_s}. \tag{8.72}$$

It was already mentioned in paragraph 6.7 that diffusive spread in phase space takes place for a sufficiently large stochasticity parameter (see also [Chirikov (1979)] and [Lichtenberg (1982)]). In what follows, we will calculate the diffusion coefficient for the standard map (8.70) in the approximation of sufficiently large value of the parameter K. For that purpose, we write the *Frobenius-Perron* operator as

$$f_{n+1}(a, I) = \widehat{U} f_n(a, I) = f_n(a - I, I + K \sin{(a - I)}). \tag{8.73}$$

In order to calculate the matrix elements of the *Frobenius-Perron* operator, it is convenient to use the quantum mechanics bra-ket notations [Sudbery (1986)]. An arbitrary function $f(a, I)$ periodic in the angle variable a can be written in the form

$$|f\rangle = f(a, I) = \sum_{k} \langle k|\, f\rangle |k\rangle, \tag{8.74}$$

where $|k\rangle$ and $\langle k|$ constitute the angle-dependent basis

$$|k\rangle = \frac{1}{\sqrt{2\pi}} e^{ika}, \qquad \langle k| = \frac{1}{\sqrt{2\pi}} e^{-ika}, \tag{8.75}$$

in *Hilbert* space. Moreover,

$$\langle k|\, f\rangle = \frac{1}{\sqrt{2\pi}} \int_{0}^{2\pi} da f(a, I) e^{-ika}. \tag{8.76}$$

Then the *Frobenius-Perron* operator (8.73) can be cast in the following alternative form

$$|f_{n+1}\rangle = \widehat{U} |f_n\rangle. \tag{8.77}$$

Next, we introduce the projected density distribution $\langle 0\,|f_{n+1}\,\rangle$ and the k-th harmonic of its angle-dependent part $\langle k\,|f_{n+1}\,\rangle$

$$\langle 0\,|f_{n+1}\,\rangle = \sum_{k} \left\langle 0 \left| \widehat{U} \right| k \right\rangle \langle k\,|f_n\,\rangle, \tag{8.78}$$

$$\langle k\,|f_{n+1}\,\rangle = \sum_{m} \left\langle k \left| \widehat{U} \right| m \right\rangle \langle m\,|f_n\,\rangle. \tag{8.79}$$

In what follows, we will need the matrix element $\left\langle k_1 \left| \widehat{U} \right| k_2 \right\rangle$. We have

$$\widehat{U} G(I) e^{ik_2 a} = G(I + K \sin(a - I)) e^{ik_2 (a - I)}$$

$$= e^{ik_2 (a-I)} \widehat{\mathbf{\Gamma}}_{inv} \exp\left[K \sin(a - I) \frac{\partial}{\partial I} \right] G(I), \tag{8.80}$$

where $G(I)$ is a generic function of the action variable I. The action of the operator $\widehat{\mathbf{\Gamma}}_{inv}$ is analogous to the action of the one introduced earlier [see equation (4.82) and the comment that follows]. It implies that all powers of

$\sin(a - I)$ must be shifted to the left, so that the powers of the differential operator $\partial/\partial I$ acts on $G(I)$ only. Using the expansion (4.62), we find

$$\widehat{U} G(I) e^{ik_2 a} = e^{ik_2(a-I)} \sum_{n=-\infty}^{\infty} e^{in(a-I)} \mathcal{J}_n\left(\frac{K}{i}\frac{\partial}{\partial I}\right) G(I). \tag{8.81}$$

Therefore, for the matrix element, we obtain

$$\left\langle k_1 \left| \widehat{U} \right| k_2 \right\rangle = e^{-ik_1 I} \mathcal{J}_{k_1 - k_2}\left(\frac{K}{i}\frac{\partial}{\partial I}\right). \tag{8.82}$$

From equation (8.77) it is easy to see that

$$|f_n\rangle = \widehat{U}^n |f_0\rangle. \tag{8.83}$$

Since, the leading contribution to the evolution of the system on large time scales is comes from the projected density distribution, it is important to calculate the quantity $\langle 0 \,|f_n \rangle$. We have

$$\langle 0 \,|f_n \rangle = \sum_{k_1, k_2, \ldots k_n} \left\langle 0 \left| \widehat{U} \right| k_1 \right\rangle \left\langle k_1 \left| \widehat{U} \right| k_2 \right\rangle \ldots \left\langle k_{n-1} \left| \widehat{U} \right| k_n \right\rangle \langle k_n \,|f_0 \rangle. \tag{8.84}$$

Suppose that

$$\langle k \,|f_0 \rangle = \mathcal{F}(I) \delta_{k0}, \tag{8.85}$$

which means that the initial density distribution depends only on the action variable I. Therefore,

$$\mathcal{F}_n = \sum_{k_1, k_2, \ldots k_{n-1}} \left\langle 0 \left| \widehat{U} \right| k_1 \right\rangle \left\langle k_1 \left| \widehat{U} \right| k_2 \right\rangle \ldots \left\langle k_{n-1} \left| \widehat{U} \right| 0 \right\rangle \mathcal{F}. \tag{8.86}$$

or

$$\mathcal{F}_n = \left\langle 0 \left| \widehat{U}^n \right| 0 \right\rangle \mathcal{F}. \tag{8.87}$$

Using the explicit form (8.82) of the matrix element, we can write

$$\mathcal{M}_n = \left\langle 0 \left| \widehat{U}^n \right| 0 \right\rangle = \sum_{k_1, k_2, \ldots k_{n-1}} \left\langle 0 \left| \widehat{U} \right| k_1 \right\rangle \left\langle k_1 \left| \widehat{U} \right| k_2 \right\rangle \ldots \left\langle k_{n-1} \left| \widehat{U} \right| 0 \right\rangle$$

$$= \sum_{k_1, k_2, \ldots k_{n-1}} \mathcal{J}_{-k_1}\left(\frac{K}{i}\frac{\partial}{\partial I}\right) e^{-ik_1 I} \mathcal{J}_{k_1 - k_2}\left(\frac{K}{i}\frac{\partial}{\partial I}\right) e^{-ik_2 I}$$

$$\times \mathcal{J}_{k_2-k_3}\left(\frac{K}{i}\frac{\partial}{\partial I}\right)\dots e^{-ik_{n-1}I}\mathcal{J}_{k_{n-1}}\left(\frac{K}{i}\frac{\partial}{\partial I}\right)$$

$$= \sum_{k_1,k_2,\dots k_{n-1}} \exp\left(-iI\sum_{m=1}^{n-1}k_m\right)\mathcal{J}_{-k_1}\left(-K(k_1+\dots+k_{n-1})+\frac{K}{i}\frac{\partial}{\partial I}\right)$$

$$\times \mathcal{J}_{k_1-k_2}\left(-K(k_2+\dots+k_{n-1})+\frac{K}{i}\frac{\partial}{\partial I}\right)$$

$$\times \mathcal{J}_{k_2-k_3}\left(-K(k_3+\dots+k_{n-1})+\frac{K}{i}\frac{\partial}{\partial I}\right)\dots$$

$$\times \mathcal{J}_{k_{n-2}-k_{n-1}}\left(-Kk_{n-1}+\frac{K}{i}\frac{\partial}{\partial I}\right)\mathcal{J}_{k_{n-1}}\left(\frac{K}{i}\frac{\partial}{\partial I}\right). \tag{8.88}$$

It becomes clear from the above expression that the leading contribution to the slow, non oscillating behaviour comes from terms for which

$$k_1+k_2+\dots+k_{n-1}=0, \tag{8.89}$$

so that

$$\exp\left(-iI\sum_{m=1}^{n-1}k_m\right)=1.$$

Hence, equation (8.88) can be rewritten as

$$\mathcal{M}_n = \sum_{k_1,k_2,\dots k_{n-1}} \mathcal{J}_{-k_1}\left(\frac{K}{i}\frac{\partial}{\partial I}\right)\mathcal{J}_{k_1-k_2}\left(Kk_1+\frac{K}{i}\frac{\partial}{\partial I}\right)$$

$$\times \mathcal{J}_{k_2-k_3}\left(K(k_1+k_2)+\frac{K}{i}\frac{\partial}{\partial I}\right)\mathcal{J}_{k_{n-2}-k_{n-1}}\left(K(k_1+\dots+k_{n-2})+\frac{K}{i}\frac{\partial}{\partial I}\right)$$

$$\times \mathcal{J}_{k_{n-1}}\left(\frac{K}{i}\frac{\partial}{\partial I}\right)\delta\left(\sum_{m=1}^{n-1}k_m\right). \tag{8.90}$$

Each term in equation (8.90) is characterized by the sequence

$$(k_1,k_2,\dots k_{n-1}).$$

The leading contribution

$$\mathcal{M}_n^{(0)} = \mathcal{J}_0^n\left(\frac{K}{i}\frac{\partial}{\partial I}\right) \approx \left(1 + \frac{K^2}{4}\frac{\partial^2}{\partial I^2}\right)^n, \tag{8.91}$$

results from the sequence where all k_i are equal to zero. Consider now a particular sequence in which some of k_i do not vanish, and let k_f be the the first non vanishing summation index and k_l be the last non vanishing one. The contribution of this sequence will result in a term

$$\mathcal{M}_n^{(1)} = \mathcal{J}_0^{f-1}\left(\frac{K}{i}\frac{\partial}{\partial I}\right)\mathcal{J}_{-k_f}\left(\frac{K}{i}\frac{\partial}{\partial I}\right)\mathcal{J}_{k_f-k_{f+1}}\left(Kk_f + \frac{K}{i}\frac{\partial}{\partial I}\right)$$

$$\times\mathcal{J}_{k_{f+1}-k_{f+2}}\left(K(k_f + k_{f+1}) + \frac{K}{i}\frac{\partial}{\partial I}\right)\cdots$$

$$\times\mathcal{J}_{k_{l-1}-k_l}\left(K(k_f + \cdots + k_{l-1}) + \frac{K}{i}\frac{\partial}{\partial I}\right)\mathcal{J}_{k_l}\left(\frac{K}{i}\frac{\partial}{\partial I}\right)\mathcal{J}_0^{n-l-1}\left(\frac{K}{i}\frac{\partial}{\partial I}\right). \tag{8.92}$$

Since, we are interested in the limit of large K, it is evident that the largest contribution to $\mathcal{M}_n^{(1)}$ results from the shortest sequence for which $l = f+1$. Because of equation (8.89), the leading contribution is from a sequence $k_f = -k_l = \pm 1$. Since

$$\mathcal{J}_{k_f-k_l}\left(Kk_f + \frac{K}{i}\frac{\partial}{\partial I}\right) = \mathcal{J}_{\pm 2}\left(\pm K + \frac{K}{i}\frac{\partial}{\partial I}\right) \approx \mathcal{J}_2(K),$$

we have

$$\mathcal{M}_n^{(1)} \approx 2(n-2)\mathcal{J}_0^{n-3}\left(\frac{K}{i}\frac{\partial}{\partial I}\right)\mathcal{J}_2(K)\mathcal{J}_1^2\left(\frac{K}{i}\frac{\partial}{\partial I}\right)$$

$$\approx -2(n-2)\left(1 + \frac{K^2}{4}\frac{\partial^2}{\partial I^2}\right)^{n-3}\mathcal{J}_2(K)\left(\frac{K}{2}\frac{\partial}{\partial I}\right)^2. \tag{8.93}$$

In obtaining equation (8.93) an account has been taken of: (i) the fact that the sequence can start at $(n-2)$ places, (ii) the relation $k_f = \pm 1$ yields a factor of 2 in the final result. Moreover, it has been assumed that the differential operator $\partial/\partial I$ when acting on what is to the right of it, produces terms much smaller in value than K. Summing up the contributions (8.91) and (8.93), we obtain

$$\mathcal{M}_n^{(0)} + \mathcal{M}_n^{(1)} \approx \left(1 + \frac{K^2}{4}\frac{\partial^2}{\partial I^2}\right)^n - 2n\left(1 + \frac{K^2}{4}\frac{\partial^2}{\partial I^2}\right)^{n-1} \mathcal{J}_2(K)$$

$$\times \left(\frac{K}{2}\frac{\partial}{\partial I}\right)^2 \left(\frac{n-2}{n}\right)\left(1 + \frac{K^2}{4}\frac{\partial^2}{\partial I^2}\right)^{-2}. \qquad (8.94)$$

In the leading order

$$\lim_{n\to\infty} \frac{n-2}{n} = 1, \qquad \left(1 + \frac{K^2}{4}\frac{\partial^2}{\partial I^2}\right)^{-1} \sim 1. \qquad (8.95)$$

Therefore, we finally obtain

$$\mathcal{M}_n^{(0)} + \mathcal{M}_n^{(1)} \sim \left[1 + \frac{K^2}{4}\frac{\partial^2}{\partial I^2} - 2\mathcal{J}_2(K)\frac{K^2}{4}\frac{\partial^2}{\partial I^2}\right]^n$$

$$= \left\{1 + \frac{K^2}{4}[1 - 2\mathcal{J}_2(K)]\frac{\partial^2}{\partial I^2}\right\}^n. \qquad (8.96)$$

In the continuous limit the map (8.87) passes into the *Fokker-Planck* equation

$$\frac{\partial \mathcal{F}}{\partial n} = \mathcal{D}(K)\frac{\partial^2 \mathcal{F}}{\partial I^2}, \qquad (8.97)$$

where

$$\mathcal{D}(K) = \frac{K^2}{4}[1 - 2\mathcal{J}_2(K)], \qquad (8.98)$$

is the diffusion coefficient.

An important comment is now in order. The standard map is an example of the so-called mixed systems as is the case for most systems of physical interest. Mixed systems are characterized by the coexistence and mutual sticking of chaotic components and regions of phase space where regular motion is being performed, such as stable islands and accelerator modes. Elimination of the effect of sticking can be achieved by formally adding a small external noise. Since the appearance of islands and the sticking is a nonperturbative effect, it is not trivial to reproduce the latter by the theory developed here. However, a more detailed analysis of equation (8.87) with due account of the oscillating terms in the matrix element (8.88) could lead to better understanding of the process of formation of regular islands and the sticking.

Chapter 9

The Vlasov Equation

9.1 Introduction

In most of the works so far, dedicated to the study of the properties of intense charged particle beams, the effect of interparticle collisions has been neglected. In many important cases this is a sensible approximation giving satisfactory results, yet one has to elucidate the limits of validity of "collisionless beam" approach and to investigate the role of collision phenomena in beam physics. Collisions are expected to bring about effects such as thermalization, resistivity, diffusion etc. that influence the long term behaviour of charged particle beams. The reasoning commonly adopted for employing the "collisionless beam" approach is that characteristic beam-plasma frequencies are much greater than collision frequencies for a large number of situations in beam physics. Such an assumption is not based on stable physical grounds as pointed in [Klimontovich (1995)].

The term "collisionless beam" implies that interactions between particles giving rise to dissipation and hence leading to establishment of equilibrium state are not taken into account. In a number of cases involving reasonable approximations it is sufficient to compute the macroscopic characteristics (charge and current densities) in relatively big volume elements containing a large number of particles. As a result interaction manifests itself in the form of a mean, self-consistent field thus preserving the reversible character of the dynamics involved, and leading to the time *reversible Vlasov equation.*

The notion of "collisional beam" usually conceived as the counterpart of "collisionless beam" implies that dissipation due to redistribution of beam particles is taken into account, resulting in additional term (in the form of *Landau* or *Balescu-Lenard*) in the kinetic equation. In a sense, *Landau* and

Vlasov approximations correspond to two limit cases: namely the *Landau* collision integral takes into account interactions that determine dissipation while the effect of the mean, self-consistent field is not included into the physical picture involved. On the contrary, the latter is the only way interactions manifest themselves in the *Vlasov* equation, leaving the question concerning the role of collisions near particle-wave resonances unanswered. The *Balescu-Lenard* approximation lies somewhat in between *Landau* and *Vlasov* limit cases. It takes into account the dynamic polarization of the beam, which constitutes a more complete inclusion of collective effects resulting from interactions between charged particles.

9.2 The Vlasov Equation for Collisionless Beams

An important parameter in beam-plasma system is the so-called beam-plasma parameter μ, which is defined as

$$\mu = \frac{1}{nr_D^3} \ll 1. \tag{9.1}$$

Note that the beam-plasma parameter is inversely proportional to the number of particles contained in the *Debye* sphere.

When addressing particular problems of practical interest, one has to deal with external parameters characteristic for the case under consideration. This could be, for instance, the length of the vacuum chamber, or its transverse dimensions. In the simplest case, we can assume that there is one such parameter L. Let us also introduce a characteristic external time scale T, correspondent to the parameter L. In paragraph 7.9, we introduced and calculated the relaxation "time" θ_{rel} [see equation (7.172)] and its corresponding physical relaxation length $l_{rel} \sim R\theta_{rel}$. One can also define the physical relaxation time τ_{rel} as

$$\tau_{rel} \sim \frac{1}{\mu\omega_p\mathcal{L}} \gg \frac{1}{\omega_p}. \tag{9.2}$$

We now consider the limit case where

$$r_D \ll L \ll l_{rel}, \qquad \frac{1}{\omega_p} \ll T \ll \tau_{rel}. \tag{9.3}$$

The zero order approximation in the parameter L/l_{rel} is equivalent to completely neglecting the collisions or in other words to neglecting the fluctuations δN_M, $\delta \mathbf{E}$ and $\delta \mathbf{B}$. In this case, one has to set the collision integrals defined by equation (7.72) to zero. As a result equation (7.71) reduces to

$$\frac{\partial f}{\partial \theta} + \left(\widehat{\mathbf{v}} \cdot \widehat{\nabla}_{\mathbf{x}}\right) f + \widehat{\nabla}_{\mathbf{p}} \cdot \left[\left(\widehat{\mathbf{F}}_0 + \left\langle \widehat{\mathbf{F}} \right\rangle\right) f\right] = 0, \tag{9.4}$$

which together with the *Maxwell* equations

$$\widehat{\nabla}_{\mathbf{x}} \times \langle \mathbf{B} \rangle = \frac{1}{c^2} \frac{\partial \langle \mathbf{E} \rangle}{\partial t} + \mu_0 e n \mathbf{j}(\widehat{\mathbf{x}}; t), \qquad \widehat{\nabla}_{\mathbf{x}} \times \langle \mathbf{E} \rangle = -\frac{\partial \langle \mathbf{B} \rangle}{\partial t}, \tag{9.5}$$

$$\widehat{\nabla}_{\mathbf{x}} \cdot \langle \mathbf{B} \rangle = 0, \qquad \widehat{\nabla}_{\mathbf{x}} \cdot \langle \mathbf{E} \rangle = \frac{en}{\epsilon_0} \rho(\widehat{\mathbf{x}}; t), \tag{9.6}$$

for the averaged electromagnetic fields comprises a closed set of equations. Here, as usual

$$\rho(\widehat{\mathbf{x}}; t) = \int \mathrm{d}^3 \widehat{\mathbf{p}}^{(k)} f\left(\widehat{\mathbf{x}}, \widehat{\mathbf{p}}^{(k)}; t\right), \qquad \mathbf{j}(\widehat{\mathbf{x}}; t) = \int \mathrm{d}^3 \widehat{\mathbf{p}}^{(k)} \mathbf{v} f\left(\widehat{\mathbf{x}}, \widehat{\mathbf{p}}^{(k)}; t\right), \tag{9.7}$$

are the charge and current densities.

Equation (9.4) is the well-known *Vlasov* equation, which was introduced in 1937 by *A.A. Vlasov* in his pioneering studies on collisionless plasmas. It is important to emphasize that the *Vlasov* equation is reversible, which put in an alternate way means that the indeterminateness in the specification of the initial state of the system does not increase. Another remarkable property of the *Vlasov-Maxwell* system (9.4)–(9.6) is the fact that it looks exactly the same as the equations for the microscopic quantities N_M, $\mathbf{E}^{(M)}$ and $\mathbf{B}^{(M)}$, which are random functions of their arguments. The latter are reversible, so are the *Vlasov-Maxwell* equations.

A second important comment is now in order. The *Vlasov-Maxwell* equations (9.4)–(9.6) should be regarded as an approximation for the deterministic functions f, $\langle \mathbf{E} \rangle$ and $\langle \mathbf{B} \rangle$. Like every approximation yielding a closed system of equations, the *Vlasov-Maxwell* approximation provides an incomplete description of the system and therefore is irreversible. The reversibility of the *Vlasov-Maxwell* equations is only illusory. Actually, for beam-plasma systems with arbitrarily specified particle density and emittance, the relaxation time and the relaxation length are finite. Since, the *Vlasov-Maxwell* equations (9.4)–(9.6) comprise a limit case of equation (7.71) if and only if the conditions (9.3) hold, provided l_{rel} is finite the characteristic length scale of the system L should be finite too.

Thus, to obtain a realistic solution of equations (9.4)–(9.6), one has to impose boundary conditions, which for real systems are always *dissipative*.

Therefore, the *Vlasov-Maxwell* equations combined with real boundary conditions which are an irrevocable attribute of each case of practical interest are dissipative equations.

9.3 The Hamiltonian Formalism for Solving the Vlasov Equation

We consider a N-dimensional dynamical system, described by the canonical conjugate pair of vector variables $(\mathbf{q}, \mathbf{p})$ with components

$$\mathbf{q} = (q_1, q_2, \ldots, q_N),$$
$$\mathbf{p} = (p_1, p_2, \ldots, p_N). \tag{9.8}$$

The Vlasov equation (9.4) for the distribution function $f(\mathbf{q}, \mathbf{p}; t)$ can be expressed equivalently as

$$\frac{\partial f}{\partial t} + [f, H]_{\mathbf{q},\mathbf{p}} = 0, \tag{9.9}$$

where

$$[F, G]_{\mathbf{q},\mathbf{p}} = \frac{\partial F}{\partial q_i}\frac{\partial G}{\partial p_i} - \frac{\partial F}{\partial p_i}\frac{\partial G}{\partial q_i} \tag{9.10}$$

is the Poisson bracket, $H(\mathbf{q}, \mathbf{p}; t)$ is the Hamiltonian of the system, and summation over repeated indices is implied. Next we define a canonical transformation via the generating function of the second type according to

$$S = S(\mathbf{q}, \mathbf{P}; t), \tag{9.11}$$

and assume that the Hessian matrix

$$\widehat{\mathcal{H}}_{ij}(\mathbf{q}, \mathbf{P}; t) = \frac{\partial^2 S}{\partial q_i \partial P_j} \tag{9.12}$$

of the generating function $S(\mathbf{q}, \mathbf{P}; t)$ is non-degenerate, i.e.,

$$\det\left(\widehat{\mathcal{H}}_{ij}\right) \neq 0. \tag{9.13}$$

This implies that the inverse matrix $\widehat{\mathcal{H}}_{ij}^{-1}$ exists. The new canonical variables $(\mathbf{Q}, \mathbf{P})$ are defined by the canonical transformation as

$$p_i = \frac{\partial S}{\partial q_i}, \qquad Q_i = \frac{\partial S}{\partial P_i}. \tag{9.14}$$

We also introduce the distribution function defined in terms of the new canonical coordinates $(\mathbf{Q}, \mathbf{P})$ and the mixed pair of canonical variables $(\mathbf{q}, \mathbf{P})$ according to

$$f_0(\mathbf{Q}, \mathbf{P}; t) = f(\mathbf{q}(\mathbf{Q}, \mathbf{P}; t), \mathbf{p}(\mathbf{Q}, \mathbf{P}; t); t), \tag{9.15}$$

$$F_0(\mathbf{q}, \mathbf{P}; t) = f(\mathbf{q}, \mathbf{p}(\mathbf{q}, \mathbf{P}; t); t). \tag{9.16}$$

In particular, in equation (9.15) the old canonical variables are expressed in terms of the new ones, which is ensured by the implicit function theorem, provided the relation (9.13) holds. As far as the function $F_0(\mathbf{q}, \mathbf{P}; t)$ is concerned, we simply replace the old momentum $\mathbf{p}$ by its counterpart taken from the first of equations (9.14). Because

$$\frac{\partial p_i}{\partial P_j} = \frac{\partial^2 S}{\partial q_i \partial P_j} = \widehat{\mathcal{H}}_{ij} \qquad \Longrightarrow \qquad \frac{\partial P_i}{\partial p_j} = \widehat{\mathcal{H}}_{ij}^{-1}, \tag{9.17}$$

we can express the Poisson bracket in terms of the mixed variables in the form

$$[F, G]_{\mathbf{q},\mathbf{P}} = \widehat{\mathcal{H}}_{ji}^{-1} \left(\frac{\partial F}{\partial q_i} \frac{\partial G}{\partial P_j} - \frac{\partial F}{\partial P_j} \frac{\partial G}{\partial q_i} \right). \tag{9.18}$$

Differentiation of equation (9.14) with respect to time t, keeping the old variables $(\mathbf{q}, \mathbf{p})$ fixed, yields

$$\frac{\partial^2 S}{\partial q_i \partial t} + \frac{\partial^2 S}{\partial q_i \partial P_j} \left(\frac{\partial P_j}{\partial t} \right)_{qp} = 0, \tag{9.19}$$

$$\left(\frac{\partial Q_i}{\partial t} \right)_{qp} = \frac{\partial^2 S}{\partial P_i \partial t} + \frac{\partial^2 S}{\partial P_i \partial P_j} \left(\frac{\partial P_j}{\partial t} \right)_{qp}, \tag{9.20}$$

or

$$\left(\frac{\partial P_j}{\partial t} \right)_{qp} = -\widehat{\mathcal{H}}_{ji}^{-1} \frac{\partial^2 S}{\partial q_i \partial t}. \tag{9.21}$$

Our goal is to express the Vlasov equation (9.9) in terms of the mixed variables $(\mathbf{q}, \mathbf{P})$. Taking into account the identities

$$\frac{\partial Q_i}{\partial q_j} = \frac{\partial^2 S}{\partial q_j \partial P_i} = \widehat{\mathcal{H}}_{ji} \quad \Longrightarrow \quad \frac{\partial q_i}{\partial Q_j} = \widehat{\mathcal{H}}_{ji}^{-1}, \tag{9.22}$$

$$\frac{\partial f_0}{\partial Q_i} = \widehat{\mathcal{H}}_{ij}^{-1} \frac{\partial F_0}{\partial q_j}, \tag{9.23}$$

$$\frac{\partial f_0}{\partial P_i} = \frac{\partial F_0}{\partial P_i} - \frac{\partial f_0}{\partial Q_j} \frac{\partial^2 S}{\partial P_i \partial P_j}, \tag{9.24}$$

we obtain

$$\left(\frac{\partial f}{\partial t}\right)_{qp} = \frac{\partial f_0}{\partial t} + \frac{\partial f_0}{\partial Q_i}\left(\frac{\partial Q_i}{\partial t}\right)_{qp} + \frac{\partial f_0}{\partial P_i}\left(\frac{\partial P_i}{\partial t}\right)_{qp}$$

$$= \frac{\partial F_0}{\partial t} + \widehat{\mathcal{H}}_{ji}^{-1}\left(\frac{\partial F_0}{\partial q_i}\frac{\partial^2 S}{\partial t \partial P_j} - \frac{\partial F_0}{\partial P_j}\frac{\partial^2 S}{\partial t \partial q_i}\right)$$

$$= \frac{\partial F_0}{\partial t} + \left[F_0, \frac{\partial S}{\partial t}\right]_{\mathbf{q},\mathbf{P}}. \tag{9.25}$$

Furthermore, using the relation

$$[f, H]_{\mathbf{q},\mathbf{P}} = [F_0, \mathcal{H}]_{\mathbf{q},\mathbf{P}}, \tag{9.26}$$

where

$$\mathcal{H}(\mathbf{q}, \mathbf{P}; t) = H(\mathbf{q}, \nabla_q S; t), \tag{9.27}$$

we express the Vlasov equation in terms of the mixed variables according to

$$\frac{\partial F_0}{\partial t} + [F_0, \mathcal{K}]_{\mathbf{q},\mathbf{P}} = 0, \tag{9.28}$$

where

$$\mathcal{K}(\mathbf{q}, \mathbf{P}; t) = \frac{\partial S}{\partial t} + H(\mathbf{q}, \nabla_q S; t) \tag{9.29}$$

is the new Hamiltonian.

For the distribution function $f_0(\mathbf{Q}, \mathbf{P}; t)$, depending on the new canonical variables, we clearly obtain

$$\frac{\partial f_0}{\partial t} + [f_0, \mathcal{K}]_{\mathbf{Q},\mathbf{P}} = 0, \tag{9.30}$$

where the new Hamiltonian $\mathcal{K}$ is a function of the new canonical pair $(\mathbf{Q}, \mathbf{P})$, such that

$$\mathcal{K}(\nabla_P S, \mathbf{P}; t) = \frac{\partial S}{\partial t} + H(\mathbf{q}, \nabla_q S; t), \tag{9.31}$$

and the Poisson bracket entering equation (9.30) has the same form as equation (9.10), expressed in the new canonical variables.

9.4 Propagation of an Intense Beam Through a Periodic Focusing Lattice

As a first application of the Hamiltonian formalism, we consider the propagation of a continuous beam through a periodic focusing lattice in a circular ring with mean radius R. Particle motion, as we already discussed in chapter 1, is accomplished in two degrees of freedom in a plane transverse to the design orbit. The model we are going to discuss in this paragraph consists of the *Vlasov* equation

$$\frac{\partial f}{\partial \theta} + [f, H]_{\mathbf{x},\mathbf{p}} = 0, \tag{9.32}$$

coupled to the *Poisson* equation

$$\nabla_{\mathbf{x}}^2 \psi = -4\pi\varrho = -4\pi \int d\mathbf{p} f(\mathbf{x}, \mathbf{p}; \theta). \tag{9.33}$$

Here

$$H(\mathbf{x}, \mathbf{p}; \theta) = \frac{R}{2}\left(\widehat{p}_x^2 + \widehat{p}_z^2\right) + \frac{1}{2R}\left(G_x \widehat{x}^2 + G_z \widehat{z}^2\right) + \lambda\psi(\mathbf{x}; \theta) \tag{9.34}$$

is the Hamiltonian (1.77) with an additional term $\lambda\psi(\mathbf{x}; \theta)$ due to self-consistent potential, and $\mathbf{x} = (\widehat{x}, \widehat{z})$. The transverse canonical momenta $\mathbf{p} = (\widehat{p}_x, \widehat{p}_z)$ entering the Hamiltonian (9.34) are dimensionless variables. For simplicity, in what follows the "hat" above the x and p variables will be omitted. The case of a beam propagation in a straight focusing channel is most appropriately described in terms of the path length $s = R\theta$ chosen as an independent variable. Then the Hamiltonian (9.34) should be divided by the mean radius R and the coefficients $G_{x,z}/R^2$ redefined accordingly.

In addition, ψ is a normalized self-consistent potential related to the actual electric potential φ according to

$$\psi = \frac{4\pi\varepsilon_0}{Ne}\varphi, \tag{9.35}$$

where N is the total number of particles in the beam, e is the particle charge, and ε_0 is the electric susceptibility of vacuum. Moreover, the parameter λ is defined by

$$\lambda = \frac{NRr_c}{\beta_s^2\gamma_s}, \tag{9.36}$$

where

$$r_c = \frac{e^2}{4\pi\varepsilon_0 m_0 c^2} \tag{9.37}$$

is the classical radius of a beam particle with charge e and rest mass m_0. The coefficients $G_{x,z}(\theta)$ determining the focusing strength in both transverse directions are periodic functions of θ

$$G_{x,z}(\theta + \Theta) = G_{x,z}(\theta), \tag{9.38}$$

with period Θ.

Following the procedure outlined in the preceding paragraph, we transform equations (9.32) – (9.34) according to

$$[F_0, \mathcal{K}]_{\mathbf{q},\mathbf{P}} \equiv 0, \tag{9.39}$$

$$\frac{\partial S}{\partial \theta} + \epsilon H(\mathbf{q}, \nabla_{\mathbf{q}} S; \theta) = \mathcal{K}(\mathbf{q}, \mathbf{P}), \tag{9.40}$$

$$\nabla_{\mathbf{q}}^2 \psi = -4\pi \int d^2\mathbf{P} F_0(\mathbf{q}, \mathbf{P}) \det\left(\nabla_{\mathbf{q}} \nabla_{\mathbf{P}} S\right), \tag{9.41}$$

where ϵ is formally a small parameter, which will be set equal to unity at the end of the calculation. Note that all the contributions to the original Hamiltonian (9.34) are allowed to be of the same order of magnitude, where the small parameter ϵ multiplies $H(\mathbf{q}, \nabla_{\mathbf{q}} S; \theta)$ in equation (9.40). The next step is to expand the quantities S, $\mathcal{K}$ and ψ in a power series in ϵ according to

$$S = S_0 + \epsilon S_1 + \epsilon^2 S_2 + \epsilon^3 S_3 + \dots, \tag{9.42}$$

$$\mathcal{K} = \mathcal{K}_0 + \epsilon \mathcal{K}_1 + \epsilon^2 \mathcal{K}_2 + \epsilon^3 \mathcal{K}_3 + \dots, \tag{9.43}$$

$$\psi = \psi_0 + \epsilon \psi_1 + \epsilon^2 \psi_2 + \epsilon^3 \psi_3 + \dots. \tag{9.44}$$

We now substitute the expansions (9.42) – (9.44) into equations (9.40) and (9.41) and obtain perturbation equations that can be solved order by order.

The lowest order solution is evident and has the form

$$S_0 = \mathbf{q} \cdot \mathbf{P}, \qquad \mathcal{K}_0 \equiv 0, \tag{9.45}$$

$$\nabla_{\mathbf{q}}^2 \psi_0 = -4\pi \int d^2\mathbf{P} F_0(\mathbf{q}, \mathbf{P}). \tag{9.46}$$

First order $O(\epsilon)$: Taking into account the already obtained lowest order solutions (9.45) and (9.46), the Hamilton-Jacobi equation (9.40) to first order in ϵ can be expressed as

$$\frac{\partial S_1}{\partial \theta} + \frac{R}{2}\left(P_x^2 + P_z^2\right) + \frac{1}{2R}\left(G_x x^2 + G_z z^2\right) + \lambda \psi_0 = \mathcal{K}_1(\mathbf{q}, \mathbf{P}). \tag{9.47}$$

Imposing the condition that the first order Hamiltonian $\mathcal{K}_1$ be equal to

$$\mathcal{K}_1(\mathbf{q}, \mathbf{P}) = \frac{R}{2}\left(P_x^2 + P_z^2\right) + \frac{1}{2R}\left(\overline{G}_x x^2 + \overline{G}_z z^2\right) + \lambda \psi_0(\mathbf{q}), \tag{9.48}$$

we obtain immediately

$$S_1 = -\frac{1}{2R}\left[\widetilde{G}_x(\theta) x^2 + \widetilde{G}_z(\theta) z^2\right], \tag{9.49}$$

$$\psi_1 \equiv 0. \tag{9.50}$$

Here we have introduced the notation

$$\overline{G}_{x,z} = \frac{1}{\Theta} \int\limits_{\theta_0}^{\theta_0 + \Theta} d\theta G_{x,z}(\theta), \qquad \widetilde{G}_{x,z}(\theta) = \int\limits_{\theta_0}^{\theta_0 + \theta} d\tau \left[G_{x,z}(\tau) - \overline{G}_{x,z}\right]. \tag{9.51}$$

Note that since the focusing coefficients are periodic functions of θ they can be expanded in a Fourier series

$$G_{x,z}(\theta) = \sum_{n=-\infty}^{\infty} G_{x,z}^{(n)} \exp(in\Omega\theta), \tag{9.52}$$

where

$$G_{x,z}^{(n)} = \frac{1}{\Theta} \int\limits_0^{\Theta} d\theta G_{x,z}(\theta) \exp(-in\Omega\theta), \tag{9.53}$$

and $\Omega = 2\pi/\Theta$. Therefore for the quantities $\overline{G}_{x,z}$ and $\widetilde{G}_{x,z}(\theta)$ expressed in terms of the Fourier amplitudes, we obtain

$$\overline{G}_{x,z} = G_{x,z}^{(0)}, \qquad \widetilde{G}_{x,z}(\theta) = -\frac{i}{\Omega} \sum_{n \neq 0} \frac{G_{x,z}^{(n)}}{n} \exp(in\Omega\theta). \tag{9.54}$$

Second order $O(\epsilon^2)$: To this order, the Hamilton-Jacobi equation (9.40) takes the form

$$\frac{\partial S_2}{\partial \theta} - \left(xP_x\widetilde{G}_x + zP_z\widetilde{G}_z\right) = \mathcal{K}_2(\mathbf{q}, \mathbf{P}). \tag{9.55}$$

It is straightforward to solve equation (9.55), yielding the obvious result

$$S_2 = xP_x\widetilde{\widetilde{G}}_x(\theta) + zP_z\widetilde{\widetilde{G}}_z(\theta), \qquad \mathcal{K}_2(\mathbf{q}, \mathbf{P}) \equiv 0. \tag{9.56}$$

For the second order potential ψ_2 we obtain the equation

$$\nabla_{\mathbf{q}}^2\psi_2 = -4\pi\left(\widetilde{\widetilde{G}}_x + \widetilde{\widetilde{G}}_z\right)\int d^2\mathbf{P}F_0(\mathbf{q}, \mathbf{P}), \tag{9.57}$$

or, making use of equation (9.46),

$$\psi_2(\mathbf{q}; \theta) = \left[\widetilde{\widetilde{G}}_x(\theta) + \widetilde{\widetilde{G}}_z(\theta)\right]\psi_0(\mathbf{q}). \tag{9.58}$$

In equations (9.56) – (9.58), $\widetilde{\widetilde{G}}_{x,z}(\theta)$ denotes application of the integral operation in equation (9.51) to $\widetilde{G}_{x,z}(\theta)$, i.e.,

$$\widetilde{\widetilde{G}}_{x,z}(\theta) = \int_{\theta_0}^{\theta_0+\theta} d\tau\widetilde{G}_{x,z}(\tau), \tag{9.59}$$

because $\overline{\widetilde{G}}_{x,z} = 0$.

Third order $O(\epsilon^3)$: To third order in ϵ, the Hamilton-Jacobi equation (9.40) can be written as

$$\frac{\partial S_3}{\partial \theta} + R\left(P_x^2\widetilde{\widetilde{G}}_x + P_z^2\widetilde{\widetilde{G}}_z\right) + \frac{1}{2R}\left(\widetilde{G}_x^2x^2 + \widetilde{G}_z^2z^2\right)$$

$$+\lambda\left(\widetilde{\widetilde{G}}_x + \widetilde{\widetilde{G}}_z\right)\psi_0 = \mathcal{K}_3(\mathbf{q}, \mathbf{P}). \tag{9.60}$$

The third-order Hamiltonian $\mathcal{K}_3$ is given by the expression

$$\mathcal{K}_3(\mathbf{q}, \mathbf{P}) = \frac{1}{2R}\left(\overline{\widetilde{G}_x^2}x^2 + \overline{\widetilde{G}_z^2}z^2\right). \tag{9.61}$$

Equation (9.60) can be easily solved for the third-order generating function S_3. The result is

$$S_3 = -R\left(P_x^2\tilde{\tilde{\tilde{G}}}_x + P_z^2\tilde{\tilde{\tilde{G}}}_z\right) - \frac{1}{2R}\left(\widetilde{\tilde{G}_x^2}x^2 + \widetilde{\tilde{G}_z^2}z^2\right) - \lambda\left(\tilde{\tilde{\tilde{G}}}_x + \tilde{\tilde{\tilde{G}}}_z\right)\psi_0. \tag{9.62}$$

For the third-order electric potential ψ_3 we obtain simply

$$\psi_3 \equiv 0. \tag{9.63}$$

Fourth order $O(\epsilon^4)$: To the fourth order in the expansion parameter ϵ the Hamilton-Jacobi equation (9.40) can be expressed as

$$\frac{\partial S_4}{\partial \theta} - xP_x\left(\widetilde{\tilde{G}_x^2} + \tilde{G}_x\tilde{\tilde{G}}_x\right) - zP_z\left(\widetilde{\tilde{G}_z^2} + \tilde{G}_z\tilde{\tilde{G}}_z\right)$$

$$-\lambda R\left(\tilde{\tilde{\tilde{G}}}_x + \tilde{\tilde{\tilde{G}}}_z\right)\left(P_x\frac{\partial \psi_0}{\partial x} + P_z\frac{\partial \psi_0}{\partial z}\right) = \mathcal{K}_4(\mathbf{q}, \mathbf{P}). \tag{9.64}$$

The obvious condition to impose is that the fourth-order Hamiltonian $\mathcal{K}_4$ be equal to

$$\mathcal{K}_4(\mathbf{q}, \mathbf{P}) = -xP_x\overline{\tilde{G}_x\tilde{\tilde{G}}_x} - zP_z\overline{\tilde{G}_z\tilde{\tilde{G}}_z}. \tag{9.65}$$

With equation (9.65) in hand, it is straightforward to solve the fourth-order Hamilton-Jacobi equation (9.64) for S_4. We obtain

$$S_4 = xP_x\left(\widetilde{\widetilde{\widetilde{\tilde{G}_x^2}}} + \widetilde{\tilde{G}_x\tilde{\tilde{G}}_x}\right) + zP_z\left(\widetilde{\widetilde{\widetilde{\tilde{G}_z^2}}} + \widetilde{\tilde{G}_z\tilde{\tilde{G}}_z}\right)$$

$$+\lambda R\left(\tilde{\tilde{\tilde{G}}}_x + \tilde{\tilde{\tilde{G}}}_z\right)\left(P_x\frac{\partial \psi_0}{\partial x} + P_z\frac{\partial \psi_0}{\partial z}\right). \tag{9.66}$$

For the fourth-order electric potential ψ_4, we obtain the Poisson equation

$$\nabla_{\mathbf{q}}^2\psi_4 = \Bigg[\tilde{\tilde{\tilde{G}}}_x\tilde{\tilde{\tilde{G}}}_z + \widetilde{\widetilde{\widetilde{\tilde{G}_x^2}}} + \widetilde{\widetilde{\widetilde{\tilde{G}_z^2}}} + \widetilde{\tilde{G}_x\tilde{\tilde{G}}_x}$$

$$+\widetilde{\tilde{G}_z\tilde{\tilde{G}}_z} + \lambda R\left(\tilde{\tilde{\tilde{G}}}_x + \tilde{\tilde{\tilde{G}}}_z\right)\nabla_{\mathbf{q}}^2\psi_0\Bigg]\nabla_{\mathbf{q}}^2\psi_0. \tag{9.67}$$

Fifth order $O(\epsilon^5)$: In fifth order, we are interested in the Hamiltonian $\mathcal{K}_5$. Omitting algebraic details, we find

$$\mathcal{K}_5(\mathbf{q},\mathbf{P}) = \frac{R}{2}\left(\overline{\widetilde{\widetilde{G}}_x^2}P_x^2 + \overline{\widetilde{\widetilde{G}}_z^2}P_z^2\right) + \frac{1}{R}\left(\overline{\widetilde{G}_x\widetilde{\widetilde{G}_x^2}}x^2 + \overline{\widetilde{G}_z\widetilde{\widetilde{G}_z^2}}z^2\right)$$

$$+\lambda\left[\overline{\widetilde{G}_x\left(\widetilde{\widetilde{\widetilde{G}}}_x + \widetilde{\widetilde{\widetilde{G}}}_z\right)}x\frac{\partial\psi_0}{\partial x} + \overline{\widetilde{G}_z\left(\widetilde{\widetilde{\widetilde{G}}}_x + \widetilde{\widetilde{\widetilde{G}}}_z\right)}z\frac{\partial\psi_0}{\partial z}\right]. \tag{9.68}$$

In concluding this section, we collect terms up to fifth order in ϵ in the new Hamiltonian $\mathcal{K} = \mathcal{K}_0 + \epsilon\mathcal{K}_1 + \epsilon^2\mathcal{K}_2 + \ldots$ and set $\epsilon = 1$. This gives

$$\mathcal{K}(\mathbf{q},\mathbf{P}) = \sum_{u=(x,z)}\left(\frac{R\mathcal{A}_u}{2}P_u^2 + \mathcal{B}_u u P_u + \frac{\mathcal{C}_u}{2R}u^2\right) + \lambda\psi_0(\mathbf{q})$$

$$+\lambda\left[\overline{\widetilde{G}_x\left(\widetilde{\widetilde{\widetilde{G}}}_x + \widetilde{\widetilde{\widetilde{G}}}_z\right)}x\frac{\partial\psi_0}{\partial x} + \overline{\widetilde{G}_z\left(\widetilde{\widetilde{\widetilde{G}}}_x + \widetilde{\widetilde{\widetilde{G}}}_z\right)}z\frac{\partial\psi_0}{\partial z}\right], \tag{9.69}$$

where the coefficients $\mathcal{A}_u$, $\mathcal{B}_u$ and $\mathcal{C}_u$ are defined by the expressions

$$\mathcal{A}_u = 1 + \epsilon^4\overline{\widetilde{\widetilde{G}}_u^2}, \tag{9.70}$$

$$\mathcal{B}_u = -\epsilon^3\overline{\widetilde{G}_u\widetilde{\widetilde{G}}_u}, \tag{9.71}$$

$$\mathcal{C}_u = \overline{G}_u + \epsilon^2\overline{\widetilde{G}_u^2} + 2\epsilon^4\overline{\widetilde{G}_u\widetilde{\widetilde{G}_u^2}}. \tag{9.72}$$

The Hamiltonian (9.69), neglecting the contribution from the self-field ψ_0, describes the unperturbed betatron oscillations in both horizontal and vertical directions.

It is useful to compute the unperturbed betatron tunes $\nu_{x,z}$ in terms of averages over the focusing field-strengths. For a Hamiltonian system governed by a quadratic form in the canonical variables of the type in equation (9.69), it is well-known that the characteristic frequencies $\nu_{x,z}$ can be expressed as

$$\nu_u^2 = \mathcal{A}_u\mathcal{C}_u - \mathcal{B}_u^2, \qquad (u = x, z). \tag{9.73}$$

Keeping terms up to sixth order in the perturbation parameter ϵ, we obtain

$$\nu_u^2 = \overline{G}_u + \epsilon^2 \overline{\widetilde{G}_u^2} + \epsilon^4 \left(\overline{G}_u \overline{\widetilde{\widetilde{G}}_u^2} + 2 \overline{\widetilde{G}_u \widetilde{\widetilde{G}_u^2}} \right) + \epsilon^6 \left[\overline{\widetilde{G}_u^2 \overline{\widetilde{\widetilde{G}}_u^2}} - \left(\overline{\widetilde{G}_u \widetilde{\widetilde{G}}_u} \right)^2 \right]. \quad (9.74)$$

In terms of Fourier amplitudes of the focusing coefficients, equation (9.74) can be expressed as

$$\nu_u^2 = G_u^{(0)} + \frac{2\epsilon^2}{\Omega^2} \sum_{n=1}^{\infty} \frac{\left| G_u^{(n)} \right|^2}{n^2} + \frac{2\epsilon^4}{\Omega^4} \left[G_u^{(0)} \sum_{n=1}^{\infty} \frac{\left| G_u^{(n)} \right|^2}{n^4} \right.$$

$$\left. + 2 \sum_{\substack{m,n=1 \\ m \neq n}}^{\infty} \frac{\mathrm{Re}\left(G_u^{(m)*} G_u^{(n)} G_u^{(m-n)} \right)}{mn(m-n)^2} - 2 \sum_{m,n=1}^{\infty} \frac{\mathrm{Re}\left(G_u^{(m)} G_u^{(n)} G_u^{(m+n)*} \right)}{mn(m+n)^2} \right]$$

$$+ \frac{4\epsilon^6}{\Omega^6} \sum_{m,n=1}^{\infty} \frac{\left| G_u^{(m)} \right|^2 \left| G_u^{(n)} \right|^2}{m^2 n^4}. \quad (9.75)$$

For purposes of illustration, we consider a simple *FODO* lattice with equal focusing and defocusing strengths $+G$ and $-G$, and period Θ. We also assume that the longitudinal dimensions θ_f of the focusing and defocusing lenses are equal; the longitudinal dimensions θ_d of the corresponding drift spaces are assumed to be equal as well. Moreover,

$$2(\theta_f + \theta_d) = \Theta. \quad (9.76)$$

For simplicity we consider the horizontal degree of freedom only (the vertical one can be treated in analogous manner). The Fourier amplitudes of the focusing coefficients are

$$G_x^{(2n+1)} = \frac{iG}{(2n+1)\pi} \{ \exp\left[-i(2n+1)\Omega\theta_f \right] - 1 \}, \qquad G_x^{(2n)} = 0, \quad (9.77)$$

where $n = 0, 1, 2, \ldots$. To second order in ϵ, we obtain for the horizontal betatron tune

$$\nu_x^2 = \frac{2\epsilon^2 \Theta^2 G^2}{\pi^4} \sum_{m=1}^{\infty} \frac{1}{(2m-1)^4} \sin^2 \frac{(2m-1)\pi\theta_f}{\Theta}. \quad (9.78)$$

In the limit of infinitely thin lenses, $\theta_f \to 0$, equation (9.78) reduces to the well-known expression

$$\nu_x^2 = \frac{\epsilon^2 \theta_f^2 G^2}{4}, \tag{9.79}$$

where use of the identity

$$\sum_{m=1}^{\infty} \frac{1}{(2m-1)^2} = \frac{\pi^2}{8} \tag{9.80}$$

has been made.

It is evident from equations (9.75) and (9.78), that the Hamiltonian averaging technique developed here represents a powerful formalism for evaluating the betatron tunes in terms of averages over the focusing field strength. The analysis in this section has been carried out to sixth order to demonstrate the ease and flexibility of the mixed-variable approach. In some practical applications, a description to lower order (e.g. fourth order) may be adequate.

9.5 Propagation of an Intense Beam with a Uniform Phase-Space Density

In this paragraph, we consider a special case where considerable analytical simplification occurs in the analysis of the *Vlasov-Poisson* equations (9.32) and (9.33). First of all, we would like to put the basic equations into a form more convenient to handle in the subsequent exposition. We begin with the Hamiltonian (9.34) describing the one-dimensional betatron motion in the presence of space-charge field

$$\widehat{H} = \frac{R}{2}\widehat{p}^2 + \frac{G(\theta)\widehat{x}^2}{2R} + V(\widehat{x};\theta) + \frac{eR}{E_s \beta_s^2}\varphi_{sc}(\widehat{x};\theta), \tag{9.81}$$

where by $V(\widehat{x};\theta)$, we have denoted the contribution coming from nonlinear lattice elements (sextupoles, octupoles, etc.). In addition, $\varphi_{sc}(\widehat{x};\theta)$ is the self-field potential due to space charge, which satisfies the one-dimensional Poisson equation

$$\frac{\partial^2 \varphi_{sc}}{\partial \widehat{x}^2} = -\frac{en}{\epsilon_0} \int d\widehat{p} f(\widehat{x}, \widehat{p};\theta). \tag{9.82}$$

Here, as usual $n = N_p/V_t$ is the density of beam particles (N_p is the number of particles in the beam, while V_t is the volume of the area occupied by the beam in the transverse direction), ϵ_0 is the dielectric susceptibility and $f(\widehat{x}, \widehat{p}; \theta)$ is the distribution function in phase space. Equations (9.81) and (9.82) can be written in an alternate form as

$$\widehat{H} = \frac{R}{2}\widehat{p}^2 + \frac{G(\theta)\widehat{x}^2}{2R} + V(\widehat{x}; \theta) + \lambda\widetilde{\varphi}(\widehat{x}; \theta), \tag{9.83}$$

$$\frac{\partial^2 \widetilde{\varphi}}{\partial \widehat{x}^2} = -\int \mathrm{d}\widehat{p} f(\widehat{x}, \widehat{p}; \theta), \tag{9.84}$$

where

$$\lambda = \frac{Re^2 N_p}{\epsilon_0 E_s \beta_s^2}, \tag{9.85}$$

is the beam perveance, and $\varphi_{sc} = eN_p\widetilde{\varphi}/\epsilon_0$. Note that the beam perveance λ is a quantity with dimension $meter^{-1}$, while the dimension of $\widetilde{\varphi}$ is $meter^2$. In addition, the one-particle distribution function $f(\widehat{x}, \widehat{p}; \theta)$ entering the integral on the right-hand-side of equation (9.82) is substituted with the normalized one $\widetilde{f}(\widehat{x}, \widehat{p}; \theta) = (1/V_t) f(\widehat{x}, \widehat{p}; \theta)$ possessing the formal normalization property

$$\int \mathrm{d}\widehat{x}\mathrm{d}\widehat{p}\widetilde{f}(\widehat{x}, \widehat{p}; \theta) = 1. \tag{9.86}$$

For simplicity, we will omit the tilde sign in what follows.

Next, we perform the canonical transformation (2.38)

$$\widehat{x} = x\sqrt{\beta}, \qquad \widehat{p} = \frac{1}{\sqrt{\beta}}(p - \alpha x), \tag{9.87}$$

defined by the generating function [see equation (2.37)]

$$F_2(\widehat{x}, p; \theta) = \frac{\widehat{x}p}{\sqrt{\beta}} - \frac{\alpha\widehat{x}^2}{2\beta}, \tag{9.88}$$

As a result, we obtain the new Hamiltonian

$$H = \frac{\dot{\chi}}{2}\left(p^2 + x^2\right) + V(x; \theta) + \lambda\sqrt{\beta}U(x; \theta), \tag{9.89}$$

where the self-field potential $U(x; \theta)$ $\left(\widetilde{\varphi} = \sqrt{\beta}U\right)$ satisfies the equation

$$\frac{\partial^2 U}{\partial x^2} = -\int \mathrm{d}p f(x, p; \theta), \tag{9.90}$$

and

$$\dot{\chi} = \frac{\mathrm{d}\chi}{\mathrm{d}\theta} = \frac{R}{\beta}, \tag{9.91}$$

is the derivative of the phase advance with respect to θ [compare with equation (2.39)].

We are now ready to write the *Vlasov* equation for the one-particle distribution function $f(x, p; \theta)$ in the two-dimensional normalized phase space (x, p). It reads as

$$\frac{\partial f}{\partial \theta} + \dot{\chi} p \frac{\partial f}{\partial x} - \left(\dot{\chi} x + \frac{\partial V}{\partial x} + \lambda \sqrt{\beta} \frac{\partial U}{\partial x} \right) \frac{\partial f}{\partial p} = 0, \tag{9.92}$$

and should be solved self-consistently with the Poisson equation (9.90). Following Davidson et al. [Davidson (2002)], we consider the case where the distribution function $f(x, p; \theta)$ is constant (independent of x, p and θ) inside a region in phase space confined by the simply connected boundary curves $p_{(+)}(x; \theta)$ and $p_{(-)}(x; \theta)$, and zero outside. In other words,

$$f(x, p; \theta) = \mathcal{C}, \qquad \text{for} \qquad p_{(-)}(x; \theta) < p < p_{(+)}(x; \theta), \tag{9.93}$$

and $f(x, p; \theta) = 0$ otherwise.

In order to prove that the expression (9.100) represents the solution of the Vlasov equation (9.92), we write the equations for the characteristics of the latter in the form

$$\frac{\mathrm{d}\theta}{1} = \frac{\mathrm{d}x}{\dot{\chi} p} = -\mathrm{d}p \left(\dot{\chi} x + \frac{\partial V}{\partial x} + \lambda \sqrt{\beta} \frac{\partial U}{\partial x} \right)^{-1} = \frac{\mathrm{d}f}{0}. \tag{9.94}$$

Let us also assume that the distribution function $f(x, p; \theta_0)$ at some initial time θ_0 is given by

$$f(x, p; \theta_0) = f_0(x, p) = \mathcal{C} \left[\mathcal{H} \left(p - p_{(-)}^{(0)}(x) \right) - \mathcal{H} \left(p - p_{(+)}^{(0)}(x) \right) \right]. \tag{9.95}$$

Here $\mathcal{H}(z)$ is the well-known Heaviside function. Suppose now that we have been able to solve the equations for the characteristics (9.94), and we have expressed the solution subsequently according to

$$p = P(x; \theta), \tag{9.96}$$

where $P(x; \theta)$ is an appropriate function of its arguments. At each instant of time θ equation (9.96) defines two curves $p_{(+)}(x; \theta)$ and $p_{(-)}(x; \theta)$ in the phase space (x, p), such that

$$p_{(+)}(x;\theta_0) = p_{(+)}^{(0)}(x), \qquad p_{(-)}(x;\theta_0) = p_{(-)}^{(0)}(x). \tag{9.97}$$

Therefore, if initially the distribution function is given by equation (9.95), then the solution of the Vlasov equation (9.92) at every subsequent instant of time θ is represented by the expression (9.100).

It is interesting to note that the chain of equations (excluding the last equation) for the characteristics (9.94) of the Vlasov equation formally coincide with those for the equation

$$\frac{\partial P}{\partial \theta} + \dot{\chi} P \frac{\partial P}{\partial x} = -\dot{\chi} x - \frac{\partial V}{\partial x} - \lambda\sqrt{\beta}\frac{\partial U}{\partial x}. \tag{9.98}$$

As a matter of fact, equation (9.98) is equivalent to the two equations

$$\frac{\partial p_{(\pm)}}{\partial \theta} + \dot{\chi} p_{(\pm)} \frac{\partial p_{(\pm)}}{\partial x} = -\dot{\chi} x - \frac{\partial V}{\partial x} - \lambda\sqrt{\beta}\frac{\partial U}{\partial x}, \tag{9.99}$$

with the initial conditions (9.97), provided U is defined as a solution to equation (9.104).

The equations for the boundary curves $p_{(\pm)}(x;\theta)$ can be derived as follows. We substitute the explicit form

$$f(x,p;\theta) = \mathcal{C}\big[\mathcal{H}\big(p - p_{(-)}(x;\theta)\big) - \mathcal{H}\big(p - p_{(+)}(x;\theta)\big)\big] \tag{9.100}$$

of the distribution function into the Vlasov equation (9.92). Using the fact that the derivative of the Heaviside function with respect to its argument equals the Dirac δ-function, we obtain

$$\begin{aligned} &-\delta\big(p - p_{(-)}\big)\frac{\partial p_{(-)}}{\partial \theta} + \delta\big(p - p_{(+)}\big)\frac{\partial p_{(+)}}{\partial \theta} \\ &+\dot{\chi} p\left[-\delta\big(p - p_{(-)}\big)\frac{\partial p_{(-)}}{\partial x} + \delta\big(p - p_{(+)}\big)\frac{\partial p_{(+)}}{\partial x}\right] \\ &-\left(\dot{\chi} x + \frac{\partial V}{\partial x} + \lambda\sqrt{\beta}\frac{\partial U}{\partial x}\right)\big[\delta\big(p - p_{(-)}\big) - \delta\big(p - p_{(+)}\big)\big] = 0. \end{aligned} \tag{9.101}$$

Multiplying equation (9.101) first by 1 and then by p, and integrating the result over p, we readily obtain

$$\frac{\partial}{\partial \theta}\big(p_{(+)} - p_{(-)}\big) + \frac{\dot{\chi}}{2}\frac{\partial}{\partial x}\left(p_{(+)}^2 - p_{(-)}^2\right) = 0, \tag{9.102}$$

$$\frac{1}{2}\frac{\partial}{\partial\theta}\left(p_{(+)}^2 - p_{(-)}^2\right) + \frac{\dot{\chi}}{3}\frac{\partial}{\partial x}\left(p_{(+)}^3 - p_{(-)}^3\right)$$

$$= -\left(p_{(+)} - p_{(-)}\right)\left(\dot{\chi}x + \frac{\partial V}{\partial x} + \lambda\sqrt{\beta}\frac{\partial U}{\partial x}\right), \tag{9.103}$$

$$\frac{\partial^2 U}{\partial x^2} = -\mathcal{C}\left(p_{(+)} - p_{(-)}\right), \tag{9.104}$$

It is convenient to recast equations (9.102) - (9.104) in a more familiar form, widely used in hydrodynamics. Let us define

$$\rho = \int_{-\infty}^{\infty} \mathrm{d}p f(x, p; \theta) = \mathcal{C}\left(p_{(+)} - p_{(-)}\right), \tag{9.105}$$

$$\rho v = \int_{-\infty}^{\infty} \mathrm{d}p p f(x, p; \theta) = \frac{\mathcal{C}}{2}\left(p_{(+)}^2 - p_{(-)}^2\right), \tag{9.106}$$

where $\rho(x;\theta)$ and $v(x;\theta)$ are the density and the current velocity, respectively. From the two definitions above, it follows that

$$v = \frac{1}{2}\left(p_{(+)} + p_{(-)}\right). \tag{9.107}$$

Defining further the pressure $\mathcal{P}(x;\theta)$ and the heat flow $\mathcal{Q}(x;\theta)$ and using equation (9.107), we have

$$\mathcal{P} = \int_{-\infty}^{\infty} \mathrm{d}p(p - v)^2 f(x, p; \theta) = \frac{1}{12\mathcal{C}^2}\left[\mathcal{C}\left(p_{(+)} - p_{(-)}\right)\right]^3, \tag{9.108}$$

$$\mathcal{Q} = \int_{-\infty}^{\infty} \mathrm{d}p(p - v)^3 f(x, p; \theta) = 0. \tag{9.109}$$

From equation (9.105), it follows that the pressure can be expressed as

$$\mathcal{P} = \frac{\mathcal{P}_0}{\widehat{\rho}_0^3}\rho^3, \qquad \frac{\mathcal{P}_0}{\widehat{\rho}_0^3} = \frac{1}{12\mathcal{C}^2}. \tag{9.110}$$

In addition, it is straightforward to verify that

$$\frac{\mathcal{C}}{3}\left(p_{(+)}^3 - p_{(-)}^3\right) = \mathcal{P} + \rho v^2, \tag{9.111}$$

which provides a closure for the equations (9.102) and (9.103) governing the evolution of the moments (boundary curves). With all the above definitions and relations in hand, we can write the system of hydrodynamic equations (supplemented by the Poisson equation) in the form

$$\frac{\partial \rho}{\partial \theta} + \dot{\chi}\frac{\partial}{\partial x}(\rho v) = 0, \tag{9.112}$$

$$\frac{\partial}{\partial \theta}(\rho v) + \dot{\chi}\frac{\partial}{\partial x}(\mathcal{P} + \rho v^2) = -\rho\left(\dot{\chi}x + \frac{\partial V}{\partial x} + \lambda\sqrt{\beta}\frac{\partial U}{\partial x}\right), \tag{9.113}$$

$$\frac{\partial^2 U}{\partial x^2} = -\rho. \tag{9.114}$$

Using the continuity equation (9.112), we cast the system of equations (9.112) - (9.114) in its final form

$$\frac{\partial \rho}{\partial \tau} + \frac{\partial}{\partial x}(\rho v) = 0, \tag{9.115}$$

$$\frac{\partial v}{\partial \tau} + v\frac{\partial v}{\partial x} + v_T^2\frac{\partial}{\partial x}(\rho^2) = -x - \frac{\beta}{R}\frac{\partial V}{\partial x} - \lambda\frac{\beta\sqrt{\beta}}{R}\frac{\partial U}{\partial x}, \tag{9.116}$$

$$\frac{\partial^2 U}{\partial x^2} = -\rho, \tag{9.117}$$

which will be the starting point for the subsequent analysis. Here

$$v_T^2 = \frac{3\mathcal{P}_0}{2\widehat{\rho}_0^3}, \tag{9.118}$$

is the normalized thermal speed-squared, and $\tau = \chi(\theta) + \tau_0$ is the new independent "time"-variable.

9.6 Dynamical Equations for the Beam Envelope and for the Mean Emittance

The full solutions for the distribution function $f(x, p; \tau)$ and for the self-field potential $U(x; \tau)$ are of special interest, however, it is extremely difficult to obtain a realistic solution to the *Vlasov-Maxwell* equations for problems of practical interest. In paragraph 9.3, we developed the Hamiltonian averaging technique capable to handle a number of cases only approximately.

We will address now the derivation of exact equations for the evolution of the statistical averages of certain dynamical variables of special interest. Let $A(x, p; \tau)$ be a dynamical variable in the normalized two-dimensional phase space (x, p). The statistical average over the one-particle distribution function $f(x, p; \tau)$ is defined by

$$\langle A \rangle(\tau) = \frac{1}{V} \int \mathrm{d}x \mathrm{d}p A(x, p; \tau) f(x, p; \tau). \tag{9.119}$$

Let us write the *Vlasov* equation (9.92) in the form

$$\frac{\partial f}{\partial \tau} + p \frac{\partial f}{\partial x} - \left(x + \frac{\beta}{R} \frac{\partial V}{\partial x} + \lambda \frac{\beta \sqrt{\beta}}{R} \frac{\partial U}{\partial x} \right) \frac{\partial f}{\partial p} = 0. \tag{9.120}$$

We further assume the boundary conditions

$$f(x, p = \pm\infty; \tau) = 0, \qquad f(|x| > |x_L|, p; \tau) = 0. \tag{9.121}$$

This means that there are no beam particles beyond some transverse dimension $\pm x_L$. For $A = x$ and $A = p$, taking the appropriate moments of equation (9.120) readily gives

$$\frac{\mathrm{d}\langle x \rangle}{\mathrm{d}\tau} = \langle p \rangle, \tag{9.122}$$

$$\frac{\mathrm{d}\langle p \rangle}{\mathrm{d}\tau} = -\langle x \rangle - \frac{\beta}{R} \left\langle \frac{\partial V}{\partial x} \right\rangle - \lambda \frac{\beta \sqrt{\beta}}{R} \left\langle \frac{\partial U}{\partial x} \right\rangle, \tag{9.123}$$

where use has been made of the boundary condition (9.121). The mean value of the particle momentum $\langle p \rangle$ can be eliminated and instead of the two equations (9.122) and (9.123) one can obtain a single equation governing the evolution of the beam centroid $\langle x \rangle(\tau)$

$$\frac{\mathrm{d}^2 \langle x \rangle}{\mathrm{d}\tau^2} + \langle x \rangle = -\frac{\beta}{R} \left\langle \frac{\partial V}{\partial x} \right\rangle - \lambda \frac{\beta \sqrt{\beta}}{R} \left\langle \frac{\partial U}{\partial x} \right\rangle, \tag{9.124}$$

In equation (9.124) the average self-field force, $-\langle \partial U / \partial x \rangle$, is determined self-consistently by solving the *Vlasov-Poisson* equations, while to calculate the average nonlinear external force, $-\langle \partial V / \partial x \rangle$, one has to know the distribution function $f(x, p; \tau)$ only. In the usual case where the potential V is a polynomial in x of order equal or higher than three, $-\langle \partial V / \partial x \rangle$ involves higher moments of the normalized coordinate x. Similar manipulations yield

$$\frac{\mathrm{d}\langle x^2\rangle}{\mathrm{d}\tau} = 2\langle xp\rangle, \tag{9.125}$$

and

$$\frac{\mathrm{d}\langle xp\rangle}{\mathrm{d}\tau} = \langle p^2\rangle - \langle x^2\rangle - \frac{\beta}{R}\left\langle x\frac{\partial V}{\partial x}\right\rangle - \lambda\frac{\beta\sqrt{\beta}}{R}\left\langle x\frac{\partial U}{\partial x}\right\rangle, \tag{9.126}$$

or equivalently,

$$\frac{1}{2}\frac{\mathrm{d}^2\langle x^2\rangle}{\mathrm{d}\tau^2} + \langle x^2\rangle = \langle p^2\rangle - \frac{\beta}{R}\left\langle x\frac{\partial V}{\partial x}\right\rangle - \lambda\frac{\beta\sqrt{\beta}}{R}\left\langle x\frac{\partial U}{\partial x}\right\rangle. \tag{9.127}$$

Taking the second moment in the normalized momentum p, it can be further shown that

$$\frac{\mathrm{d}\langle p^2\rangle}{\mathrm{d}\tau} + 2\langle xp\rangle = -\frac{2\beta}{R}\left\langle p\frac{\partial V}{\partial x}\right\rangle - 2\lambda\frac{\beta\sqrt{\beta}}{R}\left\langle p\frac{\partial U}{\partial x}\right\rangle, \tag{9.128}$$

or equivalently,

$$\frac{1}{2}\frac{\mathrm{d}}{\mathrm{d}\tau}\left(\langle p^2\rangle + \langle x^2\rangle\right) = -\frac{\beta}{R}\left\langle p\frac{\partial V}{\partial x}\right\rangle - \lambda\frac{\beta\sqrt{\beta}}{R}\left\langle p\frac{\partial U}{\partial x}\right\rangle. \tag{9.129}$$

In going over from equation (9.128) to equation (9.129) use has been made of equation (9.125). Note that equations (9.124), (9.127) and (9.129) describing the nonlinear evolution of the moments $\langle x\rangle(\tau)$, $\langle x^2\rangle(\tau)$ and $\langle p^2\rangle(\tau)$ are exact. The dynamics of the moments of the one-particle distribution function is a direct consequence of the underlying evolution in phase space governed by the *Vlasov* equation (9.120). In a similar manner, manipulating the *Vlasov* equation (9.120), dynamical equations can be derived for the evolution of the statistical averages for higher moments $\langle x^m\rangle(\tau)$, $\langle p^n\rangle(\tau)$ and $\langle x^k p^l\rangle(\tau)$, for $m > 2$, $n > 2$ and $k, l > 1$.

The mean-square beam dimension $X_e^2(\tau)$ (or the so-called *beam envelope*) is defined according to the relation

$$X_e^2 = \left\langle (x - \langle x\rangle)^2\right\rangle. \tag{9.130}$$

The obvious identities

$$X_e^2 = \langle x^2\rangle - \langle x\rangle^2, \qquad \left\langle (p - \langle p\rangle)^2\right\rangle = \langle p^2\rangle - \langle p\rangle^2, \tag{9.131}$$

$$\langle (x - \langle x\rangle)(p - \langle p\rangle)\rangle = \frac{1}{2}\frac{\mathrm{d}X_e^2}{\mathrm{d}\tau}, \tag{9.132}$$

hold and will be used in what follows. Differentiating equation (9.132) with respect to τ and making use of equations (9.122), (9.123) and (9.126), we readily obtain

$$\frac{1}{2}\frac{\mathrm{d}^2 X_e^2}{\mathrm{d}\tau^2} + X_e^2 = \left\langle (p - \langle p \rangle)^2 \right\rangle$$

$$-\frac{\beta}{R}\left\langle (x - \langle x \rangle)\frac{\partial V}{\partial x} \right\rangle - \lambda\frac{\beta\sqrt{\beta}}{R}\left\langle (x - \langle x \rangle)\frac{\partial U}{\partial x} \right\rangle. \tag{9.133}$$

Next, we introduce the *average beam emittance* $\widehat{\epsilon}$ by [compare with equation (2.76)]

$$\widehat{\epsilon}^2(\tau) = X_e^2\left\langle (p - \langle p \rangle)^2 \right\rangle - \langle (x - \langle x \rangle)(p - \langle p \rangle) \rangle^2$$

$$= X_e^2\left\langle (p - \langle p \rangle)^2 \right\rangle - X_e^2\left(\frac{\mathrm{d}X_e}{\mathrm{d}\tau}\right)^2. \tag{9.134}$$

Eliminating $\left\langle (p - \langle p \rangle)^2 \right\rangle$ in favour of $\widehat{\epsilon}^2(\tau)$, we obtain the dynamical equation

$$\frac{\mathrm{d}^2 X_e}{\mathrm{d}\tau^2} + X_e = \frac{\widehat{\epsilon}^2(\tau)}{X_e^3} - \frac{\beta}{RX_e}\left\langle (x - \langle x \rangle)\frac{\partial V}{\partial x} \right\rangle - \lambda\frac{\beta\sqrt{\beta}}{RX_e}\left\langle (x - \langle x \rangle)\frac{\partial U}{\partial x} \right\rangle. \tag{9.135}$$

Equation (9.135) is equivalent to equation (9.129) and describes the nonlinear evolution of the beam envelope function X_e.

In a similar manner, one can derive a dynamical equation for the average beam emittance $\widehat{\epsilon}(\tau)$. Differentiating equation (9.134) with respect to τ and taking into account equations (9.123), (9.128), (9.132) and (9.133), we obtain

$$\frac{1}{2}\frac{\mathrm{d}\widehat{\epsilon}^2}{\mathrm{d}\tau} = -X_e^2\frac{\beta}{R}\left\langle (p - \langle p \rangle)\left(\frac{\partial V}{\partial x} + \lambda\sqrt{\beta}\frac{\partial U}{\partial x}\right) \right\rangle$$

$$+X_e\frac{\mathrm{d}X_e}{\mathrm{d}\tau}\frac{\beta}{R}\left\langle (x - \langle x \rangle)\left(\frac{\partial V}{\partial x} + \lambda\sqrt{\beta}\frac{\partial U}{\partial x}\right) \right\rangle. \tag{9.136}$$

In previous chapters, we demonstrated on a number of examples that nonlinear terms due to higher order magnetic multipoles lead to the destruction of the adiabatic action invariant J. This implies that the emittance ϵ defined by equation (2.76) is no longer a constant. Equation (9.136) shows clearly that in addition to the nonlinear external force proportional to $\partial V/\partial x$, space-charge effects proportional to $\partial U/\partial x$ generally cause a variation in the beam emittance $\widehat{\epsilon}(\tau)$ as well.

9.7 Solution of the Equations for the Boundary Curves

As an application of the analysis in paragraph 9.5, we now consider the uniform phase-space distribution, where the boundary curves $p_{(\pm)}(x;\theta)$ are the following functions

$$p_{(\pm)}(x;\tau) = -\frac{A(\tau)}{B(\tau)}x \pm \frac{1}{B(\tau)}\sqrt{2JB(\tau) - x^2} \tag{9.137}$$

of the coordinate x and the time τ. Here $A(\tau)$ and $B(\tau)$ are functions which are related to the *Twiss* parameters α and β [see equations (2.40) and (2.41)], respectively and will be determined later. The parameter J remains to be specified as well. The example we will consider in the present paragraph concerns the case where the external nonlinear force proportional to $-\partial V/\partial x$ is zero. For the density $\rho(x;\tau)$ and for the current velocity $v(x;\tau)$, we obtain

$$\rho(x;\tau) = \frac{2\mathcal{C}}{B(\tau)}\sqrt{2JB(\tau) - x^2}, \tag{9.138}$$

$$v(x;\tau) = -\frac{A(\tau)}{B(\tau)}x. \tag{9.139}$$

Let us first consider the case without the self-field $[U(x;\tau) = 0]$. Substitution of the expressions (9.138) and (9.139) into the continuity equation (9.115) readily gives

$$\frac{\mathrm{d}B}{\mathrm{d}\tau} = -2A. \tag{9.140}$$

Further substitution of the expressions (9.138) and (9.139) into the equation for the current velocity (9.116) yields a second equation for the unknown functions $A(\tau)$ and $B(\tau)$, that is

$$B\frac{\mathrm{d}A}{\mathrm{d}\tau} + A^2 + 8\mathcal{C}^2 v_T^2 = B^2. \tag{9.141}$$

Eliminating $A(\tau)$, we can obtain a single equation for $B(\tau)$

$$\frac{B}{2}\frac{\mathrm{d}^2 B}{\mathrm{d}\tau^2} - \frac{1}{4}\left(\frac{\mathrm{d}B}{\mathrm{d}\tau}\right)^2 + B^2 = 8\mathcal{C}^2 v_T^2. \tag{9.142}$$

It is important to note that equation (9.142) has a form similar to that of equation (2.43), governing the evolution of the beta-function. In analogy with the the *rms* beam size σ [see equation (2.75)], we may define the density envelope function $\mathcal{R}$ by

$$\mathcal{R}(\tau) = \sqrt{\varepsilon B(\tau)}, \tag{9.143}$$

and derive from (9.142) the following equation

$$\frac{\mathrm{d}^2\mathcal{R}}{\mathrm{d}\tau^2} + \mathcal{R} = \frac{8\mathcal{C}^2 v_T^2 \varepsilon^2}{\mathcal{R}^3} \tag{9.144}$$

it satisfies. Here, ε is yet unspecified formal parameter. We also assume that the coordinates of beam particles belong to a finite interval $x \in \left(-x_{(-)}, x_{(+)}\right)$ (see also the next chapter), where without loss of generality

$$x_{(-)}(\tau) = x_{(+)}(\tau) = x_b(\tau), \qquad\qquad x_b(\tau) = \sqrt{2JB(\tau)}. \tag{9.145}$$

Furthermore, from the normalization condition

$$\int\limits_{-x_b}^{x_b} \mathrm{d}x\rho(x;\tau) = 1, \tag{9.146}$$

it is straightforward to obtain

$$J = \frac{1}{2\pi\mathcal{C}}. \tag{9.147}$$

Evidently, the density distribution (9.138) is centered, that is

$$\langle x \rangle = 0. \tag{9.148}$$

For the beam envelope X_e, we obtain the expression

$$X_e^2(\tau) = \frac{B(\tau)}{4\pi\mathcal{C}}. \tag{9.149}$$

We are now ready to identify

$$\mathcal{R}(\tau) = X_e(\tau), \qquad\qquad \varepsilon = \frac{1}{4\pi\mathcal{C}}. \tag{9.150}$$

Hence, the envelope equation (9.144) can be written as

$$\frac{\mathrm{d}^2 X_e}{\mathrm{d}\tau^2} + X_e = \frac{\widehat{\epsilon}^2}{X_e^3}, \qquad\qquad \widehat{\epsilon}^2 = \frac{v_T^2}{2\pi^2}. \tag{9.151}$$

Equation (9.151) possesses a stationary solution

$$X_{e0}^2 = \widehat{\epsilon}, \tag{9.152}$$

so that

$$B_0 = 4\pi\mathcal{C}\widehat{\epsilon}, \qquad A_0 = 0. \tag{9.153}$$

Then the stationary density acquires the form

$$\rho_0(x) = \frac{1}{2\pi\widehat{\epsilon}}\sqrt{4\widehat{\epsilon} - x^2}, \tag{9.154}$$

while the stationary current velocity is zero ($v_0 = 0$). The stationary boundary curves are given by the expressions

$$p_{(\pm)}^{(0)}(x) = \pm\frac{1}{4\pi\widehat{\epsilon}\mathcal{C}}\sqrt{4\widehat{\epsilon} - x^2}. \tag{9.155}$$

It is interesting and instructive to calculate perturbatively the stationary deformation of the boundary curves caused by the action of the space charge. Assuming that the beam perveance is small, we can write the first order perturbation equations

$$\frac{\partial \rho_1}{\partial \tau} + \frac{\partial}{\partial x}(\rho_0 v_1) = 0, \tag{9.156}$$

$$\frac{\partial v_1}{\partial \tau} + 2v_T^2\frac{\partial}{\partial x}(\rho_0\rho_1) = -g(\tau)\frac{\partial U_0}{\partial x}, \tag{9.157}$$

where

$$\rho = \rho_0 + \lambda\rho_1 + \ldots, \qquad v = \lambda v_1 + \ldots, \tag{9.158}$$

the function $g(\tau)$ is given by equation (10.9), and $U_0(x)$ is the zero-order self-potential determined with the stationary density ρ_0. The latter has the explicit form

$$U_0(x) = \widehat{U}_0 + \frac{1}{12\pi\widehat{\epsilon}}\left(4\widehat{\epsilon} - x^2\right)^{3/2} - \frac{1}{\pi}\left(x\arcsin\frac{x}{2X_{e0}} + \sqrt{4\widehat{\epsilon} - x^2}\right). \tag{9.159}$$

Here $\widehat{U}_0$ is an integration constant to be determined. In the smooth focusing approximation ($g(\tau) = a_0$) stationarity is preserved, so that we may again set $v_1 = 0$. Thus, for the first-order correction to the density distribution, we have

$$\rho_1(x) = \frac{a_0}{2v_T^2}\left[\frac{2\widehat{\epsilon}x\arcsin\left(x/2X_{e0}\right)}{\sqrt{4\widehat{\epsilon} - x^2}} + \frac{1}{6}\left(x^2 + 8\widehat{\epsilon}\right) - \frac{2\pi\widehat{\epsilon}\widehat{U}_0}{\sqrt{4\widehat{\epsilon} - x^2}}\right]. \tag{9.160}$$

The constant $\widehat{U}_0$ can be determined from the condition that $\rho_1(x)$ does not change the normalization condition (9.146), that is

$$\int_{-x_b}^{x_b} \mathrm{d}x \rho_1(x) = 0, \tag{9.161}$$

This gives

$$\widehat{U}_0 = \frac{64 X_{e0}}{9\pi^2}. \tag{9.162}$$

Although the first-order density distribution $\rho_1(x)$ is singular at the edges $\pm x_b$ ($x_b = 2X_{e0}$) of the interval of possible locations of the beam particles, it clearly shows the tendency of flattening of the beam profile due the the effect of the space charge.

9.8 Description of Beam Dynamics in Terms of Lagrangian Variables

It is convenient to rewrite the hydrodynamic equations (9.115)–(9.117) as follows

$$\left(\frac{\partial}{\partial \tau} + v\frac{\partial}{\partial x}\right)\rho + \rho\frac{\partial v}{\partial x} = 0, \tag{9.163}$$

$$\left(\frac{\partial}{\partial \tau} + v\frac{\partial}{\partial x}\right)v + v_T^2\frac{\partial}{\partial x}(\rho^2) = -x + \lambda g(\tau) E, \tag{9.164}$$

$$\frac{\partial E}{\partial x} = \rho, \tag{9.165}$$

where $E = -\partial U/\partial x$ is the self-consistent electric field. Equations (9.163)–(9.165) are particularly amenable to analysis in *Lagrangian variables* following the motion of a fluid element [Landau (1987)]. We will show that they acquire a remarkably simple form and reduce to a single (highly nonlinear) equation for the beam density. The Lagrangian variables (z, T) following a fluid element are introduced according to

$$T = \tau, \qquad x = z + \int_0^T \mathrm{d}\tau v(z; \tau). \tag{9.166}$$

It follows from the definition (9.166) that derivatives transform according to

$$\frac{\partial}{\partial x} = \frac{\partial z}{\partial x}\frac{\partial}{\partial z} = \left[1 + \int_0^T \mathrm{d}\tau \frac{\partial v(z;\tau)}{\partial z}\right]^{-1} \frac{\partial}{\partial z}, \tag{9.167}$$

$$\frac{\partial}{\partial \tau} = \frac{\partial}{\partial T} + \frac{\partial z}{\partial T}\frac{\partial}{\partial z} = \frac{\partial}{\partial T} - v(z;T)\left[1 + \int_0^T \mathrm{d}\tau \frac{\partial v(z;\tau)}{\partial z}\right]^{-1} \frac{\partial}{\partial z}. \tag{9.168}$$

Therefore, the substantial derivative entering the continuity equation and the equation for the balance of momentum acquires a remarkably simple form in Lagrangian variables. It is simply the partial derivative with respect to the new time variable

$$\frac{\partial}{\partial \tau} + v\frac{\partial}{\partial x} = \frac{\partial}{\partial T}. \tag{9.169}$$

Hence, the continuity equation in Lagrangian variables becomes

$$\frac{\partial \rho}{\partial T} + \frac{\partial v(z;T)}{\partial z}\left[1 + \int_0^T \mathrm{d}\tau \frac{\partial v(z;\tau)}{\partial z}\right]^{-1} \rho = 0. \tag{9.170}$$

Since, equation (9.170) is linear in ρ, it can be integrated exactly to give

$$\rho(z;T) = \rho_0(z)\left[1 + \int_0^T \mathrm{d}\tau \frac{\partial v(z;\tau)}{\partial z}\right]^{-1}, \tag{9.171}$$

where $\rho_0(z)$ is the initial density profile. *Poisson*'s equation also simplifies in Lagrangian variables. We have

$$\left[1 + \int_0^T \mathrm{d}\tau \frac{\partial v(z;\tau)}{\partial z}\right]^{-1} \frac{\partial E}{\partial z} = \rho(z;T). \tag{9.172}$$

Making use of the explicit solution (9.171), we obtain

$$\frac{\partial E}{\partial z} = \rho_0(z). \tag{9.173}$$

A very important consequence of the transformation to Lagrangian variables is the fact that the self-field is *independent* of T, and depends only on the initial density $\rho_0(z)$.

Let us now see how the equation for the current velocity transforms in Lagrangian variables. We have

$$\frac{\partial v}{\partial T} + v_T^2 \left[1 + \int_0^T \mathrm{d}\tau \frac{\partial v(z;\tau)}{\partial z} \right]^{-1} \frac{\partial}{\partial z} \left\{ \rho_0^2(z) \left[1 + \int_0^T \mathrm{d}\tau \frac{\partial v(z;\tau)}{\partial z} \right]^{-2} \right\}$$

$$= \lambda g(T) E - z - \int_0^T \mathrm{d}\tau v(z;\tau). \tag{9.174}$$

Operating on equation (9.174) with $\partial/\partial z$ and rearranging terms, we readily obtain

$$\frac{\partial^2}{\partial T^2} \left[1 + \int_0^T \mathrm{d}\tau \frac{\partial v(z;\tau)}{\partial z} \right] + \left[1 + \int_0^T \mathrm{d}\tau \frac{\partial v(z;\tau)}{\partial z} \right]$$

$$+ \frac{2v_T^2}{3} \frac{\partial}{\partial z} \left\{ \frac{1}{\rho_0} \frac{\partial}{\partial z} \left(\rho_0^3 \left[1 + \int_0^T \mathrm{d}\tau \frac{\partial v(z;\tau)}{\partial z} \right]^{-3} \right) \right\} = \lambda g(T) \rho_0. \tag{9.175}$$

Equation (9.175) is a partial differential equation for the Jacobian of the transformation (9.166) to Lagrangian variables, which on the other hand represents a density compression factor

$$1 + \int_0^T \mathrm{d}\tau \frac{\partial v(z;\tau)}{\partial z} = \frac{\rho_0(z)}{\rho(z;T)}. \tag{9.176}$$

It is important to mention that as a result of the passing over to Lagrangian description of nonlinear intense beam propagation, we obtained a single equation (9.175) which is fully equivalent to the original hydrodynamic equations (9.163)–(9.165) in laboratory frame. Substitution of the relation (9.176) into equation (9.175) readily gives

$$\frac{\partial^2}{\partial T^2} \left(\frac{1}{\rho} \right) + \frac{1}{\rho} + \frac{2v_T^2}{3\rho_0} \frac{\partial}{\partial z} \left[\frac{1}{\rho_0} \frac{\partial}{\partial z} \left(\rho^3 \right) \right] = \lambda g(T), \tag{9.177}$$

in the region where $\rho_0(z)$ is nonzero.

Nevertheless highly nonlinear and therefore difficult to solve analytically, equation (9.177) constitutes a convenient and powerful tool to study the nonlinear evolution and formation of patterns and coherent structures in intense beams for a wide variety of lattice functions $g(T)$, beam intensity λ and beam emittance v_T^2. The only input characteristics needed to be specified are the initial density profile $\rho_0(z)$ and the initial current velocity $v_0(z)$. Note that equation (9.177) is a well posed initial value problem, since for any initial $\rho_0(z)$ and $v_0(z)$ it follows from equation (9.176) that

$$\left. \frac{\partial \rho(z;T)}{\partial T} \right|_{T=0} = -\rho_0(z) \frac{\partial v_0(z)}{\partial z}. \tag{9.178}$$

Once equation (9.177) is solved for $\rho(z;T)$, the expression on the right-hand-side of equation (9.176) can be used to determine the inverse transformation to laboratory-frame variables (x, τ) defined in equation (9.166).

9.9 Landau Damping

Neglecting the contribution of the lattice nonlinearities (that is, setting $V = 0$), we write once again the *Vlasov* equation (9.120) in the form

$$\frac{\partial f}{\partial \tau} + p \frac{\partial f}{\partial x} - \left(x + \lambda a_0 \frac{\partial U}{\partial x} \right) \frac{\partial f}{\partial p} = 0. \tag{9.179}$$

Here only the first term a_o in the expansion (10.9) is retained, which corresponds to the smooth focusing approximation. As usual, the self-potential satisfies the *Poisson* equation

$$\frac{\partial^2 U}{\partial x^2} = - \int \mathrm{d}p f(x, p; \tau), \tag{9.180}$$

Consider further small perturbation $g_1(x, p; \tau)$ from the equilibrium condition

$$f(x, p; \tau) = \frac{f_0(p)}{\lambda a_0} + g_1(x, p; \tau) \qquad U(x; \tau) = -\frac{x^2}{2\lambda a_0} + U_1(x; \tau). \tag{9.181}$$

The linearized *Vlasov-Poisson* equations read as

$$\frac{\partial g_1}{\partial \tau} + p \frac{\partial g_1}{\partial x} - \frac{\partial U_1}{\partial x} \frac{\partial f_0}{\partial p} = 0. \tag{9.182}$$

$$\frac{\partial^2 U_1}{\partial x^2} = - \int \mathrm{d}p g_1(x, p; \tau), \tag{9.183}$$

Following the original paper of Landau [Landau (1946)], we suppose that at some time $\tau = 0$ the initial value of $g_1(x, p; \tau)$ is $g_{10}(x, p)$. Then the evolution of the beam is determined by equation (9.182) for $\tau > 0$. In order to use the *Fourier* transform, it is convenient to modify equation (9.182) so that the complete range of τ from $-\infty$ to $+\infty$ can be considered, rather than the restricted range $\tau > 0$. This is achieved by replacing equation (9.182) with

$$\frac{\partial g_1}{\partial \tau} + p\frac{\partial g_1}{\partial x} - \frac{\partial U_1}{\partial x}\frac{\partial f_0}{\partial p} = g_{10}(x,p)\delta(\tau), \tag{9.184}$$

where

$$g_1 = 0, \qquad U_1 = 0, \qquad \text{for} \qquad \tau < 0. \tag{9.185}$$

We can now introduce the *Fourier* transforms defined by equations (7.117) and (7.118)

$$\mathcal{F}(x;\tau) = \frac{1}{(2\pi)^2}\int \mathrm{d}\omega \mathrm{d}k \widetilde{\mathcal{F}}(k;\omega)e^{i(kx-\omega\tau)}, \tag{9.186}$$

$$\widetilde{\mathcal{F}}(k;\omega) = \int \mathrm{d}\tau \mathrm{d}x \mathcal{F}(x;\tau)e^{i(\omega\tau - kx)}, \tag{9.187}$$

where $\mathcal{F}(x;\tau)$ is a generic function of its arguments. Then equations (9.184) and (9.183) become

$$(\omega - kp)\widetilde{g}_1 + k\widetilde{U}_1\frac{\partial f_0}{\partial p} = i\widetilde{g}_{10}, \tag{9.188}$$

$$k^2\widetilde{U}_1 = \int \mathrm{d}p\widetilde{g}_1(k,p;\omega). \tag{9.189}$$

On eliminating $\widetilde{U}_1$ and integrating over momentum p, we obtain

$$\int \mathrm{d}p\widetilde{g}_1(p) = -\frac{i}{k\varepsilon(k,\omega)}\int_{\Gamma_p} \frac{\mathrm{d}p\widetilde{g}_{10}(p)}{p - \omega/k}, \tag{9.190}$$

where for present purposes, it is convenient to show explicitly only the argument p. Here Γ_p is a contour, yet to be determined, in the complex p-plane, and $\varepsilon(k,\omega)$ is the *dielectric function* defined as

$$\varepsilon(k,\omega) = 1 - \frac{1}{k^2}\int \frac{\mathrm{d}p f_0'(p)}{p - \omega/k}, \tag{9.191}$$

where $f_0'(p)$ is the derivative of $f_0(p)$. Once the integral on the right-hand-side of equation (9.190) has been evaluated, we can determine $\widetilde{U}_1$

from equation (9.189) and subsequently calculate $\widetilde{g}_1$ from equation (9.188). The evolution in time of a wave with given wave number k can then be determined from the equations

$$g_1(k,p;\tau) = \frac{1}{2\pi}\int_{\Gamma_\omega} \mathrm{d}\omega \widetilde{g}_1(k,p;\omega)e^{-i\omega\tau}, \tag{9.192}$$

$$U_1(k;\tau) = \frac{1}{2\pi}\int_{\Gamma_\omega} \mathrm{d}\omega \widetilde{U}_1(k;\omega)e^{-i\omega\tau}, \tag{9.193}$$

where the contour Γ_ω in the complex ω plane is yet to be determined.

The form of the contour Γ_ω can be specified in accordance with the causality condition (9.185). If it lies above all singularities then it can be displaced in the direction $\omega = +i\infty$, so that for $\tau < 0$ equations (9.192) and (9.193) clearly yield $g_1 = 0$ and $U_1 = 0$. An equivalent contour can be chosen as the one that runs along the real ω axis except for indentations running above all singularities in the upper-half complex ω-plane. This contour may be deformed in the form of a rectangle in the upper-half complex ω-plane. In the limit that it becomes arbitrarily large, the contribution from the vertical segments are zero since the integrand oscillates with an arbitrarily large frequency. The contribution from the upper horizontal segment tends to zero because the integrand now contains the factor $e^{\omega_I\tau}$, where $\tau < 0$ and $\omega_I \to \infty$. This implies that $U_1(k;\tau) = 0$ for $\tau < 0$.

For $\tau > 0$, we can evaluate the integrals by deforming the contour Γ_ω such that it surrounds the poles in the complex ω-plane. If there are poles in the lower-half complex ω-plane, it is necessary that the function $\varepsilon(k,\omega)$ that enters the right-hand-side of equation (9.190), should be extended analytically from the upper half plane to the lower half plane. We assume that k is real and moreover, positive. We also assume that $f_0(p)$ is well behaved and $f_0'(p)$ has no poles on the real p-axis. Since the real p-axis is the original path of integration in the integral in equation (9.191), this integral is well-defined for $\omega_I > 0$, but it cannot be used to evaluate $\varepsilon(k,\omega)$ if ω is real. However, it is possible to extend the definition of $\varepsilon(k,\omega)$ to real values of ω and into the lower half ω-plane by analytic continuation. This can be done by deforming the contour Γ_p from initially the real axis for the case that $\omega_I > 0$ to a form in the p-plane, as ω_I takes negative values, so that no singularity is allowed to cross the contour in the p-plane. Three different cases arise, which we treat separately.

(1) $\omega_I > 0$.
We are assuming that k is real and positive. The integration can be performed with the original path of integration Γ_p that runs along the real p-axis. The result is

$$\varepsilon(k,\omega) = 1 - \frac{1}{k^2}\int_{-\infty}^{\infty} \frac{\mathrm{d}p f_0'(p)}{p - \omega/k}, \qquad \text{for} \qquad \omega_I > 0, \tag{9.194}$$

(2) $\omega_I = 0$.
This case is characteristic with a real pole $p = \omega/k$. The function $\varepsilon(k,\omega)$ can be continued analytically by deforming the original path Γ_p into a new one that runs below the real pole $p = \omega/k$. The direction of integration is positive (counterclockwise) along the semi-circle indentation. Hence, the contribution from this part of the contour is half of the contribution from a path that encloses the pole completely. Therefore,

$$\varepsilon(k,\omega) = 1 - \frac{1}{k^2}\mathcal{P}\int \frac{\mathrm{d}p f_0'(p)}{p - \omega/k} - \frac{i\pi}{k^2} f_0'(\omega/k), \qquad \text{for} \quad \omega_I = 0. \tag{9.195}$$

In the above expression, the integral denotes the *Cauchy* principal value, defined by

$$\mathcal{P}\int \mathrm{d}p\mathcal{F}(p) = \lim_{\Delta\to 0}\left[\int_{-\infty}^{p_0-\Delta} \mathrm{d}p\mathcal{F}(p) + \int_{p_0+\Delta}^{\infty} \mathrm{d}p\mathcal{F}(p)\right], \tag{9.196}$$

where $p = p_0$ is the location of the singularity on the real p-axis.

(3) $\omega_I < 0$.
As ω_I takes on negative values, the singularity shifts to the lower half p-plane. The contour Γ_p is now deformed so that it completely encircles the pole from below. This means that we obtain a full contribution from the circular integration around the pole

$$\varepsilon(k,\omega) = 1 - \frac{1}{k^2}\int \frac{\mathrm{d}p f_0'(p)}{p - \omega/k} - \frac{2i\pi}{k^2} f_0'(\omega/k), \qquad \text{for} \quad \omega_I < 0. \tag{9.197}$$

We can now write the dispersion relation for the three cases considered above

$$\varepsilon(k,\omega) = 1 - \frac{1}{k^2}\int_{-\infty}^{\infty} \frac{\mathrm{d}p f_0'(p)}{p - \omega/k}, \qquad \text{for} \qquad \omega_I > 0, \tag{9.198}$$

$$\varepsilon(k,\omega) = 1 - \frac{1}{k^2}\mathcal{P}\int \frac{\mathrm{d}p f_0'(p)}{p-\omega/k} - \frac{i\pi}{k^2} f_0'(\omega/k), \qquad \text{for} \quad \omega_I = 0, \tag{9.199}$$

and finally

$$\varepsilon(k,\omega) = 1 - \frac{1}{k^2}\int \frac{\mathrm{d}p f_0'(p)}{p-\omega/k} - \frac{2i\pi}{k^2} f_0'(\omega/k), \qquad \text{for} \quad \omega_I < 0. \tag{9.200}$$

If ω lies in the upper-half plane, the wave grows in time the the corresponding mode is unstable. If ω is located on the real ω-axis, the wave has constant in time amplitude and the mode is stable. If ω lies in the lower-half ω-plane, the wave decays in time, and therefore the mode is once again stable.

Let us now analyze in detail the case where the imaginary part of the wave frequency is much smaller than the real part ($\omega_i \ll \omega_r$). We have

$$\varepsilon(k,\omega) = 1 - \frac{1}{k^2}\int_{\Gamma_p} \frac{\mathrm{d}p f_0'(p)}{p-(\omega_r/k) - i(\omega_i/k)}.$$

On expanding to the lowest order in ω_i/k, we re-express $\varepsilon(k,\omega)$ as

$$\varepsilon(k,\omega) = 1 - \frac{1}{k^2}\int \frac{\mathrm{d}p f_0'(p)}{p-\omega_r/k} - i\frac{\omega_r}{k^3}\int \frac{\mathrm{d}p f_0''(p)}{p-\omega_r/k}. \tag{9.201}$$

In both integrals on the right-hand-side of equation (9.201) the pole is located on the real p-axis so that it is necessary to adopt the contour used to obtain expression (9.195). Hence, we obtain

$$\varepsilon(k,\omega) = 1 - \frac{1}{k^2}\mathcal{P}\int \frac{\mathrm{d}p f_0'(p)}{p-\omega_r/k} - i\frac{\pi}{k^2} f_0'\left(\frac{\omega_r}{k}\right)$$

$$-i\frac{\omega_r}{k^3}\mathcal{P}\int \frac{\mathrm{d}p f_0''(p)}{p-\omega_r/k} + \frac{\pi\omega_i}{k^3} f_0''\left(\frac{\omega_r}{k}\right). \tag{9.202}$$

The real and imaginary part of the dispersion equation

$$\varepsilon(k,\omega) = 0$$

can be written as

$$1 - \frac{1}{k^2}\mathcal{P}\int \frac{\mathrm{d}p f_0'(p)}{p-\omega_r/k} + \frac{\pi\omega_i}{k^3} f_0''\left(\frac{\omega_r}{k}\right) = 0, \tag{9.203}$$

and

$$\pi k f_0'\left(\frac{\omega_r}{k}\right) + \omega_i\mathcal{P}\int \frac{\mathrm{d}p f_0''(p)}{p-\omega_r/k} = 0. \tag{9.204}$$

Therefore, to first order, ω_i is given by

$$\omega_i = -\pi k f_0'\Big(\frac{\omega_r}{k}\Big)\Big[\mathcal{P}\int \frac{\mathrm{d}p f_0''(p)}{p-\omega_r/k}\Big]^{-1}. \tag{9.205}$$

The integral in the above expression can be evaluated as follows. To the lowest order in ω_i/ω_r, equation (9.203) reduces to

$$\varepsilon_r(k;\omega_r) = 1 - \frac{1}{k^2}\mathcal{P}\int \frac{\mathrm{d}p f_0'(p)}{p-\omega_r/k} = 0. \tag{9.206}$$

Allowing a variation Δk in k and a variation $\Delta\omega_r$ in ω_r, we obtain

$$\varepsilon_r(k+\Delta k;\omega_r+\Delta\omega_r) = \frac{\partial\varepsilon_r}{\partial k}\Delta k + \frac{\partial\varepsilon_r}{\partial\omega_r}\Delta\omega_r = 0. \tag{9.207}$$

The group velocity Ω_g of waves is defined as

$$\Omega_g = \frac{\Delta\omega_r}{\Delta k} = \frac{\mathrm{d}\omega_r}{\mathrm{d}k} = -\frac{\partial\varepsilon_r}{\partial k}\Big(\frac{\partial\varepsilon_r}{\partial\omega_r}\Big)^{-1}. \tag{9.208}$$

Calculating the derivatives in the above expression explicitly, we can express the integral entering the right-hand-side of equation (9.205) according to

$$\mathcal{P}\int \frac{\mathrm{d}p f_0''(p)}{p-\omega_r/k} = \frac{2k^3}{k\Omega_g - \omega_r}.$$

On substituting this expression into equation (9.205), we finally arrive at

$$\omega_i = \frac{\pi}{2k^2} f_0'\Big(\frac{\omega_r}{k}\Big)(\omega_r - k\Omega_g). \tag{9.209}$$

Suppose now that the equilibrium one-particle distribution function $f_0(p)$ has a form for which

$$f_0(p) = 0 \quad \text{for} \quad |p| > p_0.$$

Integrating by parts in expression (9.191) and expanding the denominator for $|\omega/k| \gg p_0$, we obtain the dispersion equation

$$1 - \frac{1}{\omega^2}\int \mathrm{d}p f_0(p)\Big(1 + \frac{2kp}{\omega} + \frac{3k^2p^2}{\omega^2} + \ldots\Big) = 0. \tag{9.210}$$

If $f_0(p)$ is a symmetric function in p, the second term in the integral vanishes, so that

$$\omega^2 - 1 - \frac{3k^2}{\omega^2}\langle p^2\rangle = 0. \tag{9.211}$$

In the limit $\langle p^2 \rangle \ll \omega^2/k^2$, the dispersion relation becomes

$$\omega^2 = 1 + 3\langle p^2 \rangle k^2. \tag{9.212}$$

Note that the above derivation is approximately correct provided the phase velocity of the wave is much higher than the thermal velocities of the beam particles so that none of the particles is in resonance with the wave.

Chapter 10

Nonlinear Waves and Turbulence in Intense Beams

10.1 Introduction

One of the main goals in the commissioning and operation of modern high energy accelerators and storage rings is the achievement of higher and higher beam currents and charge densities. That is why, of particular importance are the effects of intense self-fields due to space charge and current, influencing the beam propagation, its stability and transport properties. In the previous chapter, we showed that, in general, a complete description of collective processes in intense charged particle beams is provided by the Vlasov-Maxwell equations for the self-consistent evolution of the beam distribution function and the electromagnetic fields. Although the analytical basis for modeling the dynamics and behaviour of space-charge dominated beams is well established, a thorough and satisfactory understanding of collective processes, detailed equilibrium and formation of patterns and coherent structures is far from being complete.

In the present chapter, we will make an attempt to take a view at the description of the evolution and the collective behaviour of intense charged particle beams from an entirely different perspective, as compared to the ones available in the literature. We will be mainly interested in describing the slow evolution of some coarse-grained quantities that are easily measurable, such as the amplitudes of density waves. Due to the nonlinear wave interaction contingent on the nonlinear coupling between the Vlasov and Maxwell equations, one can expect a formation of nontrivial coherent structure that might be fairly stable in space and time. In the subsequent exposition, we will show that solitary wave patterns in the beam density distribution are an irrevocable feature, characteristic of intense beams. Moreover, density condensates in the special case where a parametric resonance in terms of wave frequency between a particular mode of the fundamental density waves and the external focusing occurs, can be formed.

In particular, in the non-resonant case, we will show that the renormalized solution (by means of the renormalization group approach) for the beam density describes the process of formation of *holes (cavitons)* in intense particle beams. The case where a parametric resonance is present is more interesting. In the cold-beam limit it will be demonstrated that the evolution of the forward and the backward resonant wave amplitudes can be well described by a system of two coupled Gross-Pitaevskii equations. Finally, the spectrum of fluctuations of the microscopic phase space density will be analyzed and a scaling law for the envelope of the fluctuation spectrum will be calculated.

10.2 Renormalization Group Reduction of the Hydrodynamic Equations

Before proceeding further, we make an important remark. Let $F(\theta)$ be a periodic function of θ with period 2π. Noting that the phase advance $\chi(\theta)$ can be represented in the form

$$\chi(\theta) = \nu\theta + \chi_p(\theta) - \chi_p(\pi), \tag{10.1}$$

where ν is the betatron tune, and $\chi_p(\theta + 2\pi) = \chi_p(\theta)$, we can expand $F(\theta)$ regarded as a function of χ (respectively τ) in a Fourier series in the new variable χ (respectively τ) as follows

$$F(\theta) = \sum_{n=-\infty}^{\infty} \mathcal{A}_n \exp\left(in\frac{\chi}{\nu}\right). \tag{10.2}$$

The expansion coefficients $\mathcal{A}_n$ are given by the expressions

$$\mathcal{A}_n = \frac{1}{2\pi\nu} \int\limits_{-\pi\nu}^{\pi\nu} \mathrm{d}\chi F(\theta) \exp\left(-in\frac{\chi}{\nu}\right). \tag{10.3}$$

Using the definition of the phase advance (10.1), and choosing $\tau_0 = \chi_p(\pi)$, we can rewrite the Fourier expansion (10.2) as

$$F(\theta) = \sum_{n=-\infty}^{\infty} \mathcal{B}_n \exp\left(in\frac{\tau}{\nu}\right), \tag{10.4}$$

where the expansion coefficients $\mathcal{B}_n$ are given by the expressions

$$\mathcal{B}_n = \frac{R}{2\pi\nu} \int\limits_{-\pi}^{\pi} \mathrm{d}\theta \frac{F(\theta)}{\beta(\theta)} e^{-in\theta} \exp\left[-in\frac{\chi_P(\theta)}{\nu}\right]. \tag{10.5}$$

In what follows, we consider the case where the external potential V is zero, so that the set of equations to be analyzed acquires the form

$$\frac{\partial \rho}{\partial \tau} + \frac{\partial}{\partial x}(\rho v) = 0, \tag{10.6}$$

$$\frac{\partial v}{\partial \tau} + v\frac{\partial v}{\partial x} + v_T^2 \frac{\partial}{\partial x}(\rho^2) = -x - \lambda g(\tau)\frac{\partial U}{\partial x}, \tag{10.7}$$

$$\frac{\partial^2 U}{\partial x^2} = -\rho, \tag{10.8}$$

where

$$g(\tau) = \sum_{n=-\infty}^{\infty} a_n \exp\left(in\frac{\tau}{\nu}\right), \tag{10.9}$$

$$a_n = \frac{1}{2\pi\nu} \int\limits_{-\pi}^{\pi} \mathrm{d}\theta \sqrt{\beta(\theta)} e^{-in\theta} \exp\left[-in\frac{\chi_P(\theta)}{\nu}\right]. \tag{10.10}$$

Further, we assume that there exists a nontrivial solution to equations (10.6) - (10.8) in the interval $x \in \left(-x_{(-)}, x_{(+)}\right)$, and that the sheet beam density is zero ($\varrho = 0$) outside of the interval. Let us introduce the ansatz

$$\rho(x;\tau) = \frac{1}{\mathcal{E}} + \epsilon R(x;\tau), \qquad v(x;\tau) = \frac{x}{\mathcal{E}}\frac{\mathrm{d}\mathcal{E}}{\mathrm{d}\tau} + \epsilon u(x;\tau), \tag{10.11}$$

$$U(x;\tau) = -\frac{x^2}{2\mathcal{E}} + \epsilon \mathcal{U}(x;\tau), \tag{10.12}$$

where the envelope function $\mathcal{E}(\tau)$ is a solution to the equation

$$\frac{\mathrm{d}^2\mathcal{E}}{\mathrm{d}\tau^2} + \mathcal{E} = \lambda g(\tau), \tag{10.13}$$

and therefore can be represented in explicit form as

$$\mathcal{E}(\tau) = \lambda a_0 + \lambda \sum_{n\neq 0} b_n e^{in\tau/\nu}, \qquad b_n = \frac{a_n}{1 - n^2/\nu^2}. \tag{10.14}$$

It enables us to rewrite equations (10.6) - (10.8) as follows

$$\frac{\partial R}{\partial \tau} + \frac{1}{\mathcal{E}} \frac{\partial u}{\partial x} + \frac{\dot{\mathcal{E}}}{\mathcal{E}} \frac{\partial}{\partial x}(xR) + \epsilon \frac{\partial}{\partial x}(Ru) = 0, \tag{10.15}$$

$$\frac{\partial u}{\partial \tau} + \frac{\dot{\mathcal{E}}}{\mathcal{E}} \frac{\partial}{\partial x}(xu) + \frac{2v_T^2}{\mathcal{E}} \frac{\partial R}{\partial x} + \epsilon u \frac{\partial u}{\partial x} + \epsilon v_T^2 \frac{\partial}{\partial x}\left(R^2\right) = -\lambda g \frac{\partial \mathcal{U}}{\partial x}, \tag{10.16}$$

$$\frac{\partial^2 \mathcal{U}}{\partial x^2} = -R.$$

Next, we differentiate the first equation (10.15) of the above system with respect to τ and the second equation (10.16) with respect to x. After summing these up, we obtain

$$\mathcal{E} \frac{\partial}{\partial \tau}\left(\mathcal{E} \frac{\partial R}{\partial \tau}\right) - 2v_T^2 \frac{\partial^2 R}{\partial x^2} + \lambda g \mathcal{E} R$$

$$= \dot{\mathcal{E}} \frac{\partial^2}{\partial x^2}(xu) - \mathcal{E} \frac{\partial}{\partial \tau}\left[\dot{\mathcal{E}} \frac{\partial}{\partial x}(xR)\right]$$

$$+\epsilon \mathcal{E} \frac{\partial^2}{\partial x^2}\left(\frac{u^2}{2} + v_T^2 R^2\right) - \epsilon \mathcal{E} \frac{\partial}{\partial \tau}\left[\mathcal{E} \frac{\partial}{\partial x}(Ru)\right]. \tag{10.17}$$

Equation (10.17) supplemented with equation (10.15) rewritten as

$$\mathcal{E} \frac{\partial R}{\partial \tau} + \frac{\partial u}{\partial x} + \dot{\mathcal{E}} \frac{\partial}{\partial x}(xR) + \epsilon \mathcal{E} \frac{\partial}{\partial x}(Ru) = 0, \tag{10.18}$$

comprises the basic system of equations for the analysis in the subsequent exposition.

In what follows, we consider the case of smooth focusing, where the time variation of $g(\tau)$ [$\mathcal{E}(\tau)$, respectively] can be neglected. In the next section, we will show that when the frequencies of all fundamental modes are sufficiently far from a parametric resonance, this assumption holds true to second order in the perturbation parameter ϵ even in the case when such time variation is present. Thus, the basic equations (10.17) and (10.18) acquire the form

$$\mathcal{E}_0^2 \frac{\partial^2 R}{\partial \tau^2} - 2v_T^2 \frac{\partial^2 R}{\partial x^2} + \mathcal{E}_0^2 R = \epsilon \mathcal{E}_0 \frac{\partial^2}{\partial x^2}\left(\frac{u^2}{2} + v_T^2 R^2\right) - \epsilon \mathcal{E}_0^2 \frac{\partial^2}{\partial \tau \partial x}(Ru), \tag{10.19}$$

$$\mathcal{E}_0 \frac{\partial R}{\partial \tau} + \frac{\partial u}{\partial x} + \epsilon \mathcal{E}_0 \frac{\partial}{\partial x}(Ru) = 0, \tag{10.20}$$

where

$$\mathcal{E}_0 = \lambda a_0. \tag{10.21}$$

Let us further assume that the actual dependence of R and u on the independent variables is given by

$$R = R(x, \xi; \tau), \qquad u = u(x, \xi; \tau), \tag{10.22}$$

where

$$\xi = \epsilon x \tag{10.23}$$

is a slow spatial variable. Therefore, the basic equations (10.19) and (10.20) can be rewritten as

$$\mathcal{E}_0^2 \frac{\partial^2 R}{\partial \tau^2} - 2v_T^2 \left(\frac{\partial^2}{\partial x^2} + 2\epsilon \frac{\partial^2}{\partial x \partial \xi} + \epsilon^2 \frac{\partial^2}{\partial \xi^2} \right) R + \mathcal{E}_0^2 R$$

$$= \epsilon \mathcal{E}_0 \left(\frac{\partial^2}{\partial x^2} + 2\epsilon \frac{\partial^2}{\partial x \partial \xi} + \epsilon^2 \frac{\partial^2}{\partial \xi^2} \right) \left(\frac{u^2}{2} + v_T^2 R^2 \right)$$

$$- \epsilon \mathcal{E}_0^2 \left(\frac{\partial^2}{\partial \tau \partial x} + \epsilon \frac{\partial^2}{\partial \tau \partial \xi} \right) (Ru), \tag{10.24}$$

$$\mathcal{E}_0 \frac{\partial R}{\partial \tau} + \frac{\partial u}{\partial x} + \epsilon \frac{\partial u}{\partial \xi} + \epsilon \mathcal{E}_0 \left(\frac{\partial}{\partial x} + \epsilon \frac{\partial}{\partial \xi} \right) (Ru) = 0, \tag{10.25}$$

Following the basic idea of the RG method, we represent the solution to equations (10.24) and (10.25) in the form of a standard perturbation expansion [Nayfeh (1981)] in the formal small parameter ϵ as

$$R = \sum_{k=0}^{\infty} \epsilon^k R_k, \qquad u = \sum_{k=0}^{\infty} \epsilon^k u_k, \tag{10.26}$$

The zeroth-order equations (10.24) and (10.25) read as

$$\lambda^2 a_0^2 \frac{\partial^2 R_0}{\partial \tau^2} - 2v_T^2 \frac{\partial^2 R_0}{\partial x^2} + \lambda^2 a_0^2 R_0 = 0, \tag{10.27}$$

$$\lambda a_0 \frac{\partial R_0}{\partial \tau} + \frac{\partial u_0}{\partial x} = 0. \tag{10.28}$$

Their solutions can be found in a straightforward manner to be

$$R_0(x,\xi;\tau) = \sum_{m\neq 0}\left[A_m(\xi)e^{iz_m^{(+)}} + B_m(\xi)e^{iz_m^{(-)}}\right], \tag{10.29}$$

$$u_0(x,\xi;\tau) = -\frac{2\pi}{\sigma^2}\sum_{m\neq 0}\frac{\omega_m}{m}\left[A_m(\xi)e^{iz_m^{(+)}} - B_m(\xi)e^{iz_m^{(-)}}\right]. \tag{10.30}$$

Here A_m and B_m are constant complex amplitudes (to this end dependent on the slow variable ξ only) of the backward and the forward wave solution to equation (10.27), respectively. These will be the subject of renormalization at the final step of the renormalization group procedure resulting in RG equations governing their slow evolution. Furthermore,

$$z_m^{(\pm)}(x;\tau) = \omega_m\tau \pm m\sigma x, \qquad \sigma = \frac{2\pi}{x_{(+)} + x_{(-)}}, \qquad \lambda a_0 = \frac{2\pi}{\sigma}. \tag{10.31}$$

The discrete mode frequencies ω_m are determined from the dispersion relation

$$\omega_m^2 = 1 + K^2m^2, \qquad K^2 = \frac{v_T^2\sigma^4}{2\pi^2}. \tag{10.32}$$

It can be easily verified that the above choice of parameters leads to

$$\int_{-x_{(-)}}^{x_{(+)}} \mathrm{d}x R_0(x;\theta) = 0, \tag{10.33}$$

which means that linear perturbation to the uniform density $\rho_0 = \mathcal{E}_0^{-1}$ average to zero and do not affect the normalization properties on the interval $\left(x^{(-)}, x^{(+)}\right)$. In addition, the following conventions and notations

$$\omega_{-m} = -\omega_m, \qquad A_{-m} = A_m^*, \qquad B_{-m} = B_m^* \tag{10.34}$$

have been introduced.

The first order perturbation equation for R_1 reads as

$$\frac{\partial^2 R_1}{\partial\tau^2} - \frac{K^2}{\sigma^2}\frac{\partial^2 R_1}{\partial x^2} + R_1 = \frac{2iK^2}{\sigma}\sum_{m\neq 0} m\left(\frac{\partial A_m}{\partial\xi}e^{iz_m^{(+)}} - \frac{\partial B_m}{\partial\xi}e^{iz_m^{(-)}}\right)$$

$$-\frac{\pi}{\sigma}\sum_{m,n\neq 0}\left[\gamma_{mn}^{(+)}A_mA_ne^{i\left(z_m^{(+)}+z_n^{(+)}\right)} + 2\gamma_{mn}^{(-)}A_mB_ne^{i\left(z_m^{(+)}+z_n^{(-)}\right)}\right.$$

$$+\gamma_{mn}^{(+)} B_m B_n e^{i\left(z_m^{(-)}+z_n^{(-)}\right)}\Big], \tag{10.35}$$

where

$$\gamma_{mn}^{(\pm)} = (m \pm n)\Big[(\omega_m + \omega_n)\Big(\frac{\omega_m}{m} \pm \frac{\omega_n}{n}\Big) + (m \pm n)\Big(K^2 \pm \frac{\omega_m \omega_n}{mn}\Big)\Big]. \tag{10.36}$$

The solution for R_1 is readily obtained to be

$$R_1(x, \xi; \tau) = \frac{K^2 \tau}{\sigma} \sum_{m \neq 0} \frac{m}{\omega_m} \left(\frac{\partial A_m}{\partial \xi} e^{iz_m^{(+)}} - \frac{\partial B_m}{\partial \xi} e^{iz_m^{(-)}} \right)$$

$$-\frac{\pi}{\sigma} \sum_{m,n \neq 0} \Big[\alpha_{mn}^{(+)} A_m A_n e^{i\left(z_m^{(+)}+z_n^{(+)}\right)} + 2\alpha_{mn}^{(-)} A_m B_n e^{i\left(z_m^{(+)}+z_n^{(-)}\right)}$$

$$+\alpha_{mn}^{(+)} B_m B_n e^{i\left(z_m^{(-)}+z_n^{(-)}\right)}\Big], \tag{10.37}$$

where

$$\alpha_{mn}^{(\pm)} = \frac{\gamma_{mn}^{(\pm)}}{\mathcal{D}_{mn}^{(\pm)}}, \tag{10.38}$$

$$\mathcal{D}_{mn}^{(\pm)} = 1 - (\omega_m + \omega_n)^2 + K^2 (m \pm n)^2. \tag{10.39}$$

Note that the (infinite dimensional) matrices $\widehat{\alpha}^{(\pm)}$ are symmetric, i.e.

$$\alpha_{mn}^{(\pm)} = \alpha_{nm}^{(\pm)}, \qquad \alpha_{m,\mp m}^{(\pm)} = 0. \tag{10.40}$$

Having determined $R_1(x; \tau)$, the first-order current velocity $u_1(x; \tau)$ can be found in a straightforward manner. The result is

$$u_1(x, \xi; \tau) = -\frac{2\pi K^2 \tau}{\sigma^3} \sum_{m \neq 0} \left(\frac{\partial A_m}{\partial \xi} e^{iz_m^{(+)}} + \frac{\partial B_m}{\partial \xi} e^{iz_m^{(-)}} \right)$$

$$-\frac{2\pi i}{\sigma^3} \sum_{m \neq 0} \frac{1}{m^2 \omega_m} \left(\frac{\partial A_m}{\partial \xi} e^{iz_m^{(+)}} + \frac{\partial B_m}{\partial \xi} e^{iz_m^{(-)}} \right)$$

$$+\frac{2\pi^2}{\sigma^3} \sum_{m,n \neq 0} \Big[\beta_{mn}^{(+)} A_m A_n e^{i\left(z_m^{(+)}+z_n^{(+)}\right)} + 2\beta_{mn}^{(-)} A_m B_n e^{i\left(z_m^{(+)}+z_n^{(-)}\right)}$$

$$-\beta_{mn}^{(+)} B_m B_n e^{i\left(z_m^{(-)}+z_n^{(-)}\right)}\Big], \tag{10.41}$$

where

$$\beta_{mn}^{(\pm)} = \frac{\omega_m}{m} \pm \frac{\omega_n}{n} + \alpha_{mn}^{(\pm)} \frac{\omega_m + \omega_n}{m \pm n}, \qquad \beta_{m,-m}^{(+)} = 0. \tag{10.42}$$

Note that the (infinite dimensional) matrix $\widehat{\beta}^{(+)}$ possesses the same symmetry properties as those displayed by equation (10.40) possessed by the matrix $\widehat{\alpha}^{(+)}$, while $\widehat{\beta}^{(-)}$ is antisymmetric (and evidently $\beta_{mm}^{(-)} = 0$).

In second order, the equation for $R_2(x, \xi; \tau)$ acquires a form similar to that of equation (10.35). It is important to emphasize that this assertion holds true in every subsequent order. Each entry on the right-hand-sides of the corresponding equations can be calculated explicitly utilizing the already determined quantities from the previous orders. The right-hand-side of the equation for $R_2(x, \xi; \tau)$ contains terms which yield oscillating contributions with constant amplitudes to the solution for $R_2(x, \xi; \tau)$. Apart from these, there are resonant terms (proportional to $e^{iz_m^{(\pm)}(x;\tau)}$) leading to a secular contribution. To complete the renormalization group reduction of the hydrodynamic equations, we select these particular resonant second-order terms on the right-hand-side of the equation determining $R_2(x, \xi; \tau)$. The latter can be written as

$$\frac{\partial^2 R_2}{\partial \tau^2} - \frac{K^2}{\sigma^2} \frac{\partial^2 R_2}{\partial x^2} + R_2 = \frac{K^2}{\sigma^2} \sum_{m \neq 0} \left(\frac{\partial^2 A_m}{\partial \xi^2} e^{iz_m^{(+)}} + \frac{\partial^2 B_m}{\partial \xi^2} e^{iz_m^{(-)}} \right)$$

$$+ \frac{2iK^4 \tau}{\sigma^2} \sum_{m \neq 0} \frac{m^2}{\omega_m} \left(\frac{\partial^2 A_m}{\partial \xi^2} e^{iz_m^{(+)}} + \frac{\partial^2 B_m}{\partial \xi^2} e^{iz_m^{(-)}} \right)$$

$$+ \frac{4\pi^2}{\sigma^2} \sum_{m,n \neq 0} \left[\left(\Gamma_{mn}^{(+)} |A_n|^2 + \Gamma_{mn}^{(-)} |B_n|^2 \right) A_m e^{iz_m^{(+)}} \right.$$

$$\left. + \left(\Gamma_{mn}^{(-)} |A_n|^2 + \Gamma_{mn}^{(+)} |B_n|^2 \right) B_m e^{iz_m^{(-)}} \right], \tag{10.43}$$

where

$$\Gamma_{mn}^{(+)} = m^2 \left[\beta_{mn}^{(+)} \left(\frac{\omega_m}{m} + \frac{\omega_n}{n} \right) + \alpha_{mn}^{(+)} \left(K^2 + \frac{\omega_m \omega_n}{mn} \right) \right] \left(1 - \frac{\delta_{mn}}{2} \right), \tag{10.44}$$

$$\Gamma_{mn}^{(-)} = m^2 \left[\beta_{mn}^{(-)} \left(\frac{\omega_m}{m} - \frac{\omega_n}{n} \right) + \alpha_{mn}^{(-)} \left(K^2 - \frac{\omega_m \omega_n}{mn} \right) \right], \tag{10.45}$$

Some straightforward algebra yields the solution for $R_2(x, \xi; \tau)$ in the form

$$R_2(x, \xi; \tau) = \frac{K^4 \tau^2}{2\sigma^2} \sum_{m \neq 0} \frac{m^2}{\omega_m^2} \left(\frac{\partial^2 A_m}{\partial \xi^2} e^{iz_m^{(+)}} + \frac{\partial^2 B_m}{\partial \xi^2} e^{iz_m^{(-)}} \right)$$

$$+ \frac{K^2 \tau}{2i\sigma^2} \sum_{m \neq 0} \frac{1}{\omega_m^3} \left(\frac{\partial^2 A_m}{\partial \xi^2} e^{iz_m^{(+)}} + \frac{\partial^2 B_m}{\partial \xi^2} e^{iz_m^{(-)}} \right)$$

$$+ \frac{2\pi^2 \tau}{i\sigma^2} \sum_{m,n \neq 0} \frac{1}{\omega_m} \left[\left(\Gamma_{mn}^{(+)} |A_n|^2 + \Gamma_{mn}^{(-)} |B_n|^2 \right) A_m e^{iz_m^{(+)}} \right.$$

$$\left. + \left(\Gamma_{mn}^{(-)} |A_n|^2 + \Gamma_{mn}^{(+)} |B_n|^2 \right) B_m e^{iz_m^{(-)}} \right], \tag{10.46}$$

where non-secular oscillating terms have not been written out in full explicitly. The final step is to collect the terms proportional to the fundamental modes $e^{iz_m^{(+)}}$ and $e^{iz_m^{(-)}}$ in all orders and renormalize the amplitudes A_m and B_m. As a result one obtains the following RG equations

$$2i\omega_m \frac{\partial A_m}{\partial \tau} - 2im \frac{K^2}{\sigma} \frac{\partial A_m}{\partial x} = \frac{K^2}{\sigma^2 \omega_m^2} \frac{\partial^2 A_m}{\partial x^2}$$

$$+ \frac{4\pi^2}{\sigma^2} A_m \sum_{n \neq 0} \left(\Gamma_{mn}^{(+)} |A_n|^2 + \Gamma_{mn}^{(-)} |B_n|^2 \right), \tag{10.47}$$

$$2i\omega_m \frac{\partial B_m}{\partial \tau} + 2im \frac{K^2}{\sigma} \frac{\partial B_m}{\partial x} = \frac{K^2}{\sigma^2 \omega_m^2} \frac{\partial^2 B_m}{\partial x^2}$$

$$+ \frac{4\pi^2}{\sigma^2} B_m \sum_{n \neq 0} \left(\Gamma_{mn}^{(-)} |A_n|^2 + \Gamma_{mn}^{(+)} |B_n|^2 \right), \tag{10.48}$$

10.3 The Parametric Wave-Particle Resonance

Let us now address the system of equations (10.17) and (10.18). Without loss of generality, we assume that the time variation of $g(\tau)$ can be treated as a second-order perturbation (which is usually the case), that is

$$g(\tau) = a_0 + \epsilon^2 \sum_{n \neq 0} a_n e^{in\tau/\nu}. \tag{10.49}$$

The same holds true for the envelope function $\mathcal{E}(\tau)$

$$\mathcal{E}(\tau) = \lambda a_0 + \epsilon^2 \lambda \sum_{n \neq 0} b_n e^{in\tau/\nu}, \qquad b_n = \frac{a_n}{1 - n^2/\nu^2}. \tag{10.50}$$

Thus, the basic equations (10.17) and (10.18) can be rewritten in the form

$$\mathcal{E}\frac{\partial}{\partial \tau}\left(\mathcal{E}\frac{\partial R}{\partial \tau}\right) - 2v_T^2 \frac{\partial^2 R}{\partial x^2} + \lambda g \mathcal{E} R$$

$$= \epsilon^2 \dot{\mathcal{E}}_2 \frac{\partial^2}{\partial x^2}(xu) - \epsilon^2 \mathcal{E} \frac{\partial}{\partial \tau}\left[\dot{\mathcal{E}}_2 \frac{\partial}{\partial x}(xR)\right]$$

$$+ \epsilon \mathcal{E} \frac{\partial^2}{\partial x^2}\left(\frac{u^2}{2} + v_T^2 R^2\right) - \epsilon \mathcal{E} \frac{\partial}{\partial \tau}\left[\mathcal{E} \frac{\partial}{\partial x}(Ru)\right]. \tag{10.51}$$

$$\mathcal{E}\frac{\partial R}{\partial \tau} + \frac{\partial u}{\partial x} + \epsilon^2 \dot{\mathcal{E}}_2 \frac{\partial}{\partial x}(xR) + \epsilon \mathcal{E} \frac{\partial}{\partial x}(Ru) = 0, \tag{10.52}$$

Let us reiterate that the assumption concerning the second order of magnitude in ϵ of the time variation of $g(\tau)$ and $\mathcal{E}(\tau)$ does not restrict the generality of the subsequent analysis. If this variation were of first order in ϵ, the proper perturbation parameter to use would be $\sqrt{\epsilon}$ instead of ϵ. In addition, the variables R and u have to be rescaled accordingly

$$R \longrightarrow \frac{R}{\sqrt{\epsilon}}, \qquad u \longrightarrow \frac{u}{\sqrt{\epsilon}}. \tag{10.53}$$

In terms of the ansatz (10.22) and (10.23), equations (10.51) and (10.52) become

$$\mathcal{E}\frac{\partial}{\partial \tau}\left(\mathcal{E}\frac{\partial R}{\partial \tau}\right) - 2v_T^2 \left(\frac{\partial^2}{\partial x^2} + 2\epsilon \frac{\partial^2}{\partial x \partial \xi} + \epsilon^2 \frac{\partial^2}{\partial \xi^2}\right) R + \lambda g \mathcal{E} R$$

$$= \epsilon \dot{\mathcal{E}}_2 \left(\frac{\partial^2}{\partial x^2} + 2\epsilon \frac{\partial^2}{\partial x \partial \xi} + \epsilon^2 \frac{\partial^2}{\partial \xi^2}\right)(\xi u) - \epsilon \mathcal{E} \frac{\partial}{\partial \tau}\left[\dot{\mathcal{E}}_2 \left(\frac{\partial}{\partial x} + \epsilon \frac{\partial}{\partial \xi}\right)(\xi R)\right]$$

$$+ \epsilon \mathcal{E} \left(\frac{\partial^2}{\partial x^2} + 2\epsilon \frac{\partial^2}{\partial x \partial \xi} + \epsilon^2 \frac{\partial^2}{\partial \xi^2}\right)\left(\frac{u^2}{2} + v_T^2 R^2\right)$$

$$- \epsilon \mathcal{E} \frac{\partial}{\partial \tau}\left[\mathcal{E}\left(\frac{\partial}{\partial x} + \epsilon \frac{\partial}{\partial \xi}\right)(Ru)\right], \tag{10.54}$$

$$\mathcal{E}\frac{\partial R}{\partial \tau} + \frac{\partial u}{\partial x} + \epsilon \frac{\partial u}{\partial \xi} + \epsilon \dot{\mathcal{E}}_2 \left(\frac{\partial}{\partial x} + \epsilon \frac{\partial}{\partial \xi} \right) (\xi R) + \epsilon \mathcal{E} \left(\frac{\partial}{\partial x} + \epsilon \frac{\partial}{\partial \xi} \right) (Ru) = 0. \tag{10.55}$$

Although the general case can be in principle treated through more labour-intensive manipulations, for the sake of simplicity in what follows, we select a particular mode with mode number m. The zeroth-order solution can be written in the form

$$R_0(x, \xi; \tau) = A(\xi) e^{iz^{(+)}(x,\tau)} + B(\xi) e^{iz^{(-)}(x,\tau)} + c.c., \tag{10.56}$$

$$u_0(x, \xi; \tau) = -\frac{2\pi\omega}{m\sigma^2} \left[A(\xi) e^{iz^{(+)}(x,\tau)} - B(\xi) e^{iz^{(-)}(x,\tau)} \right] + c.c., \tag{10.57}$$

where again

$$z^{(\pm)}(x, \tau) = \omega\tau \pm m\sigma x, \qquad \omega^2 = 1 + K^2 m^2.$$

The particular mode is chosen such that an exact parametric resonance (if possible)

$$2\omega = \frac{n_0}{\nu} \tag{10.58}$$

occurs for some integer n_0. The first-order perturbation equation for R_1 can be written in the form

$$\frac{\partial^2 R_1}{\partial \tau^2} - \frac{K^2}{\sigma^2} \frac{\partial^2 R_1}{\partial x^2} + R_1 = \frac{2imK^2}{\sigma} \left(\frac{\partial A}{\partial \xi} e^{iz^{(+)}} - \frac{\partial B}{\partial \xi} e^{iz^{(-)}} \right)$$

$$+ \frac{im\sigma^2}{2\pi} \xi \lambda \sum_{n \neq 0}{}' \frac{n}{\nu} \left(2\omega + \frac{n}{\nu} \right) b_n e^{in\tau/\nu} \left(A e^{iz^{(+)}} - B e^{iz^{(-)}} \right)$$

$$- \frac{4\pi}{\sigma} (4\omega^2 - 1) \left(A^2 e^{2iz^{(+)}} + B^2 e^{2iz^{(-)}} \right) + \frac{8\pi}{\sigma} AB^* e^{i\left(z^{(+)} - z^{(-)}\right)} + c.c.. \tag{10.59}$$

Here $\sum'$ implies exclusion of the harmonic with $n = n_0$ from the sum. It is important to note that terms giving rise to secular contribution due to the parametric resonance vanish identically in the first order. The solution for R_1 can be written explicitly as

$$R_1(x, \xi; \tau) = \frac{mK^2}{\sigma\omega} \tau \left(\frac{\partial A}{\partial \xi} e^{iz^{(+)}} - \frac{\partial B}{\partial \xi} e^{iz^{(-)}} \right)$$

$$-\frac{im\sigma^2}{2\pi}\xi\lambda\sum_{n\neq 0}{}' b_n e^{in\tau/\nu}\Big(Ae^{iz^{(+)}} - Be^{iz^{(-)}}\Big)$$

$$+\frac{4\pi}{3\sigma}(4\omega^2-1)\Big(A^2e^{2iz^{(+)}} + B^2e^{2iz^{(-)}}\Big) + \frac{8\pi}{\sigma(4\omega^2-3)}AB^* e^{i\left(z^{(+)}-z^{(-)}\right)} + c.c.. \tag{10.60}$$

Straightforward calculations yield the solution for u_1

$$u_1(x,\xi;\tau) = -\frac{2\pi i}{m^2\sigma^3\omega}\left(\frac{\partial A}{\partial \xi}e^{iz^{(+)}} + \frac{\partial B}{\partial \xi}e^{iz^{(-)}}\right)$$

$$-\frac{2\pi K^2}{\sigma^3}\tau\left(\frac{\partial A}{\partial \xi}e^{iz^{(+)}} + \frac{\partial B}{\partial \xi}e^{iz^{(-)}}\right)$$

$$-\frac{i\lambda n_0}{\nu}\xi\Big(b_{n_0}e^{in_0\tau/\nu} - b_{n_0}^* e^{-in_0\tau/\nu}\Big)\Big(Ae^{iz^{(+)}} + Be^{iz^{(-)}}\Big)$$

$$+i\omega\xi\lambda\sum_{n\neq 0}{}' b_n e^{in\tau/\nu}\Big(Ae^{iz^{(+)}} + Be^{iz^{(-)}}\Big)$$

$$-\frac{8\pi^2}{3m\sigma^3}\omega(4\omega^2-5/2)\Big(A^2e^{2iz^{(+)}} - B^2e^{2iz^{(-)}}\Big) + c.c.. \tag{10.61}$$

In second order, we take into account terms providing secular contribution to the solution only. Thus, we write the equation for R_2 as

$$\frac{\partial^2 R_2}{\partial \tau^2} - \frac{K^2}{\sigma^2}\frac{\partial^2 R_2}{\partial x^2} + R_2 = \frac{2iK^4m^2}{\omega\sigma^2}\tau\left(\frac{\partial^2 A}{\partial \xi^2}e^{iz^{(+)}} + \frac{\partial^2 B}{\partial \xi^2}e^{iz^{(-)}}\right)$$

$$+\frac{\lambda\sigma}{\pi}b_{n_0}(1-\omega^2)\left[\left(\xi\frac{\partial}{\partial \xi} - 1\right)B^* e^{iz^{(+)}} + \left(\xi\frac{\partial}{\partial \xi} - 1\right)A^* e^{iz^{(-)}}\right]$$

$$+\frac{K^2}{\sigma^2}\left(\frac{\partial^2 A}{\partial \xi^2}e^{iz^{(+)}} + \frac{\partial^2 B}{\partial \xi^2}e^{iz^{(-)}}\right)$$

$$+\frac{2\lambda^2\sigma^2}{\pi^2}\omega^2 m^2\sigma^2 |b_{n_0}|^2\xi^2\Big(Ae^{iz^{(+)}} + Be^{iz^{(-)}}\Big)$$

$$-\frac{8\pi^2}{3\sigma^2}(16\omega^4 - 11\omega^2 + 1)\Big(|A|^2 Ae^{iz^{(+)}} + |B|^2 Be^{iz^{(-)}}\Big)$$

$$+\frac{16\pi^2}{\sigma^2(4\omega^2-3)}\left(|B|^2 Ae^{iz^{(+)}}+|A|^2 Be^{iz^{(-)}}\right)+c.c.. \tag{10.62}$$

Similar to equation (10.46), we obtain in a straightforward manner

$$R_2(x,\xi;\tau)=\frac{K^4m^2}{2\omega^2\sigma^2}\tau^2\left(\frac{\partial^2 A}{\partial\xi^2}e^{iz^{(+)}}+\frac{\partial^2 B}{\partial\xi^2}e^{iz^{(-)}}\right)$$

$$+\frac{K^2\tau}{2i\omega^3\sigma^2}\left(\frac{\partial^2 A}{\partial\xi^2}e^{iz^{(+)}}+\frac{\partial^2 B}{\partial\xi^2}e^{iz^{(-)}}\right)$$

$$+\frac{\lambda\sigma\tau}{2\pi i\omega}b_{n_0}(1-\omega^2)\left[\left(\xi\frac{\partial}{\partial\xi}-1\right)B^*e^{iz^{(+)}}+\left(\xi\frac{\partial}{\partial\xi}-1\right)A^*e^{iz^{(-)}}\right]$$

$$+\frac{2\lambda^2\sigma^2\tau}{2i\pi^2\omega}\omega^2m^2\sigma^2|b_{n_0}|^2\xi^2\left(Ae^{iz^{(+)}}+Be^{iz^{(-)}}\right)$$

$$-\frac{\tau}{2i\omega}\frac{8\pi^2}{3\sigma^2}(16\omega^4-11\omega^2+1)\left(|A|^2 Ae^{iz^{(+)}}+|B|^2 Be^{iz^{(-)}}\right)$$

$$+\frac{\tau}{2i\omega}\frac{16\pi^2}{\sigma^2(4\omega^2-3)}\left(|B|^2 Ae^{iz^{(+)}}+|A|^2 Be^{iz^{(-)}}\right)+c.c.. \tag{10.63}$$

Collecting once again the zeroth-order term together with all secular terms in higher orders and renormalizing the amplitudes A and B, we obtain the RG equations

$$2i\omega\frac{\partial A}{\partial\tau}-2im\frac{K^2}{\sigma}\frac{\partial A}{\partial x}=\frac{K^2}{\sigma^2\omega^2}\frac{\partial^2 A}{\partial x^2}+\frac{\lambda\sigma}{\pi}b_{n_0}(1-\omega^2)\left(x\frac{\partial}{\partial x}-1\right)B^*$$

$$+\frac{2\lambda^2\sigma^4}{\pi^2}m^2\omega^2|b_{n_0}|^2x^2A-\frac{8\pi^2}{3\sigma^2}(16\omega^4-11\omega^2+1)|A|^2A+\frac{16\pi^2}{\sigma^2(4\omega^2-3)}|B|^2A, \tag{10.64}$$

$$2i\omega\frac{\partial B}{\partial\tau}+2im\frac{K^2}{\sigma}\frac{\partial B}{\partial x}=\frac{K^2}{\sigma^2\omega^2}\frac{\partial^2 B}{\partial x^2}+\frac{\lambda\sigma}{\pi}b_{n_0}(1-\omega^2)\left(x\frac{\partial}{\partial x}-1\right)A^*$$

$$+\frac{2\lambda^2\sigma^4}{\pi^2}m^2\omega^2|b_{n_0}|^2x^2B-\frac{8\pi^2}{3\sigma^2}(16\omega^4-11\omega^2+1)|B|^2B+\frac{16\pi^2}{\sigma^2(4\omega^2-3)}|A|^2B. \tag{10.65}$$

10.4 The Nonlinear Schrodinger Equation for a Single Mode

Equations (10.47) and (10.48) represent a system of coupled nonlinear Schrodinger equations for the mode amplitudes A_m and B_m. Neglecting the contribution from modes with $n \neq \pm m$ and introducing a new amplitude $\widetilde{B}_m(x;\tau) = B_m(-x;\tau)$, for the single mode amplitudes A_m and B_m, we obtain the equations

$$i\frac{\partial A}{\partial \tau} - \frac{\partial^2 A}{\partial \zeta^2} - \frac{8\pi^2}{\omega\sigma^2(4\omega^2-3)}\left(-G|A|^2 + |B|^2\right)A = 0, \tag{10.66}$$

$$i\frac{\partial B}{\partial \tau} - \frac{\partial^2 B}{\partial \zeta^2} - \frac{8\pi^2}{\omega\sigma^2(4\omega^2-3)}\left(|A|^2 - G|B|^2\right)B = 0, \tag{10.67}$$

where

$$\zeta = \sqrt{2\omega}\left(mK\tau + \frac{\omega\sigma}{K}\right), \qquad G = \frac{4\omega^2-3}{6}\left(16\omega^4 - 11\omega^2 + 1\right), \tag{10.68}$$

the index m and the tilde sign over the new amplitude $\widetilde{B}$ has been omitted. Clearly enough, equations (10.66) and (10.67) follow directly from the system (10.64) - (10.65) if b_{n_0} is set to zero. This implies that in the case where a parametric resonance does not occur in the original system, the smooth approximation is valid to second order in the formal perturbation parameter. Moreover, it is straightforward to verify that G is always positive. The system of two coupled nonlinear Schrodinger equations (10.66) and (10.67) is in general non integrable. Its integrability is proven by Manakov [Manakov (1973)] only in the simplest case of $G = -1$.

If one of the amplitudes (B or A) in its capacity of being a particular solution to the corresponding nonlinear Schrodinger equation is identically zero, the equation for the other amplitude (say A) becomes

$$i\frac{\partial A}{\partial \tau} - \frac{\partial^2 A}{\partial \zeta^2} + \Gamma|A|^2 A = 0, \tag{10.69}$$

where

$$\Gamma = \frac{4\pi^2}{3\omega\sigma^2}\left(16\omega^4 - 11\omega^2 + 1\right). \tag{10.70}$$

In nonlinear optics equation (10.69) is known to describe the formation and evolution of the so-called *dark solitons* [Kivshar (1998)]. In the case of charged particle beams these correspond to the formation of *holes* or *cavitons* in the beam. The solution to equation (10.69) can be written as

$$A(\zeta;\tau) = \frac{r(\xi)}{\sqrt{\Gamma}} e^{i[n\tau+\theta(\xi)]}, \qquad \xi = \zeta - c\tau, \tag{10.71}$$

where

$$r(\xi) = \sqrt{n - 2a^2 \mathrm{sech}^2(a\xi)}, \qquad \theta(\xi) = \arctan\left[\frac{2a}{c}\tanh(a\xi)\right], \tag{10.72}$$

for all c and

$$a = \frac{1}{2}\sqrt{2n - c^2}, \tag{10.73}$$

provided $n > c^2/2$.

Let us now examine equations (10.64) and (10.65). In the cold-beam limit $v_T \to 0$ (or equivalently $\omega \to 1$) the second term on their right-hand-sides can be neglected as compared to the other terms. Therefore, we can write

$$2i\omega \frac{\partial \widetilde{A}}{\partial \tau} = \frac{K^2}{\sigma^2\omega^2}\frac{\partial^2 \widetilde{A}}{\partial x^2} + \frac{2\lambda^2\sigma^4}{\pi^2} m^2\omega^2 |b_{n_0}|^2 x^2 \widetilde{A}$$

$$-\frac{8\pi^2}{3\sigma^2}\left(16\omega^4 - 11\omega^2 + 1\right)\left|\widetilde{A}\right|^2 \widetilde{A} + \frac{16\pi^2}{\sigma^2(4\omega^2 - 3)}\left|\widetilde{B}\right|^2 \widetilde{A}, \tag{10.74}$$

$$2i\omega \frac{\partial \widetilde{B}}{\partial \tau} = \frac{K^2}{\sigma^2\omega^2}\frac{\partial^2 \widetilde{B}}{\partial x^2} + \frac{2\lambda^2\sigma^4}{\pi^2} m^2\omega^2 |b_{n_0}|^2 x^2 \widetilde{B}$$

$$-\frac{8\pi^2}{3\sigma^2}\left(16\omega^4 - 11\omega^2 + 1\right)\left|\widetilde{B}\right|^2 \widetilde{B} + \frac{16\pi^2}{\sigma^2(4\omega^2 - 3)}\left|\widetilde{A}\right|^2 \widetilde{B}, \tag{10.75}$$

where the local gauge transformation

$$\widetilde{A} = A \exp\left[im\omega\left(\frac{mK^2}{2}\tau + \sigma\omega x\right)\right], \tag{10.76}$$

$$\widetilde{B} = B \exp\left[im\omega\left(\frac{mK^2}{2}\tau - \sigma\omega x\right)\right], \tag{10.77}$$

has been performed on the amplitudes A and B. Equations (10.74) and (10.75) represent a system of two coupled Gross-Pitaevskii equations, known to govern the formation of condensate structures in atomic gases confined in magnetic traps. In addition, superfluidity in helium as well as patterns in the gas of paraexcitons in semiconductors are considered as a possible manifestation of Bose-Einstein condensation. It is remarkable that under certain conditions similar phenomenon can be observed in space-charge dominated beams.

It may be worth noting that an analysis in Lagrangian variables, similar to the one performed in the preceding paragraphs can be carried out adopting equation (9.177) as a starting point. Once the amplitude equations have been derived and possibly solved, the main difficulty here comes from the inverse transformation to laboratory-frame variables.

10.5 Nonlinear Damped Waves in Intense Beams

We begin with the Vlasov equation (9.9)

$$\frac{\partial f}{\partial \tau} + [f, H] = 0, \tag{10.78}$$

which describes the evolution of the one-particle distribution function $f(x, p; \tau)$ in the two-dimensional phase space (x, p). Here

$$[F, G] = \frac{\partial F}{\partial x}\frac{\partial G}{\partial p} - \frac{\partial F}{\partial p}\frac{\partial G}{\partial x}, \tag{10.79}$$

is the Poisson bracket, while τ is the betatron phase advance playing the role of the independent time variable. The Hamiltonian $H(x, p; \tau)$ [see equation (9.89)] can be written as

$$H(x, p; \tau) = \frac{1}{2}\left(p^2 + x^2\right) + \lambda\frac{\beta\sqrt{\beta}}{R}U(x; \tau), \tag{10.80}$$

where again x is the normalized particle deviation from the design orbit and p is the canonical conjugate momentum. In addition, λ is the beam perveance defined by equation (9.85), R is the mean machine radius and β is the beta-function. The Vlasov equation (10.78) should be solved self-consistently with the Poisson equation

$$\frac{\partial^2 U}{\partial x^2} = -\int \mathrm{d}p f(x, p; \tau), \tag{10.81}$$

In the subsequent analysis, we restrict ourselves to the smooth focusing approximation already defined and used in chapter 9 and in the preceding

The multivariable multiple space-time Fourier transformation is then introduced by means of the definition

$$\mathcal{G}(x_0, \tau_0; x_1, \tau_1; \ldots) = \int \prod_{n=0}^{N} \mathrm{d}k_n \mathrm{d}\omega_n \mathcal{G}(k_0, \omega_0; k_1, \omega_1; \ldots)$$

$$\times \exp\left[i \sum_{n=0}^{N} (k_n x_n - \omega_n \tau_n)\right]. \tag{10.112}$$

One can also define the inverse multiple space-time Fourier transformation according to the expression

$$\mathcal{G}(k_0, \omega_0; k_1, \omega_1; \ldots) = \frac{1}{(2\pi)^{2N+2}} \int \prod_{n=0}^{N} \mathrm{d}x_n \mathrm{d}\tau_n \mathcal{G}(x_0, \tau_0; x_1, \tau_1; \ldots)$$

$$\times \exp\left[i \sum_{n=0}^{N} (\omega_n \tau_n - k_n x_n)\right]. \tag{10.113}$$

The next step consists in transforming the equations (10.91)–(10.93). We obtain

$$(\omega - kP)S(k_0, \omega_0; k_1, \omega_1; \ldots) + i\lambda a_0 \varphi(k_0, \omega_0; k_1, \omega_1; \ldots)$$

$$-i\mathcal{C}(k_0, \omega_0; k_1, \omega_1; \ldots)$$

$$= \frac{i}{2} \int \prod_{n=0}^{N} \mathrm{d}k'_n \mathrm{d}\omega'_n k'(k - k') S(k'_0, \omega'_0; \ldots) S(k_0 - k'_0, \omega_0 - \omega'_0; \ldots), \tag{10.114}$$

$$(\omega - kP)F(k_0, \omega_0; k_1, \omega_1; \ldots) + k \frac{\mathrm{d}F_0}{\mathrm{d}P} \mathcal{C}(k_0, \omega_0; k_1, \omega_1; \ldots)$$

$$= -i \int \prod_{n=0}^{N} \mathrm{d}k'_n \mathrm{d}\omega'_n \omega'(k - k') F(k'_0, \omega'_0; \ldots) \frac{\partial}{\partial P} S(k_0 - k'_0, \omega_0 - \omega'_0; \ldots)$$

$$+ \int \prod_{n=0}^{N} \mathrm{d}k'_n \mathrm{d}\omega'_n k' \left[F(k'_0, \omega'_0; \ldots) \frac{\partial}{\partial P} \mathcal{C}(k_0 - k'_0, \omega_0 - \omega'_0; \ldots) \right.$$

$$-\mathcal{C}(k_0', \omega_0'; \dots) \frac{\partial}{\partial P} F(k_0 - k_0', \omega_0 - \omega_0'; \dots) \Big], \tag{10.115}$$

$$k^2 \varphi(k_0, \omega_0; k_1, \omega_1; \dots) = \int \mathrm{d}P F(k_0, \omega_0; \dots) + ik \int \mathrm{d}P F_0 \frac{\partial}{\partial P} S(k_0, \omega_0; \dots)$$

$$+i \int \mathrm{d}P \int \prod_{n=0}^{N} \mathrm{d}k_n' \mathrm{d}\omega_n' k' \left[\frac{\partial}{\partial P} S(k_0', \omega_0'; \dots) \right] F(k_0 - k_0', \omega_0 - \omega_0'; \dots), \tag{10.116}$$

where

$$\omega = \omega_0 + \epsilon \omega_1 + \epsilon^2 \omega_2 + \dots, \qquad k = k_0 + \epsilon k_1 + \epsilon^2 k_2 + \dots, \tag{10.117}$$

and similar relations hold for the primed quantities as well.

Let us expand the transformed quantities in a power series in the formal perturbation parameter ϵ according to

$$S = \epsilon S_1 + \epsilon^2 S_2 + \epsilon^3 S_3 + \dots, \qquad \varphi = \epsilon \varphi_1 + \epsilon^2 \varphi_2 + \epsilon^3 \varphi_3 + \dots, \tag{10.118}$$

$$\mathcal{C} = \epsilon \mathcal{C}_1 + \epsilon^2 \mathcal{C}_2 + \epsilon^3 \mathcal{C}_3 + \dots, \qquad F = \epsilon F_1 + \epsilon^2 F_2 + \epsilon^3 F_3 + \dots, \tag{10.119}$$

and proceed further with solving the transformed equations (10.114)–(10.116) order by order.

First Order $O(\epsilon)$. The first-order equations read as

$$(\omega_0 - k_0 P) S_1 + i \lambda a_0 \varphi_1 - i \mathcal{C}_1 = 0, \tag{10.120}$$

$$(\omega_0 - k_0 P) F_1 + k_0 \frac{\mathrm{d}F_0}{\mathrm{d}P} \mathcal{C}_1 = 0, \tag{10.121}$$

$$k_0^2 \varphi_1 = \int \mathrm{d}P \left(F_1 - i k_0 S_1 \frac{\mathrm{d}F_0}{\mathrm{d}P} \right). \tag{10.122}$$

From the first two equations (10.120) and (10.121), we obtain

$$F_1 - i k_0 S_1 \frac{\mathrm{d}F_0}{\mathrm{d}P} = - \frac{\lambda a_0 k_0 \varphi_1}{\omega_0 - k_0 P} \frac{\mathrm{d}F_0}{\mathrm{d}P}, \tag{10.123}$$

which after substitution in equation (10.122) yields

$$\mathcal{D}(k_0, \omega_0) \varphi_1 = 0. \tag{10.124}$$

Here

$$\mathcal{D}(k_0, \omega_0) = k_0^2 + \lambda a_0 k_0 \int \frac{\mathrm{d}P}{\omega_0 - k_0 P} \frac{\mathrm{d}F_0}{\mathrm{d}P}, \tag{10.125}$$

is the dispersion function. A straightforward observation indicates that the solution for φ_1 can be written as

$$\varphi_1(k_0, \omega_0; k_1, \omega_1; \dots) = \delta(k_0 - \kappa)\delta(\omega_0 - \Omega)E(k_1, \omega_1; k_2, \omega_2; \dots)$$

$$+\delta(k_0 + \kappa)\delta(\omega_0 + \Omega^*)E^*(k_1, \omega_1; k_2, \omega_2; \dots). \tag{10.126}$$

In addition, for a given wave number κ, the wave frequency $\Omega(\kappa)$ satisfies the dispersion equation

$$\mathcal{D}(\kappa, \Omega) = 0, \tag{10.127}$$

and is in general a complex number. Since the long-wave component of the electric field φ_1 is zero, we can set the first-order Hamiltonian $\mathcal{C}_1$ to zero, i.e,

$$\mathcal{C}_1(k_0, \omega_0; k_1, \omega_1; \dots) = 0. \tag{10.128}$$

Therefore, the first-order distribution function F_1 vanishes as well

$$F_1(k_0, \omega_0; k_1, \omega_1; \dots) = 0. \tag{10.129}$$

For the first-order generating function S_1, we obtain

$$S_1(k_0, \omega_0; k_1, \omega_1; \dots) = -\frac{i\lambda a_0}{\Omega - \kappa P}\delta(k_0 - \kappa)\delta(\omega_0 - \Omega)E(k_1, \omega_1; k_2, \omega_2; \dots)$$

$$+\frac{i\lambda a_0}{\Omega^* - \kappa P}\delta(k_0 + \kappa)\delta(\omega_0 + \Omega^*)E^*(k_1, \omega_1; k_2, \omega_2; \dots). \tag{10.130}$$

Second Order $O(\epsilon^2)$. The second-order equations read as

$$(\omega_0 - k_0 P)S_2 + (\omega_1 - k_1 P)S_1 + i\lambda a_0 \varphi_2 - i\mathcal{C}_2 =$$

$$= \frac{i}{2} \int \prod_{n=0}^{N} \mathrm{d}k'_n \mathrm{d}\omega'_n k'_0 (k_0 - k'_0) S_1(k'_0, \omega'_0; \dots) S_1(k_0 - k'_0, \omega_0 - \omega'_0; \dots), \tag{10.131}$$

$$(\omega_0 - k_0 P)F_2 + k_0 \frac{\mathrm{d}F_0}{\mathrm{d}P}\mathcal{C}_2 = 0, \tag{10.132}$$

$$k_0^2\varphi_2 + 2k_0k_1\varphi_1 = \int \mathrm{d}P\left(F_2 - ik_0S_2\frac{\mathrm{d}F_0}{\mathrm{d}P}\right) - ik_1\int \mathrm{d}PS_1\frac{\mathrm{d}F_0}{\mathrm{d}P}. \quad (10.133)$$

Equation (10.131) can be rewritten in alternate form as

$$(\omega_0 - k_0P)S_2 + (\omega_1 - k_1P)S_1 + i\lambda a_0\varphi_2 - i\mathcal{C}_2 = -\frac{i\lambda^2 a_0^2}{2}$$

$$\times\left[\frac{\kappa^2}{(\Omega-\kappa P)^2}\delta(k_0 - 2\kappa)\delta(\omega_0 - 2\Omega)\right.$$

$$\times\int\prod_{n=1}^{N}\mathrm{d}k_n'\mathrm{d}\omega_n' E(k_1',\omega_1';\ldots)E(k_1 - k_1',\omega_1 - \omega_1';\ldots)$$

$$+\frac{\kappa^2}{(\Omega^*-\kappa P)^2}\delta(k_0 + 2\kappa)\delta(\omega_0 + 2\Omega^*)$$

$$\left.\times\int\prod_{n=1}^{N}\mathrm{d}k_n'\mathrm{d}\omega_n' E^*(k_1',\omega_1';\ldots)E^*(k_1 - k_1',\omega_1 - \omega_1';\ldots)\right]$$

$$-\frac{i\lambda^2 a_0^2\kappa^2}{|\Omega-\kappa P|^2}\delta(k_0)\delta(\omega_0 - \Omega + \Omega^*)$$

$$\times\int\prod_{n=1}^{N}\mathrm{d}k_n'\mathrm{d}\omega_n' E(k_1',\omega_1';\ldots)E^*(k_1 - k_1',\omega_1 - \omega_1';\ldots). \quad (10.134)$$

Since the long-wave component represented by the last term on the right-hand-side of equation (10.134) does not vanish, we can equate the latter to the second-order Hamiltonian $\mathcal{C}_2$. Therefore,

$$\mathcal{C}_2(k_0,\omega_0;k_1,\omega_1;\ldots) = \frac{\lambda^2 a_0^2\kappa^2}{|\Omega-\kappa P|^2}\delta(k_0)\delta(\omega_0 - \Omega + \Omega^*)$$

$$\times\int\prod_{n=1}^{N}\mathrm{d}k_n'\mathrm{d}\omega_n' E(k_1',\omega_1';\ldots)E^*(k_1 - k_1',\omega_1 - \omega_1';\ldots). \quad (10.135)$$

The second term on the left-hand-side of equation (10.132) vanishes (because of $k_0\delta(k_0) = 0$), which provides an additional degree of freedom in the choice of F_2. More precisely

$$F_2(k_0, \omega_0; k_1, \omega_1; \dots) = \delta(k_0)\delta(\omega_0)\mathcal{W}(k_1, \omega_1; k_2, \omega_2; \dots). \qquad (10.136)$$

Let us examine the first harmonic [terms proportional to the factor $\delta(k_0 - \kappa)\delta(\omega_0 - \Omega)$, or to the factor $\delta(k_0 + \kappa)\delta(\omega_0 + \Omega^*)$] in the second-order perturbation equations

$$(\Omega - \kappa P)S_2^{(1)} - i\lambda a_0 \frac{\omega_1 - k_1 P}{\Omega - \kappa P} E + i\lambda a_0 \Phi = 0, \qquad (10.137)$$

$$\kappa^2 \Phi + 2\kappa k_1 E = -i\kappa \int \mathrm{d}P S_2^{(1)} \frac{\mathrm{d}F_0}{\mathrm{d}P} - \lambda a_0 k_1 E \int \frac{\mathrm{d}P}{\Omega - \kappa P} \frac{\mathrm{d}F_0}{\mathrm{d}P}. \qquad (10.138)$$

It is straightforward to verify that

$$\left(\frac{\partial \mathcal{D}}{\partial \kappa} k_1 + \frac{\partial \mathcal{D}}{\partial \Omega} \omega_1 \right) E = 0, \qquad (10.139)$$

while

$$\Phi = \Psi(k_1, \omega_1; k_2, \omega_2; \dots), \qquad (10.140)$$

where Ψ is a generic function of its arguments. Equation (10.139) can be written equivalently as

$$(\omega_1 - v_g k_1)E = 0, \qquad (10.141)$$

where

$$v_g = -\frac{\partial \mathcal{D}}{\partial \kappa} \left(\frac{\partial \mathcal{D}}{\partial \Omega} \right)^{-1} \qquad (10.142)$$

is the group velocity. Obviously

$$E(k_1, \omega_1; k_2, \omega_2; \dots) = \delta(\omega_1 - v_g k_1)\mathcal{E}(k_1; k_2, \omega_2; \dots). \qquad (10.143)$$

Next, we compute the second harmonic [terms proportional to $\delta(k_0 - 2\kappa)\delta(\omega_0 - 2\Omega)$, or to $\delta(k_0 + 2\kappa)\delta(\omega_0 + 2\Omega^*)$] in second-order perturbation equations. From the equations

$$2(\Omega - \kappa P)S_2^{(2)} = -i\lambda a_0 \varphi_2^{(2)}$$

$$-\frac{i\lambda^2 a_0^2 \kappa^2}{2(\Omega - \kappa P)^2} \int \prod_{n=1}^{N} \mathrm{d}k'_n \mathrm{d}\omega'_n E(k'_1, \omega'_1; \dots) E(k_1 - k'_1, \omega_1 - \omega'_1; \dots), \tag{10.144}$$

$$4\kappa^2 \phi_2^{(2)} = -2i\kappa \int \mathrm{d}P S_2^{(2)} \frac{\mathrm{d}F_0}{\mathrm{d}P}, \tag{10.145}$$

we readily obtain

$$\phi_2^{(2)} = -\frac{\lambda a_0 \kappa \Pi_2}{6} \int \prod_{n=1}^{N} \mathrm{d}k'_n \mathrm{d}\omega'_n E(k'_1, \omega'_1; \dots) E(k_1 - k'_1, \omega_1 - \omega'_1; \dots), \tag{10.146}$$

where

$$\Pi_n = \lambda a_0 \int \frac{\mathrm{d}P}{(\Omega - \kappa P)^{n+1}} \frac{\mathrm{d}F_0}{\mathrm{d}P}. \tag{10.147}$$

To complete the second-order calculations, let us write the second-order generating function S_2 as

$$S_2(k_0, \omega_0; k_1, \omega_1; \dots) = \delta(k_0 - \kappa)(\omega_0 - \Omega) S_2^{(1)}(k_1, \omega_1; \dots)$$

$$+\delta(k_0 - 2\kappa)(\omega_0 - 2\Omega) S_2^{(2)}(k_1, \omega_1; \dots) + c.c., \tag{10.148}$$

where

$$S_2^{(1)} = i\lambda a_0 \frac{\omega_1 - k_1 P}{(\Omega - \kappa P)^2} E - i\lambda a_0 \frac{\Psi}{\Omega - \kappa P}, \tag{10.149}$$

$$S_2^{(2)} = \frac{i\lambda^2 a_0^2 \kappa}{4} \left[\frac{\Pi_2}{3(\Omega - \kappa P)} - \frac{\kappa}{(\Omega - \kappa P)^3} \right]$$

$$\times \int \prod_{n=1}^{N} \mathrm{d}k'_n \mathrm{d}\omega'_n E(k'_1, \omega'_1; \dots) E(k_1 - k'_1, \omega_1 - \omega'_1; \dots), \tag{10.150}$$

Third Order $O(\epsilon^3)$. The third-order equations read as

$$(\omega_0 - k_0 P) S_3 + (\omega_1 - k_1 P) S_2 + (\omega_2 - k_2 P) S_1 + i\lambda a_0 \varphi_3 - i\mathcal{C}_3 =$$

$$= \frac{i}{2} \int \prod_{n=0}^{N} \mathrm{d}k'_n \mathrm{d}\omega'_n [2k'_0(k_0 - k'_0) S_1(k'_0, \omega'_0; \dots) S_2(k_0 - k'_0, \omega_0 - \omega'_0; \dots)$$

$$+(k_0'k_1 + k_0k_1' - 2k_0'k_1')S_1(k_0', \omega_0'; \ldots)S_1(k_0 - k_0', \omega_0 - \omega_0'; \ldots)], \quad (10.151)$$

$$(\omega_0 - k_0P)F_3 + (\omega_1 - k_1P)F_2 + k_0\frac{\mathrm{d}F_0}{\mathrm{d}P}\mathcal{C}_3 + k_1\frac{\mathrm{d}F_0}{\mathrm{d}P}\mathcal{C}_2 = 0, \quad (10.152)$$

$$k_0^2\varphi_3 + 2k_0k_1\varphi_2 + \left(k_1^2 + 2k_0k_2\right)\varphi_1 = \int \mathrm{d}P\left(F_3 - ik_0S_3\frac{\mathrm{d}F_0}{\mathrm{d}P}\right)$$

$$-ik_1\int \mathrm{d}PS_2\frac{\mathrm{d}F_0}{\mathrm{d}P} - ik_2\int \mathrm{d}PS_1\frac{\mathrm{d}F_0}{\mathrm{d}P}$$

$$-i\int \mathrm{d}P\int \prod_{n=0}^{N} \mathrm{d}k_n'\mathrm{d}\omega_n' k_0'S_1(k_0', \omega_0'; \ldots)\frac{\partial}{\partial P}F_2(k_0 - k_0', \omega_0 - \omega_0'; \ldots). \quad (10.153)$$

It suffices to consider the first harmonic only. We can write the equations for the first harmonic as

$$(\Omega - \kappa P)S_3 + (\omega_1 - k_1P)S_2^{(1)} - i\lambda a_0\frac{\omega_2 - k_2P}{\Omega - \kappa P}E + i\lambda a_0\varphi_3 =$$

$$\frac{2\lambda a_0\kappa^2}{\Omega^* - \kappa P}\frac{\delta(\omega_0 + \Omega^* - 2\Omega)}{\delta(\omega_0 - \Omega)}$$

$$\times\int \prod_{n=1}^{N} \mathrm{d}k_n'\mathrm{d}\omega_n' E^*(k_1', \omega_1'; \ldots)S_2^{(2)}(k_1 - k_1', \omega_1 - \omega_1'; \ldots), \quad (10.154)$$

$$\kappa^2\varphi_3 + 2\kappa k_1\Psi + \left(k_1^2 + 2\kappa k_2\right)E = -i\kappa\int \mathrm{d}PS_3\frac{\mathrm{d}F_0}{\mathrm{d}P} - ik_1\int \mathrm{d}PS_2^{(1)}\frac{\mathrm{d}F_0}{\mathrm{d}P}$$

$$-\lambda a_0k_2E\int \frac{\mathrm{d}P}{\Omega - \kappa P}\frac{\mathrm{d}F_0}{\mathrm{d}P} - \lambda a_0\kappa\int \prod_{n=1}^{N} \mathrm{d}k_n'\mathrm{d}\omega_n' E(k_1', \omega_1'; \ldots)$$

$$\times\int \frac{\mathrm{d}P}{\Omega - \kappa P}\frac{\partial}{\partial P}\mathcal{W}(k_1 - k_1', \omega_1 - \omega_1'; \ldots). \quad (10.155)$$

Straightforward algebra yields

$$\left(\frac{\partial\mathcal{D}}{\partial\kappa}k_1 + \frac{\partial\mathcal{D}}{\partial\Omega}\omega_1\right)\Psi = 0. \quad (10.156)$$

For the envelope of the self-field E, we obtain

$$\left(\frac{\partial\mathcal{D}}{\partial\kappa}k_2 + \frac{\partial\mathcal{D}}{\partial\Omega}\omega_2\right)E + \frac{1}{2}\left(\frac{\partial^2\mathcal{D}}{\partial\kappa^2}k_1^2 + 2\frac{\partial^2\mathcal{D}}{\partial\Omega\partial\kappa}\omega_1 k_1 + \frac{\partial^2\mathcal{D}}{\partial\Omega^2}\omega_1^2\right)E$$

$$= \frac{\delta(\omega_0 + \Omega^* - 2\Omega)}{2\delta(\omega_0 - \Omega)}\lambda^2 a_0^2 \kappa^4 \left(\frac{\Pi_2\Sigma_2}{3} - \kappa\Sigma_4\right)$$

$$\times \int \prod_{n=1}^{N} \mathrm{d}k'_n \mathrm{d}k''_n \mathrm{d}\omega'_n \mathrm{d}\omega''_n E^*(k'_1, \omega'_1; \ldots)$$

$$\times E(k''_1, \omega''_1; \ldots) E(k_1 - k'_1 - k''_1, \omega_1 - \omega'_1 - \omega''_1; \ldots)$$

$$-\lambda a_0 \kappa \int \prod_{n=1}^{N} \mathrm{d}k'_n \mathrm{d}\omega'_n E(k'_1, \omega'_1; \ldots) \int \frac{\mathrm{d}P}{\Omega - \kappa P}\frac{\partial}{\partial P}\mathcal{W}(k_1 - k'_1, \omega_1 - \omega'_1; \ldots), \tag{10.157}$$

where

$$\Sigma_n = \lambda a_0 \int \frac{\mathrm{d}P}{|\Omega - \kappa P|^2 (\Omega - \kappa P)^{n-1}} \frac{\partial F_0}{\partial P}. \tag{10.158}$$

For the second order distribution function $\mathcal{W}$, we have

$$(\omega_1 - k_1 P)\mathcal{W} + k_1 \frac{\partial F_0}{\partial P} \frac{\delta(\omega_0 + \Omega^* - \Omega)}{\delta(\omega_0)} \frac{\lambda^2 a_0^2 \kappa^2}{|\Omega - \kappa P|^2}$$

$$\times \int \prod_{n=1}^{N} \mathrm{d}k'_n \mathrm{d}\omega'_n E(k'_1, \omega'_1; \ldots) E^*(k_1 - k'_1, \omega_1 - \omega'_1; \ldots) = 0, \tag{10.159}$$

Performing the inverse Fourier transformation of equations (10.157) and (10.159), we obtain

$$\frac{\partial\mathcal{W}}{\partial\tau_1} + P\frac{\partial\mathcal{W}}{\partial x_1} - \frac{\lambda^2 a_0^2 \kappa^2 \mathrm{e}^{2\mathrm{Im}(\Omega)\tau}}{|\Omega - \kappa P|^2}\frac{\partial |E|^2}{\partial x_1}\frac{\partial F_0}{\partial P} = 0, \tag{10.160}$$

$$i\left(\frac{\partial}{\partial\tau_2} + v_g\frac{\partial}{\partial x_2}\right)E + \frac{1}{2}\frac{\mathrm{d}v_g}{\mathrm{d}\kappa}\frac{\partial^2 E}{\partial x_1^2}$$

$$= \frac{\lambda^2 a_0^2 \kappa^4 e^{2\mathrm{Im}(\Omega)\tau}}{2(\partial\mathcal{D}/\partial\Omega)} \left(\frac{\Pi_2\Sigma_2}{3} - \kappa\Sigma_4\right)|E|^2 E - \frac{\lambda a_0 \kappa}{(\partial\mathcal{D}/\partial\Omega)} E \int \frac{\mathrm{d}P}{\Omega - \kappa P} \frac{\partial\mathcal{W}}{\partial P}. \tag{10.161}$$

Equations (10.160) and (10.161) constitute a system of coupled equations for the coarse-grained one-particle distribution function $\mathcal{W}$ and for the slowly varying amplitude E of the self-field, and provide a complete description of the nonlinear particle-wave interaction. This system governs the slow processes of beam pattern dynamics through the evolution of the amplitude functions $\mathcal{W}$ and E. In equation (10.160) one can immediately recognize the Vlasov equation for the envelope distribution function $\mathcal{W}$ with the ponderomotive force proportional to $-\partial|E|^2/\partial x_1$, which is due to fast oscillations at frequency close to the resonant wave frequency. It is worth noting that the system (10.160) and (10.161) intrinsically contains the so-called *nonlinear Landau damping* mechanism.

It is possible to solve formally equation (10.160) for $\mathcal{W}$ in terms of $|E|^2$ and substitute the result into equation (10.161). As a result, one obtains a *generalized Ginzburg-Landau equation* for the amplitude function E [Tzenov (1998b)]. The generalized *Ginzburg-Landau* equation is known [Cross (1993)] to provide the basic framework for the study of many properties of non equilibrium systems, such as existence and interaction of coherent structures, generic onset of traveling wave disturbance in continuous media, appearance of chaos. Recent experimental and numerical evidence (see e.g. [Colestock (1998)], and the references therein) shows that similar behaviour is consistent with the propagation of charged particle beams, and the generalized Ginzburg-Landau equation could represent the appropriate analytical model to study the above mentioned phenomena.

10.6 Fluctuation Spectrum and Turbulence

The study of turbulence in beams is valuable primarily because it may be a universal phenomenon, at least at low levels, which plays a role in determining the limiting phase-space density in any machine. The effect has likely been small in machines well below their stability thresholds, however, as intensities have been pushed closer to stability limits, nonlinear wave interactions can occur which lead to a marginally stable equilibrium. It is our aim in the present paragraph to outline a theoretical approach capable to describe the formation of an equilibrium spectrum of fluctuations and as a consequence to understand the spectral amplitude dependence on the character of the interactions between particles within the beam directly and/or through the environment.

The approach taken is to develop a statistical description of fluctuations for an ensemble of coupled modes. This may be viewed as a development of the amplitude equations associated with coupled modes, as in equation (10.157). The interaction between modes may be three-wave, as given in equation (10.157), or higher-order. We will outline a general procedure, but keep only interactions up to the three-wave level. The result will be a scaling law for the envelope of the fluctuation spectrum.

The starting point for our analysis is the system of equations for the microscopic phase space density $N_M(x, p; \tau)$ and for the microscopic self-consistent potential $U_M(x; \tau)$. In paragraph 9.2 it was mentioned that the *Vlasov-Maxwell* system of equations (9.179) and (9.180) for the one-particle distribution function $f(x, p; \tau)$ and for the self-consistent potential $U(x; \tau)$ looks exactly the same as the system of equations for the microscopic quantities. Therefore, we can write

$$\frac{\partial N_M}{\partial \tau} + p\frac{\partial N_M}{\partial x} - \lambda a_0 \frac{\partial V_M}{\partial x}\frac{\partial N_M}{\partial p} = 0. \tag{10.162}$$

$$\frac{\partial^2 V_M}{\partial x^2} = \frac{1}{\lambda a_0} - \int \mathrm{d}p N_M(x, p; \tau), \tag{10.163}$$

Following the familiar procedure, we extract the average part and the fluctuation according to

$$N_M = f + \delta N, \qquad V_M = V + \delta V, \qquad \langle N_M V_M \rangle = fV + \langle \delta N \delta V \rangle. \tag{10.164}$$

As a result, we obtain the following equation

$$\left(\frac{\partial}{\partial \tau} + p\frac{\partial}{\partial x} - \lambda a_0 \frac{\partial V}{\partial x}\frac{\partial}{\partial p}\right)\delta N$$

$$= \lambda a_0 \frac{\partial \delta V}{\partial x}\frac{\partial f}{\partial p} + \lambda a_0 \frac{\partial}{\partial p}\left(\delta N \frac{\partial \delta V}{\partial x} - \left\langle \delta N \frac{\partial \delta V}{\partial x}\right\rangle\right), \tag{10.165}$$

for the fluctuating part $\delta N(x, p; \tau)$ of the microscopic phase space density. A judicious approximation consists in neglecting the last term in the brackets on the left-hand-side of equation (10.165) for reasons analogous to the ones taken into account in the derivation of the *Fokker-Planck* equation in paragraph 8.5. As a matter of fact, we assume that the fluctuation $\delta N(x, p; \tau)$ being a random function of its variables varies on small scales, so that the term in question does not manage to contribute noticeably. Therefore, the system of equations to begin the subsequent analysis with can be written in the form

$$\left(\frac{\partial}{\partial \tau} + p\frac{\partial}{\partial x}\right)\delta N = \lambda a_0 \frac{\partial \delta V}{\partial x}\frac{\partial f}{\partial p} + \lambda a_0 \frac{\partial}{\partial p}\left(\delta N \frac{\partial \delta V}{\partial x} - \left\langle \delta N \frac{\partial \delta V}{\partial x}\right\rangle\right), \tag{10.166}$$

$$\frac{\partial^2 \delta V}{\partial x^2} = -\int \mathrm{d}p \delta N(x, p; \tau). \tag{10.167}$$

The next step consists in the familiar procedure of *Fourier* transformation of the equations for the fluctuations. Applying the definition

$$\mathcal{F}(x; \tau) = \frac{1}{(2\pi)^2}\int \mathrm{d}\omega \mathrm{d}k \widetilde{\mathcal{F}}(k; \omega) e^{i(kx - \omega\tau)}, \tag{10.168}$$

$$\widetilde{\mathcal{F}}(k; \omega) = \int \mathrm{d}\tau \mathrm{d}x \mathcal{F}(x; \tau) e^{i(\omega\tau - kx)}, \tag{10.169}$$

holding for an arbitrary function $\mathcal{F}(x; \tau)$ to equations (10.166) and (10.167) readily gives

$$(\omega - kp)\delta\widetilde{N}(k; \omega) = -\lambda a_0 k \frac{\partial f}{\partial p}\delta\widetilde{V}(k; \omega) - \lambda a_0 \frac{\partial}{\partial p}\sum_{k_1, \omega_1} k_1$$

$$\times\left[\delta\widetilde{V}(k_1; \omega_1)\delta\widetilde{N}(k - k_1; \omega - \omega_1) - \left\langle \delta\widetilde{V}(k_1; \omega_1)\delta\widetilde{N}(k - k_1; \omega - \omega_1)\right\rangle\right], \tag{10.170}$$

$$\delta\widetilde{V}(k; \omega) = \frac{1}{k^2}\delta\widetilde{\rho}(k; \omega), \tag{10.171}$$

where

$$\delta\widetilde{\rho}(k; \omega) = \int \mathrm{d}p \delta\widetilde{N}(k, p; \omega), \qquad \sum_{k, \omega}\cdots = \frac{1}{(2\pi)^2}\int \mathrm{d}\omega \mathrm{d}k \ldots. \tag{10.172}$$

Introducing the operator

$$\widehat{\mathcal{G}}(k; \omega) = -\frac{\lambda a_0}{\omega - kp + i0}\frac{\partial}{\partial p}, \tag{10.173}$$

we can write equation (10.170) as

$$\delta\widetilde{N}(k; \omega) = k\widehat{\mathcal{G}}(k; \omega) f \delta\widetilde{V}(k; \omega) + \sum_{k_1, \omega_1}\widehat{\mathcal{G}}(k; \omega) k_1$$

$$\times\left[\delta\widetilde{V}(k_1;\omega_1)\delta\widetilde{N}(k-k_1;\omega-\omega_1)-\left\langle\delta\widetilde{V}(k_1;\omega_1)\delta\widetilde{N}(k-k_1;\omega-\omega_1)\right\rangle\right]. \tag{10.174}$$

Equation (10.174) can be solved perturbatively taking into account that the term under the sum on the right-hand-side is of second order of magnitude[1]. We have

$$\delta\widetilde{N}^{(0)}(k;\omega)=k\widehat{\mathcal{G}}(k;\omega)f\delta\widetilde{V}(k;\omega), \tag{10.175}$$

$$\delta\widetilde{N}^{(n)}(k;\omega)=\sum_{k_1,\omega_1}\widehat{\mathcal{G}}(k;\omega)k_1\left[\delta\widetilde{V}(k_1;\omega_1)\delta\widetilde{N}^{(n-1)}(k-k_1;\omega-\omega_1)\right.$$

$$\left.-\left\langle\delta\widetilde{V}(k_1;\omega_1)\delta\widetilde{N}^{(n-1)}(k-k_1;\omega-\omega_1)\right\rangle\right]. \tag{10.176}$$

Up to second order (cubic terms in the self-field fluctuations), we obtain

$$\delta\widetilde{N}(k;\omega)=k\widehat{\mathcal{G}}(k;\omega)f\delta\widetilde{V}(k;\omega)+\sum_{\substack{k_1+k_2=k\\ \omega_1+\omega_2=\omega}}k_1k_2\widehat{\mathcal{G}}(k;\omega)\widehat{\mathcal{G}}(k_2;\omega_2)f$$

$$\times\left[\delta\widetilde{V}(k_1;\omega_1)\delta\widetilde{V}(k_2;\omega_2)-\left\langle\delta\widetilde{V}(k_1;\omega_1)\delta\widetilde{V}(k_2;\omega_2)\right\rangle\right]$$

$$+\sum_{\substack{k_1+k_2+k_3=k\\ \omega_1+\omega_2+\omega_3=\omega}}k_1k_2k_3\widehat{\mathcal{G}}(k;\omega)\widehat{\mathcal{G}}(k_2+k_3;\omega_2+\omega_3)\widehat{\mathcal{G}}(k_3;\omega_3)f$$

$$\times\left[\delta\widetilde{V}(k_1;\omega_1)\delta\widetilde{V}(k_2;\omega_2)\delta\widetilde{V}(k_3;\omega_3)-\delta\widetilde{V}(k_1;\omega_1)\right.$$

$$\left.\times\left\langle\delta\widetilde{V}(k_2;\omega_2)\delta\widetilde{V}(k_3;\omega_3)\right\rangle-\left\langle\delta\widetilde{V}(k_1;\omega_1)\delta\widetilde{V}(k_2;\omega_2)\delta\widetilde{V}(k_3;\omega_3)\right\rangle\right]. \tag{10.177}$$

Next, we integrate both sides of the above equation over p and write the equation for the fluctuation of the self-potential in the form

$$\mathcal{D}(k;\omega)\delta\widetilde{V}(k;\omega)=\sum_{\substack{k_1+k_2=k\\ \omega_1+\omega_2=\omega}}\kappa_2(k,\omega;k_2,\omega_2)$$

$$\times\left[\delta\widetilde{V}(k_1;\omega_1)\delta\widetilde{V}(k_2;\omega_2)-\left\langle\delta\widetilde{V}(k_1;\omega_1)\delta\widetilde{V}(k_2;\omega_2)\right\rangle\right]$$

[1]Note that equation (10.171) suggests that both $\delta\widetilde{N}(k;\omega)$ and $\delta\widetilde{V}(k;\omega)$ are of the same order of magnitude.

$$+\sum_{\substack{k_1+k_2+k_3=k\\ \omega_1+\omega_2+\omega_3=\omega}} \kappa_3(k,\omega;k_1,\omega_1;k_3,\omega_3)\Big[\delta\widetilde{V}(k_1;\omega_1)\delta\widetilde{V}(k_2;\omega_2)\delta\widetilde{V}(k_3;\omega_3)$$

$$-\delta\widetilde{V}(k_1;\omega_1)\Big\langle\delta\widetilde{V}(k_2;\omega_2)\delta\widetilde{V}(k_3;\omega_3)\Big\rangle - \Big\langle\delta\widetilde{V}(k_1;\omega_1)\delta\widetilde{V}(k_2;\omega_2)\delta\widetilde{V}(k_3;\omega_3)\Big\rangle\Big], \tag{10.178}$$

where $\mathcal{D}(k;\omega)$ is the dispersion function defined by equation (10.125). In addition,

$$\kappa_2(k,\omega;k_2,\omega_2) = k_2(k-k_2)\int \mathrm{d}p\widehat{\mathcal{G}}(k;\omega)\widehat{\mathcal{G}}(k_2;\omega_2)f(p), \tag{10.179}$$

$$\kappa_3(k,\omega;k_1,\omega_1;k_3,\omega_3) = k_1k_3(k-k_1-k_3)$$

$$\times\int \mathrm{d}p\widehat{\mathcal{G}}(k;\omega)\widehat{\mathcal{G}}(k-k_1;\omega-\omega_1)\widehat{\mathcal{G}}(k_3;\omega_3)f(p), \tag{10.180}$$

are the second and the third order dielectric susceptibilities, respectively.

Let us now rewrite the equation for waves (10.178) in the form

$$\mathcal{D}(k;\omega)\delta\widetilde{V}(k;\omega) = \mathcal{J}(k;\omega), \tag{10.181}$$

where $\mathcal{J}(k;\omega)$ is a short-hand notation for the right-hand-side. Next, we represent the fluctuation of the self-potential $\delta V(x;\tau)$ as a superposition of plane eigenwaves whose frequency and wave-number obey the dispersion law (10.127). We let the amplitudes of these waves be dependent on x and τ. This dependence is slow in comparison with the phase of the wave and is characterized by scales inversely proportional to the small parameter of nonlinear wave-wave interaction. We have

$$\delta V(x;\tau) = \int \mathrm{d}kV_k(x;\tau)e^{i[kx-\omega(k)\tau]}. \tag{10.182}$$

The inverse *Fourier* transformation of equation (10.181) yields

$$\frac{1}{(2\pi)^2}\int \mathrm{d}\omega\mathrm{d}k\mathcal{J}(k;\omega)e^{i(kx-\omega\tau)} = \frac{1}{(2\pi)^2}\int \mathrm{d}\omega\mathrm{d}k\mathcal{D}(k;\omega)\delta\widetilde{V}(k;\omega)e^{i(kx-\omega\tau)}$$

$$= \mathcal{D}\left(-i\frac{\partial}{\partial x};i\frac{\partial}{\partial\tau}\right)\frac{1}{(2\pi)^2}\int \mathrm{d}\omega\mathrm{d}k\delta\widetilde{V}(k;\omega)e^{i(kx-\omega\tau)}$$

$$= \mathcal{D}\left(-i\frac{\partial}{\partial x};i\frac{\partial}{\partial\tau}\right)\delta V(x;\tau). \tag{10.183}$$

Expressing $\delta V(x;\tau)$ in terms of $V_k(x;\tau)$ according to the superposition integral (10.182), we find

$$\frac{1}{(2\pi)^2}\int \mathrm{d}\omega \mathrm{d}k \mathcal{J}(k;\omega)e^{i(kx-\omega\tau)} = \int \mathrm{d}k \mathcal{D}\Big(-i\frac{\partial}{\partial x};i\frac{\partial}{\partial \tau}\Big)V_k(x;\tau)e^{i(kx-\omega_k\tau)}$$

$$= \int \mathrm{d}k e^{i(kx-\omega_k\tau)}\mathcal{D}\Big(k-i\frac{\partial}{\partial x};\omega_k+i\frac{\partial}{\partial \tau}\Big)V_k(x;\tau)$$

$$= i\int \mathrm{d}k e^{i(kx-\omega_k\tau)}\Big(\frac{\partial \mathcal{D}}{\partial \omega}\Big)_{\omega_k}\Big(\frac{\partial}{\partial \tau}+\Omega_g\frac{\partial}{\partial x}-\Gamma_k\Big)V_k(x;\tau), \tag{10.184}$$

where

$$\Omega_g = -\frac{\partial \mathrm{Re}\mathcal{D}}{\partial k}\Big(\frac{\partial \mathrm{Re}\mathcal{D}}{\partial \omega}\Big)^{-1}\Bigg|_{\omega_k}, \qquad \Gamma_k = \mathrm{Im}\mathcal{D}\Big(\frac{\partial \mathrm{Re}\mathcal{D}}{\partial \omega}\Big)^{-1}\Bigg|_{\omega_k}, \tag{10.185}$$

is the group velocity of waves [see also equation (9.208)] and the damping coefficient, respectively. We multiply both sides of equation (10.184) by $[1/(2\pi)]\exp[i(\omega_m\tau - mx)]$. Integrating over x by taking into account that the wave amplitudes V_k vary slowly in x (thus behaving as constants in the course of integration), we obtain

$$i\Big(\frac{\partial \mathcal{D}}{\partial \omega}\Big)_{\omega_k}\Big(\frac{\partial}{\partial \tau}+\Omega_g\frac{\partial}{\partial x}-\Gamma_k\Big)V_k = \frac{1}{(2\pi)^2}\int \mathrm{d}\omega \mathcal{J}(k;\omega)e^{i(\omega_k-\omega)\tau}. \tag{10.186}$$

The *Fourier* amplitudes $\delta\widetilde{V}(k;\omega)$ entering $\mathcal{J}(k;\omega)$ can be expressed according to

$$\delta\widetilde{V}(k;\omega) = \int \mathrm{d}\tau \mathrm{d}x \delta V(x;\tau)e^{i(\omega\tau-kx)}$$

$$= \int \mathrm{d}\tau \mathrm{d}x \mathrm{d}m V_m(x;\tau)e^{i(\omega\tau-kx)}e^{i(mx-\omega_m\tau)} = (2\pi)^2 V_k(x;\tau)\delta(\omega-\omega_k). \tag{10.187}$$

Thus, the equation (10.186) for the wave amplitudes can be written in its final form

$$\Big(\frac{\partial}{\partial \tau}+\Omega_g\frac{\partial}{\partial x}-\Gamma_k\Big)V_k = -\frac{i}{(\partial\mathcal{D}/\partial\omega)_{\omega_k}}\sum_{k_1+k_2=k}\kappa_2(k,\omega_{k_1}+\omega_{k_2};k_2,\omega_{k_2})$$

$$\times(V_{k_1}V_{k_2}-\langle V_{k_1}V_{k_2}\rangle)e^{i(\omega_k-\omega_{k_1}-\omega_{k_2})\tau}$$

$$-\frac{i}{(\partial \mathcal{D}/\partial \omega)_{\omega_k}} \sum_{k_1+k_2+k_3=k} \kappa_3(k, \omega_{k_1} + \omega_{k_2} + \omega_{k_3}; k_1, \omega_{k_1}; k_3, \omega_{k_3})$$

$$\times (V_{k_1} V_{k_2} V_{k_3} - V_{k_1} \langle V_{k_2} V_{k_3} \rangle - \langle V_{k_1} V_{k_2} V_{k_3} \rangle) e^{i(\omega_k - \omega_{k_1} - \omega_{k_2} - \omega_{k_3})\tau}. \quad (10.188)$$

To obtain the spectrum law of fluctuations it suffices to consider the three-wave interaction and neglect the terms under the second sum in equation (10.188). Taking into account the dispersion relation (9.212), we represent the wave amplitudes V_k as

$$V_k = a_k e^{i\omega_k \tau}, \qquad \omega_{-k} = \omega_k, \qquad V_k^* = V_{-k}, \quad (10.189)$$

so that

$$a_k^* = a_{-k}. \quad (10.190)$$

A simple and straightforward algebra yields an equation for the new amplitudes a_k, namely

$$\left(\frac{\partial}{\partial \tau} + \Omega_g \frac{\partial}{\partial x} + i\omega_k - \Gamma_k \right) a_k = -\frac{i}{(\partial \mathcal{D}/\partial \omega)_{\omega_k}}$$

$$\times \sum_{k_1+k_2=k} \kappa_2(k, \omega_{k_1} + \omega_{k_2}; k_2, \omega_{k_2})(a_{k_1} a_{k_2} - \langle a_{k_1} a_{k_2} \rangle). \quad (10.191)$$

Note that in obtaining equation (10.191) the fact that decay processes of the form

$$\omega_k = \omega_{k_1} + \omega_{k_2}, \qquad \text{for} \qquad k = k_1 + k_2, \quad (10.192)$$

are forbidden has been taken into account. We can further simplify equation (10.191) by representing a_k as

$$a_k = (A_k + G_k) e^{-is_k \tau}, \quad (10.193)$$

where

$$s_k = \text{sgn}(k). \quad (10.194)$$

Here A_k varies slowly in comparison with G_k, and as it will become clear from what follows, the faster varying part G_k is of second order with respect to A_k. In addition, we assume that A_k is a constant during the time of change of G_k. In order to obtain an expression for G_k (in terms of A_k), we substitute the decomposition of a_k as given by equation (10.193) into (10.191) and integrate the resulting equation for G_k. We obtain

$$G_k = -\frac{1}{(\partial\mathcal{D}/\partial\omega)_{\omega_k}} \sum_{k_1+k_2=k} \frac{\kappa_2(k, \omega_{k_1} + \omega_{k_2}; k_2, \omega_{k_2})}{s_k - s_{k_1} - s_{k_2}}$$

$$\times (A_{k_1} A_{k_2} - \langle A_{k_1} A_{k_2} \rangle) e^{i\left(s_k - s_{k_1} - s_{k_2}\right)\tau}. \tag{10.195}$$

Consequently, the sought-for equation for the slowly varying wave amplitudes A_k is found to be

$$\left(\frac{\partial}{\partial\tau} + \Omega_g \frac{\partial}{\partial x} + i\Omega_k - \Gamma_k\right) A_k = i \sum_{k+k_1=k_2+k_3} \mathcal{S}(k, k_1, k_2, k_3)$$

$$\times \left(A^*_{k_1} A_{k_2} A_{k_3} - A_{k_2} \left\langle A^*_{k_1} A_{k_3} \right\rangle - \left\langle A^*_{k_1} A_{k_2} A_{k_3} \right\rangle\right), \tag{10.196}$$

where

$$\mathcal{S}(k, k_1, k_2, k_3) = \frac{\kappa_2(k - k_2, \omega_{k_3} - \omega_{k_1}; k_3, \omega_{k_3})}{(\partial\mathcal{D}/\partial\omega)_{\omega_k} (\partial\mathcal{D}/\partial\omega)_{\omega_{k-k_2}}}$$

$$\times [\kappa_2(k, \omega_{k_2} + \omega_{k-k_2}; k - k_2, \omega_{k-k_2}) + \kappa_2(k - k_2, \omega_{k_2} + \omega_{k-k_2}; k_2, \omega_{k_2})], \tag{10.197}$$

$$\Omega_k = \omega_k - s_k. \tag{10.198}$$

We now proceed to the statistical description of the system of slowly varying wave amplitudes A_k described by equation (10.196). It is reasonable to assume that the oscillation phases corresponding to different wave numbers k are random, which suggests a change to a new variable - the intensity of waves defined as

$$\langle A_k A_m \rangle = \mathcal{I}_k \delta(k + m). \tag{10.199}$$

On averaging equation (10.196), we finally arrive at the *kinetic equation* for waves

$$\left(\frac{\partial}{\partial\tau} + \Omega_g \frac{\partial}{\partial x} - 2\Gamma_k\right) \mathcal{I}_k = 2\pi \sum_{k+k_1=k_2+k_3} |\mathcal{S}(k, k_1, k_2, k_3)|^2$$

$$\times \delta(\Omega_k + \Omega_{k_1} - \Omega_{k_2} - \Omega_{k_3})(\mathcal{I}_{k_1}\mathcal{I}_{k_2}\mathcal{I}_{k_3} + \mathcal{I}_k\mathcal{I}_{k_2}\mathcal{I}_{k_3} - \mathcal{I}_k\mathcal{I}_{k_1}\mathcal{I}_{k_3} - \mathcal{I}_k\mathcal{I}_{k_1}\mathcal{I}_{k_2}), \tag{10.200}$$

Consider the evolution of a wave packet with average wave number k_a. The entire space of wave numbers can be divided into three regions. The first one is the energy-containing region for waves (*plasmons*) with wave numbers $k \sim k_a$, the second region is the so-called intermediate region for $k_a < k < k_s$ and the last one is the scattering region for $k \sim k_s$. Here k_s is a characteristic wave number for which wave-particle interactions become important. In analogy with hydrodynamic turbulence [Kolmogorov (1941)], we assume that the turbulence spectrum in the region of intermediate wave numbers is determined by the flux of plasmon kinetic energy,

$$P = \sum_k \Omega_k \frac{\partial \mathcal{I}_k}{\partial \tau}, \tag{10.201}$$

in the region of large k, so that the intermediate region is a region of universal equilibrium. The flux can be expressed in terms of the characteristics of the wave packet in the energy containing region according to

$$P \sim k_a^8 (k_a \mathcal{I}_{k_a})^3. \tag{10.202}$$

In writing equation (10.202) use has been made of the fact that the matrix element $\mathcal{S}(k, k_1, k_2, k_3)$ scales as

$$\mathcal{S}(k, k_1, k_2, k_3) \sim k^4.$$

In the intermediate region the flux should be expressed equivalently in terms of $\mathcal{I}_k$ and k, so that

$$P \sim k^8 (k \mathcal{I}_k)^3. \tag{10.203}$$

Hence,

$$\mathcal{I}_k \sim k^{-11/3}. \tag{10.204}$$

We note that this power law spectrum, which in this case is due to a space-charge impedance, would indeed lead to the type of broad fluctuation spectrum seen in a number of experiments. Remarkably enough, expression (10.204) is perfectly analogous to the Kolmogorov spectrum law in ordinary hydrodynamic turbulence $\mathcal{E}_k \sim \epsilon^{2/3} k^{-11/3}$, where $\mathcal{E}_k$ is the energy of the k-mode and ϵ is the energy flux in the region of large k.

Bibliography

M. Abramowitz and I.A. Stegun (1984). *Handbook of Mathematical Functions with Formulas, Graphs, and Mathematical Tables*, Wiley, New York.

V.I. Arnold and A. Avez (1968). *Ergodic Problems of Classical Mechanics* , Benjamin, New York.

V.I. Arnold (1983). *Geometrical Methods in the Theory of Ordinary Differential Equations*, Springer-Verlag, Berlin.

R. Balescu (1988). *Transport Processes in Plasmas*, Volume 1 and 2, North-Holland, Amsterdam.

A.O. Barut and R. Raczka (1987). *Theory of Group Representations and Applications*, World Scientific Pub. Co., Singapore.

C.M. Bender and S.A. Orszag (1978). *Advanced Mathematical Methods for Scientists and Engineers*, McGraw-Hill, New York.

N.N. Bogoliubov and Y.A. Mitropolsky (1961). *Asymptotic Methods in the Theory of Nonlinear Oscillations*, Gordon and Breach, New York.

N.N. Bogoliubov (1965). *The Dynamical Theory in Statistical Physics*, Hindustan Pub. Corp., Delhi.

H. Bruck (1966). *Accelerateurs Circulaires de Particules*, Presses Universitaires de France, Paris.

P. J. Bryant and Kjell Johnsen (1993). *The Principles of Circular Accelerators and Storage Rings*, Cambridge University Press, Cambridge.

A.W. Chao (1993). *Physics of Collective Beam Instabilities in High Energy Accelerators*, Wiley, New York.

L.-Y. Chen, N. Goldenfeld and Y. Oono (1996). "Renormalization Group and Singular Perturbations: Multiple Scales, Boundary Layers, and Reductive Perturbation Theory", *Physical Review* **E 54**, 376-394.

B.V. Chirikov (1979). "A Universal Instability of Many-Dimensional Oscillator Systems", *Physics Reports* **52**, 263-379.

P.L. Colestock, L.K. Spentzouris and S.I. Tzenov (1998). "Coherent Nonlinear Phenomena in High Energy Synchrotrons: Observations and Theoretical Models", R.A. Carrigan and N.V. Mokhov eds., *International Symposium on Near Beam Physics, Fermilab, September 22-24, 1997*, 94-104, Fermilab, Batavia.

E.D. Courant and H.S. Snyder (1958). “Theory of the Alternating Gradient Synchrotron”, *Annals of Physics* **3**, 1-48.

M.C. Cross and P.C. Hohenberg (1993). "Pattern Formation Outside of Equilibrium", *Reviews of Modern Physics* **65**, 851.

R.C. Davidson, H. Qin, S.I. Tzenov, and E.A. Startsev (2002). “Kinetic Description of Intense Beam Propagation Through a Periodic Focusing Field for Uniform Phase-Space Density”, *Physical Review ST Accel. Beams* **5**, 084402.

A.J. Dragt (1982). “Lectures on Nonlinear Orbit Dynamics”, R.A. Carrigan et al. eds., *Physics of High Energy Particle Accelerators: (Fermilab Summer School, 1981)*, AIP Conference Proceedings, No. **87**, 147-313, New York.

M. Dubois-Violette (1969). “On the Functional Formalism in Quantum Field Theory”, *Nuovo Cimento* **62B**, 235-246.

M. Dubois-Violette (1970a). “Theory of Formal Series with Applications to Quantum Field Theory”, *Journal of Mathematical Physics* **11**, 2539-2546.

M. Dubois-Violette (1970b). “Methode Formelle de Solutions d'Equations Nonlineaires”, *International Journal of Nonlinear Mechanics* **5**, 311-323.

D.A. Edwards and M.J. Syphers (1993). *An Introduction to the Physics of High Energy Accelerators*, Wiley, New York.

D.F. Escande and F. Doveil (1981). “Renormalization Method for Computing the Threshold of the Large-Scale Stochastic Instability in Two Degrees of Freedom Hamiltonian Systems”, *Journal of Statistical Physics* **26**, 257.

W. Fleming (1966). *Functions of Several Variables*, Addison-Wesley, Reading, Massachusetts.

F.R. Gantmacher (1959). *The Theory of Matrices*, Chelsea Pub. Co., New York.

C.W. Gardiner (1983). *Handbook of Stochastic Methods for Physics, Chemistry and the Natural Sciences*, Springer-Verlag, Berlin.

P. Gaspard (1998). *Chaos, Scattering and Statistical Mechanics*, Cambridge University Press, Cambridge.

R. Gilmore (1981). *Catastrophe Theory for Scientists and Engineers*, Wiley, New York.

H. Goldstein (1980). *Classical Mechanics*, Addison-Wesley, Reading, Massachusetts.

I.S. Gradshteyn and I.M. Ryzhik (2000). *Table of Integrals, Series and Products*, Academic Press, San Diego.

S. Goto, Y. Masutomi and K. Nozaki (1999). “Lie-Group Approach to Perturbative Renormalization Group Method”, *Progress of Theoretical Physics* **102**, 471-497.

S. Goto and K. Nozaki (2001). “Regularized Renormalization Group Reduction of Symplectic Maps”, *Journal of the Physical Society of Japan* **70**, 49-54.

G. Guignard (1978). “A General Treatment of Resonances in Accelerators”, CERN 78-11, Geneva.

G. Guignard and J. Hagel (1986). “Sextupole Correction and Dynamic Aperture Numerical and Analytical Tools”, *Particle Accelerators* **18**, 129.

R. Hagedorn (1957). “Stability and Amplitude Ranges of Two-Dimensional Nonlinear Oscillations with Periodical Hamiltonian”, CERN 57-1, Geneva.

H. Haken (1990). *Information and Self-Organization. A Microscopic Approach to Complex Systems*, Springer-Verlag, Berlin.

M. Hammermesh (1962). *Group Theory and its Applications to Physical Problems*, Addison-Wesley, Reading, Massachusetts.

H.H. Hasegawa and W.C. Saphir (1992). "Unitarity and irreversibility in chaotic systems", *Physical Review* **A 46**, 74017423.

S. Humphries, Jr. (1986). *Principles of Charged Particle Acceleration*, Wiley, New York.

E.L. Ince (1944). *Ordinary Differential Equations*, Dover, New York.

E.T. Jaynes (1957a). "Information Theory and Statistical Mechanics", *Physical Review* **106**, 620-630.

E.T. Jaynes (1957b). "Information Theory and Statistical Mechanics. II", *Physical Review* **108**, 171-190.

A. Jeffrey and T. Kawahara (1982). *Asymptotic Methods in Nonlinear Wave Theory*, Pitman Advanced Publishing Program, Boston.

J. Kevorkian and J.D. Cole (1996). *Multiple Scale and Singular Perturbation Methods*, Springer-Verlag, Berlin.

Y.S. Kivshar and B. Luther-Davies (1998). "Dark Optical Solitons: Physics and Applications", *Physics Reports* **298**, 81-197.

Yu.L. Klimontovich (1986). *Statistical Physics*, Harwood Academic Publishers, Chur.

Yu.L. Klimontovich (1995). *Statistical Theory of Open Systems*, Kluwer Academic Publishers, Dordrecht.

A.N. Kolmogorov (1941), *DAN SSSR* **30**, 299.

A.A. Kolomensky and A.N. Lebedev (1966). *Theory of Cyclic Accelerators*, North-Holland, Amsterdam.

T. Kunihiro (1995a). "A Geometrical Formulation of the Renormalization Group Method for Global Analysis", *Progress of Theoretical Physics* **94**, 503-514.

T. Kunihiro (1995b). "A Geometrical Formulation of the Renormalization Group Method for Global Analysis II: Partial Differential Equations", Ryukoko Preprint RYUTHP-95/4 (patt-sol/9508001), Ohtsu-city.

L.D. Landau (1946). "On the Vibrations of the Electronic Plasma", *J. Phys. (USSR)* **10**, 25.

L.D. Landau and E.M. Lifshitz (1987). *Fluid Mechanics*, Pergamon Press, Oxford.

S.Y. Lee (1999). *Accelerator Physics*, World Scientific Pub. Co., Singapore.

A.J. Lichtenberg and M.A. Lieberman (1982). *Regular and Stochastic Motion*, Springer-Verlag, Berlin.

S.V. Manakov (1973). "On the Theory of Two-Dimensional Stationary Self-Focusing of Electromagnetic Waves", *Zh. Eksp. Teor. Fiz. [Sov. Phys. JETP]* **65 [38]**, 505-516 [248-253] (1973 [1974]).

H. Mori and Y. Kuramoto (1998). *Dissipative Structure and Chaos*, Springer-Verlag, Berlin.

J.A. Murdock (1991). *Perturbation Theory and Methods*, Wiley, New York.

A.H. Nayfeh (1981). *Introduction to Perturbation Techniques*, Wiley, New York.

A.H. Nayfeh (1985). "Perturbation Methods in Nonlinear Dynamics", J.M. Jowett et al. eds., *Nonlinear Dynamics Aspects of Particle Accelerators*, Springer-

Verlag, Berlin.
H. Poincare (1992). "New Methods of Celestial Mechanics", *History of Modern Physics and Astronomy, Vol. 13*, Chap. 27, Sec. 311, AIP, New York.
A.P. Prudnikov, Yu.A. Brychkov and O.I. Marichev (1986). *Integrals and Series*, Gordon and Breach Science Publishers, New York.
M. Reiser (1994). *Theory and Design of Charged Particle Beams*, Wiley, New York.
D. Ruelle (1978). *Statistical Mechanics, Thermodynamic Formalism*, Addison-Wesley, Reading, Massachusetts.
R.D. Ruth (1986). "Single Particle Dynamics in Circular Accelerators", SLAC - PUB - 4103, Stanford.
R.Z. Sagdeev, D.A. Usikov and G.M. Zaslavsky (1988). *Nonlinear Physics: From the Pendulum to Turbulence and Chaos*, Harwood Academic Publishers, Chur.
G. Sandri (1965). "A New Method of Expansion in Mathematical Physics", *Nuovo Cimento* **B36**, 67-93.
A. Schoch (1957). "Theory of Linear and Nonlinear Perturbations of Betatron Oscillations in Alternating Gradient Synchrotrons", CERN 57-21, Geneva.
S. Sternberg (1964). *Lectures on Differential Geometry*, Prentice-Hall, Englewood Cliffs, New Jersey.
J.J. Stoker (1950). *Nonlinear Vibrations in Mechanical and Electrical Systems*, Interscience, New York.
A. Sudbery (1986). *Quantum Mechanics and the Particles of Nature*, Cambridge University Press, Cambridge.
T. Suzuki (1978). "Hamiltonian Formalism for the Study of Closed Orbits and Betatron Oscillations in a Perturbed Machine", KEK Report 78-14, Tsukuba.
R. Thom (1975). *Structural Stability and Morphogenesis*, Addison-Wesley, Reading, Massachusetts.
S.I. Tzenov (1990). "Theory of Accelerated Orbits in AVF Cyclotrons", JINR Communication E9-90-92, Dubna.
S.I. Tzenov (1991). "Linear Synchro-Betatron Resonances, Driven by Beam-Beam Interaction", JINR Communication E9-91-51, Dubna.
S.I. Tzenov (1992a). "Diffusion of Particles Due to Periodic Crossing of Nonlinear Resonances", CERN SL/92-07 (AP), Geneva.
S.I. Tzenov (1992b). "The Method of Effective Potential for the Stability Analysis of Isolated Nonlinear Resonances", CERN SL/92-48 (AP), Geneva.
S.I. Tzenov (1992c). "Nonlinear Beam Dynamics Using the Method of Formal Series", CERN SL/92-35 (AP), Geneva.
S.I. Tzenov (1995). "The Method of Formal Series of Dubois-Violette as a Non-Perturbative Tool for the Analysis of Nonlinear Beam Dynamics", S. Chattopadhyay et al. eds., *Nonlinear Dynamics in Particle Accelerators: Theory and Experiments, Arcidosso, Italy, 1994*, AIP Conference Proceedings, No. **344**, 237-248, New York.
S.I. Tzenov (1997a). "Beam Dynamics in $e\bar{e}$ Storage Rings and a Stochastic Schrödinger-Like Equation ", *Physics Letters A* **232**, 260-268.
S.I. Tzenov (1997b). "On the Unified Kinetic, Hydrodynamic and Diffusion De-

scription of Particle Beam Propagation", S. Chattopadhyay et al. eds., *Nonlinear and Collective Phenomena in Beam Physics, Arcidosso, Italy, 1996*, AIP Conference Proceedings, No. **395**, 391-406, New York.

S.I. Tzenov (1998a). "Collision Integrals and the Generalized Kinetic Equation for Charged Particle Beams", Fermilab FERMILAB-Pub-98/287, Batavia.

S.I. Tzenov (1998b). "Formation of Patterns and Coherent Structures in Charged Particle Beams", Fermilab FERMILAB-Pub-98/275, Batavia.

S.I. Tzenov (2002a). "Renormalization Group Reduction of Non-Integrable Hamiltonian Systems", *New Journal of Physics* **4**, 6.1-6.12.

S.I. Tzenov (2002b). "Renormalization Group Approach to the Beam-Beam Interaction in Circular Colliders", *Proceedings of the Eighth European Particle Accelerator Conference (EPAC 2002)*, Paris, 3-7 June 2002, pp. 1422-1424.

E.T. Whittaker (1944). *A Treatise on the Analytical Dynamics of Particles and Rigid Bodies, with an Introduction to the Problem of Three Bodies*, Dover, New York.

E.T. Whittaker and G.N. Watson (1963). *A Course of Modern Analysis*, Cambridge University Press, Cambridge.

H. Wiedemann (1993). *Particle Accelerator Physics*, Springer-Verlag, Berlin.

S. Wiggins (1990). *Introduction to Applied Nonlinear Dynamical Systems and Chaos*, Springer-Verlag, Berlin.

E.J.N. Wilson (2001). *An Introduction to Particle Accelerators*, Oxford University Press, Oxford.

D.N. Zubarev and M.Yu. Novikov (1993). *Theoretical and Mathematical Physics* **13**, No. 3, 406-420 (in Russian).

R.W. Zwanzig (1961). *Lectures in Theoretical Physics*, Vol. **3**, 106-141, W.E. Brittin ed., Wiley, New York.

Index